T0139814

Current Microbiological Research in Africa

Akebe Luther King Abia • Guy R. Lanza
Editors

Current Microbiological Research in Africa

Selected Applications for Sustainable
Environmental Management

 Springer

Editors
Akebe Luther King Abia
University of KwaZulu-Natal
Westville, Durban
KwaZulu-Natal, South Africa

Guy R. Lanza
Department of Environmental and Forest
Biology (ESF)
State University of New York
Syracuse, NY, USA

ISBN 978-3-030-35298-1   ISBN 978-3-030-35296-7   (eBook)
https://doi.org/10.1007/978-3-030-35296-7

This Springer imprint is published by the registered company Springer Nature Switzerland AG
The registered company address is: Gewerbestrasse 11, 6330 Cham, Switzerland

We dedicate this book to our beloved fathers David Abia and Thomas Lanza.
They unselfishly supported our quest for knowledge.

Foreword

Environmental degradation remains among the most significant challenges faced by the African continent, fueled by a combination of numerous factors such as rapid population growth, high rates of urban migration, the need for increased food supply, lack of basic sanitation and water supply facilities, among others. The adverse consequences of this degradation include decreased land for human habitat, decreased soil fertility and associated hunger, biodiversity loss, and most importantly, increased incidence of human diseases, including those linked to antibiotic resistance.

Despite the numerous challenges, Africa is still reported to only contribute less than 1% of research findings globally. Limited or absence of funding, lack of infrastructure, low availability of modern equipment, and lack of skilled personnel are only a few of the factors that may make the situation worse. Even where the funding is available, prioritizing where the funds should be directed is also challenging. These issues, in turn, may affect policy meant to protect and preserve the environment.

We have compiled this volume on current microbiological research in Africa, with a focus on approaches for sustainable environmental management. Our aim is to stimulate areas that may need research attention on the continent, such as eco-friendly sustainable agriculture and pollution abatement, the current state of the environment's role in disease transmission on the continent, and most importantly, antibiotic resistance, which is a problem of global concern.

It is our firm conviction that this book is the beginning of a bigger initiative in which we are only playing a small part, and that the bigger picture lies on the combined efforts of the global society to ensure that we contribute to a better environment in Africa, through research, for present and future generations.

One hand cannot tie a bundle.—Cameroonian proverb

Akebe Luther King Abia
Antimicrobial Research Unit (ARU)
University of KwaZulu-Natal
Westville, Durban, KwaZulu-Natal, South Africa

Guy R. Lanza
Department of Environmental and Forest Biology (ESF)
State University of New York
Syracuse, NY, USA

Preface

The world's second largest and second most-populous continent, Africa, is achieving new levels of growth and economic development and occupies about 6% of the earth's total surface area and about 20% of earth's land area. Degradation of terrestrial and aquatic ecosystems on the continent is lowering water and soil quality, decreasing biodiversity, and is rapidly changing the African microbiome. One immediate concern with shifts in the microbial communities is the challenge of increased microbial disease transmission.

The 13 chapters in our book were chosen to illustrate interesting and informative examples of the current applied research agenda in environmental microbiology in selected countries in Africa. Chapter authors from Northern Africa to South Africa are included along with contributions from Nigeria and Tanzania. Different chapters are built on specific research objectives and provide a variety of applied methods and approaches to meet the pragmatic needs faced by environmental microbiologists in Africa. Topics provided in our book include studies of the three major groups of microorganisms: viruses, bacteria (including cyanobacteria), and protozoa. Their important roles in disease transmission, the enhancement of agriculture and aquaculture for food production, and in the bioremediation/degradation of oil and hydrocarbons are emphasized. The activities of microorganisms in different common indoor (e.g., household kitchens, latrines, and hospitals) and outdoor settings including air, soil, and water habitats are described.

The book contains chapters on the subject of the biodegradation of hydrocarbons in both halophilic and non-halophilic environments and descriptions of the application of biosurfactants for the bioremediation of oil spills in water. New information on topics relating to food production is offered in two chapters: one describing the risks to humans and fish from the use of antibiotics in aquaculture while another chapter illustrates the prospects for developing competitive strains of *Rhizobia* inoculants for agricultural applications. The important topics of the emergence and re-emergence of pathogenic, zoonotic, and hemorrhagic viruses (e.g., Dengue, Zika, Marburg, Chikungunya, Hantavirus, Monkeypox virus, Ebola, Lassa, and Influenza), bacteria (*Vibrio cholera*, and other emerging bacterial pathogens), and protozoa (*Cryptosporidium* and *Giardia*) in water and wastewater are presented in several chapters. One chapter provides new information on the important topic of ecotoxicity changes measured with indicator organisms following the biological treatment of the toxic cyanobacteria *Oscillatoria* and *Microcystis*.

Africa stands at the crossroad of rapidly changing methods due to scientific breakthroughs and the need for modern training in environmental microbiology in African universities. Modern research training in environmental microbiology is essential to the health and economic prosperity of Africa. Taken together, the 13 chapters in our book illustrate the types of applied environmental microbiology currently under study in selected African countries. However, the realm of environmental microbiology is changing rapidly, driven by the advent of powerful new genomic methods including CRISPR, next-generation sequencing, and innovative approaches to address the challenges of increasing environmental degradation, food and water security, and continuing emergence of resistance to antibiotics and other medications. Hopefully, the modern training required to prepare African microbiologists to meet the challenges they face will become widely available in African universities.

Durban, KwaZulu-Natal, South Africa   Akebe Luther King Abia
Syracuse, NY, USA   Guy R. Lanza

Contents

Chapter 1
Some Bacterial Pathogens of Public Health Concern in Water and Wastewater: An African Perspective

Mohamed Azab El-Liethy and Akebe Luther King Abia

1.1 The Primary Water Sources in Africa

It is well known that freshwater resources are always renewed via a continuous cycle of evaporation, precipitation, and runoff. The African continent has about 9% of the world's freshwater resources and also contains 11% of the global population (CIWA (Cooperation in International Water in Africa) 2016; UNESCO 2016). The primary water sources in Africa are rivers, fresh lakes, groundwater, and rainfall. Africa contains 17 rivers with catchment areas greater than 100,000 km², and it has more than 160 lakes larger than 27 km². The majority of these are located in the equatorial region and the sub-humid East African highlands within the Rift Valley (African Union 2017). The four major rivers of Africa (Fig. 1.1) are the Nile which is considered the world's longest and is approximately 6670 km long, followed by the Congo (4700 km), Niger (4180 km), and the Zambezi (2574 km) Rivers. River Nile runs across ten different African countries with Egypt, Sudan, and Eritrea contributing the lowest annual amount of rainfall within the basin. However, Sudan represents 63.6% of the total area of the basin, while Egypt covers only 10.5%.

River Nile is produced by the meeting of the Blue River Nile, which comes from the Tana Lake located in the highlands of Ethiopia, and the White River Nile, which begins in the headwaters of Lake Victoria. Both rivers join at Khartoum, Sudan, before draining northward to the Mediterranean Sea. The Congo River comes from the highlands of East Africa with many tributaries. The Congo River flows in the north direction, then drifts to the west before draining into the Atlantic Ocean.

M. A. El-Liethy (✉)
Environmental Microbiology Laboratory, Water Pollution Research Department, National Research Centre, Giza, Egypt
e-mail: ma.el-liethy@nrc.sci.eg

A. L. K. Abia
Antimicrobial Research Unit, College of Health Sciences, University of KwaZulu-Natal, Durban, South Africa

© Springer Nature Switzerland AG 2020
A. L. K. Abia, G. R. Lanza (eds.), *Current Microbiological Research in Africa*,
https://doi.org/10.1007/978-3-030-35296-7_1

1

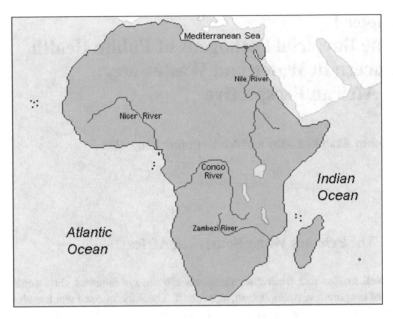

Fig. 1.1 The four major rivers in Africa (http://sayregeographyclass.weebly.com/uploads/1/4/8/8/14883058/major_rivers_in_africa.pdf)

The Niger River comes from the Guinea hills to Mali in the northeast, then to Niger and Nigeria in the southwest. The Zambezi River originates from Central Africa and flows southeast to the Indian Ocean.

Groundwater is a vital water source for drinking and irrigation in Africa. This resource was estimated at 0.66 million km³ (0.36–1.75 million km³) in 2012 and is the primary source of water for drinking and other purposes, especially in the northern and southern countries (MacDonald et al. 2012). According to the African Union (2017), although a substantial percentage of people living in urban areas had been covered by piped drinking water, boreholes are still the most essential water source for over 24% of the people in these areas. Northern African countries, including Libya, Algeria, Sudan, Egypt, and Chad, have the largest groundwater reserves (MacDonald et al. 2012). On the other hand, many of the Saharan aquifers are not, however, actively recharged, but were recharged in wet climate since more than 5000 years ago (Scanlon et al. 2006; Edmunds 2008). The storage volumes of these aquifers are more than 5×10^6 m³ km² (equivalent to 75 m water depth), and the volume of groundwater in Africa is higher than freshwater by 100 times and also more than the storage water in fresh African lakes by 20 times (MacDonald et al. 2012). Although the mean annual rainfall on the Africa continent is similar to that of other continents, the evaporation rates are high in Africa. The annual amounts of rainfall in Africa are about 670 mm and vary from one place to another and from time to time by about 40% (African Union 2017).

It is estimated that achieving universal access to safe and affordable drinking water for all by 2030 would require immense efforts to supply over 844 million people still living without even a basic water service, the majority of who are in Africa (United Nations 2018). Those lacking these services are among the poorest and most vulnerable in the world (Hunter et al. 2010). It is well known that life quality, agriculture, and industry are enhanced by ready and steady access to clean water and good sanitation. Thus, moving these people out of their poverty levels through the provision of safe potable water for drinking and other purposes is an undisputable need (Grey and Sadoff 2007; Hunter et al. 2010). Meeting this need is, however, challenged by the projected water shortages which are predictable to be potential triggers for new wars in the future as about 25 African countries will be entering into water stress by 2025 compared to 13 in 1995 (African Union 2017).

1.2 The Relationship Between Bacterial Indicators and Pathogens

Added to the quantity-related stress, there are ever-increasing reports on pollution of the already stressed water resources. These pollution events introduce undesirable and hazardous substances, including microbial pathogens, into the water bodies, with potentials to cause severe adverse human, animal, and environmental health effects. Thus, to provide water that is microbiologically safe for human consumption, the microbial quality of these water bodies needs to be monitored constantly for the presence of microbes, including pathogenic ones. However, given the excessively large number and significantly diverse nature of microbes that can be present in any water body at any given time, an attempt to obtain a complete picture of the microbial quality of such water bodies would not only be technically but also be financially unrealistic. Thus, it has been suggested that checking for the presence of microbial indicators could provide a rapid indication of the potential dangers associated with exposure to a water source.

The scientist van Fritsch in 1880 suggested the expression "indicators" depending on his observations of a bacterium called *Klebsiella* in human feces that was also detected in water (Hendricks 1978). In 1885, Escherich recommended *Bacillus coli* that was later named *Escherichia coli* as a proper indicator for fecal contamination of water since it was found in high densities in feces (Leiter 1929). In the twentieth century, certain bacterial groups were recognized as indicators of the presence of potentially pathogenic microbes in natural and treated waters. The bacterial groups most frequently used as indicators are heterotrophic bacteria (HB), also known as total bacterial counts or total viable bacterial counts. Other bacterial group indicators of fecal contamination are total coliforms (TC), fecal coliforms (FC), *E. coli*, and fecal streptococci (FS) recently named Enterococci (Fig. 1.2). *E. coli* is a subgroup of fecal coliforms, and fecal coliforms are a subgroup of the total coliforms. Therefore, many investigators have considered that *E. coli* is the most suitable fecal indicator for the microbial quality of water (Sunny et al. 2007; Haller et al. 2009; Odonkor et al. 2013).

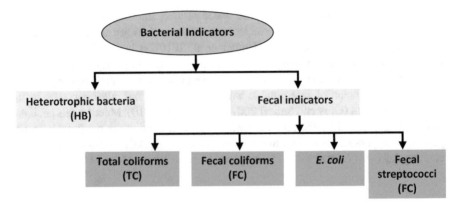

Fig. 1.2 A schematic diagram representing the bacterial indicator groups of water pollution

Table 1.1 Heterotrophic bacteria limits for drinking water in some African countries

Country	Heterotrophic bacteria at 20 or 22 °C	35 or 37 °C	Units	Standard guideline	References
Egypt	<50	<50	CFU/mL	Egyptian Standard for drinking water	Egyptian Standard (2007)
Nigeria	100	100	CFU/mL	Environmental Protection Agency Safe Drinking Water Act. EPA 816 – F – 03 – 016	EPA (2003)
South Africa	100	100	CFU/mL	South African Bureau of Standards	SABS (2001)
Sudan	100	100	CFU/mL	Sudanese Standards and Metrology Organization	SSMO (Sudanese Standards and Metrology Organization) (2002)
International limit (WHO)	100	100	CFU/mL	International Standard	WHO (2008)

Bacterial groups recovered through the HB test generally include normal flora bacteria (non-hazardous) but may also include pathogenic bacteria in polluted water (Bartram et al. 2003). Generally, HB level in drinking water should not exceed the defined limits as such excess would generally signal a deterioration of water quality. Some of the African countries use the same permissible limits of the WHO. For example, the South African Bureau of Standards allows HB in drinking water up to 100 colony forming units (CFU)/mL (SABS 2001). In Egypt, the HB should not exceed 50 CFU/mL according to the Egyptian Standard (2007) (Table 1.1). Moreover, drinking water should be free from fecal bacterial indicators, including *E. coli* (that is 0 CFU/100 mL).

The use of indicator bacteria has been suggested based on the results of previous studies that showed a positive correlation between the bacterial indicators and the presence of some pathogens (Odonkor et al. 2013; Gruber et al. 2014; El-Liethy et al. 2016). However, the study carried out by Grabow et al. (1995) found that there was no direct correlation between the numbers of any indicator bacteria and total pathogens. Numerous epidemiological studies have also failed to illustrate any relationship between microbial indicators and pathogens; this has been related to factors like improper study design, the extensively fluctuating ratio of pathogens to fecal indicators, and the presence of varying virulence factors of the pathogens (Fleisher 1990; Oliveira et al. 2018). This temporal lack of correlation has led to conclusions that there is no single indicator able to assure that water is free from pathogens (Horman et al. 2004; Abia et al. 2015; Hassard et al. 2017). Although it may not be feasible to check for the presence of all microbial pathogens in water during routine monitoring programs, checking for their presence is encouraged, especially in cases such as during disease outbreaks.

1.3 Pathogenic *E. coli* and *Salmonella* in Africa Waters

1.3.1 *Pathogenic* Escherichia coli

1.3.1.1 Brief History

Escherichia coli is a normal flora of both humans and other animals (Nontongana et al. 2014). Some strains of this organism were only associated with disease in the early 1940s (Bray and Beavan 1948). The history of the various pathogenic forms has previously been reviewed (Nataro and Kaper 1998; Deborah Chen et al. 2005). Some key dates associated with the major diarrheagenic forms include:

- 1885—Discovery of *Escherichia coli* (Eseherich 1885).
- 1955—Coining of the term EPEC (Neter et al. 1955).
- 1956—First test of ETEC in mice (De et al. 1956).
- 1971—First confirmation of ETEC in human volunteers (DuPont et al. 1971).
- 1971—First demonstration of EIEC pathogenic potential in humans (DuPont et al. 1971).
- 1978—Confirmation of EPEC's involvement in human disease (Levine et al. 1978).
- 1982—Recognition of EHEC O157:H7 as a human pathogen (Riley et al. 1983).
- 1987—First description of EAEC (Nataro et al. 1987).

Since their discovery, these pathogenic *E. coli* strains have been implicated in numerous disease outbreaks both in developed and in developing countries.

1.3.1.2 Taxonomy and Microbiological Characteristics

The pathogenic strains of *E. coli* are a subgroup of *E. coli* that is in the family *Enterobacteriaceae*. The bacteria are facultative anaerobes, Gram-negative bacilli often testing positive for catalase, and can reduce nitrate. Motility is through peritrichous flagella while some strains are not motile (Don et al. 1986; Gyles 2007). Strains of *Escherichia coli* have been placed into different serogroups and serotypes depending on the antigenic features expressed on the bacterial cell surface, that is, the outer membrane (O), flagella (H), and capsule (k) antigens (DebRoy et al. 2016). Over 170 distinct serogroups based on the O, >50 based on the k, and more than 100 based on the H antigens have been reported and have been employed to further categorize the bacterium into different serotypes (Gyles 2007).

Different pathogenic *E. coli* pathotypes have been identified based on their virulence potentials and the mode through which they initiate infection. These pathogenic strains can further be divided into intestinal or diarrheagenic *E. coli* (DEC), which are associated with diarrhea (Gomes et al. 2016), and extraintestinal pathogenic *E. coli* (ExPEC) (Luna et al. 2010). The DEC pathotypes include:

1. Shiga toxin- or verotoxin-producing *E. coli* (STEC or VTEC), which harbor genes that encode either Stx1 or Stx2 or both and are among the leading food-borne pathogens globally (Ranjbar et al. 2018).
2. Members of the EHEC group are considered a subset of STEC strains and are known to cause hemorrhagic colitis (HC) and HUS (Viazis and Diez-Gonzalez 2011).
3. Enteropathogenic *E. coli* (EPEC), responsible for severe infantile diarrhea especially in developing countries (Trabulsi et al. 2002b).
4. Enterotoxigenic *E. coli* (ETEC) include strains that are the leading cause of travelers' diarrhea in developing countries (Mudrak and Kuehn 2010).
5. Enteroinvasive *E. coli* (EIEC), which gain access to epithelial cells, become intracellular, and cause dysentery-like diarrhea similar to *Shigella* spp. (Pasqua et al. 2017).
6. Recently identified diffuse-adhering *E. coli* (DAEC), with a mechanism of infection that has not yet been fully understood but are believed to cause diarrhea by exhibiting a diffuse attachment pattern to intestinal cells (Mansan-Almeida et al. 2013).
7. Entero-aggregative *E. coli* (EAggEC), for its aggregative or "stacked-brick"-like adherence to cultured mammalian cells (APHA 2012), hence its name, and producing toxins which lead to the initiation of infection (Kaper et al. 2004).

The ExPEC members include uropathogenic *E. coli* (UPEC), which is responsible for most urinary tract infections globally (Welch et al. 2002), and neonatal meningitis *E. coli* (NMEC), which is among the primary causes of meningitis in neonates worldwide (Logue et al. 2012). Also, with the increase in implication of some *E. coli* strains in human bacteremia, another ExPEC pathotype known as sepsis-associated *E. coli* (SEPEC) has been reported (Sarowska et al. 2019).

1.3.1.3 Virulence Factors and Antibiotic Resistance of Major Human-Associated Pathotypes

The pathogenic potentials of disease-causing bacteria are determined by virulence factors which allow them to replicate and spread within a susceptible host, sometimes overcoming or avoiding the host's defense system (Cross 2008). These virulence determinants are regulated by a set of genes, the products of which allow the bacteria to adapt, survive, and cause infection in the environment in which they find themselves either intentionally or accidentally (Thomas and Wigneshweraraj 2014). This set of genes for a particular organism is known as the virulome of that organisms (Lebughe et al. 2017) and may vary from as little as 100 to as many as several thousand, depending on the prevailing environmental conditions in which the organism is and the methods used for the detection of the gene (Thomas and Wigneshweraraj 2014). The different *E. coli* pathotypes possess different virulence factors that allow them to cause disease (Mainil 2013; Hazen et al. 2017). Although numerous genes may be necessary to initiate infection by these pathogenic strains in humans, the main virulence factors involved and mostly used for characterizing these strains are summarized in Table 1.2.

With advanced techniques such as whole-genome sequencing, the virulence factors aiding in pathogenesis in the different pathogenic *E. coli* strains could be more than is presented in the current write-up and previous literature.

Several studies in Africa reported that pathogenic *E. coli* isolated from diverse sources were resistant to different antibiotics. In Ismailia City, Egypt, El-Alfy et al. (2013) found multiresistant antibiotic EHEC serotypes isolated from humans, animals, and environmental samples. In a separate study in Egypt carried out by El-Shatoury et al. (2015), all STEC O157:H7 isolated from El-Rahawy Drain and untreated hospital wastewater samples were resistant to amoxicillin, and 77% were resistant to clarithromycin. In the Eastern Cape, South Africa, multi-antibiotic-resistant *E. coli* O157 strains were reported in feces of dairy cattle by Iweriebor et al. (2015). Also, in Nigeria, *E. coli* O157 and other *E. coli* isolated from diarrheal stools and surface waters were resistant to different types of antibiotics (Chigor et al. 2010). A study conducted on wastewater and a river in Cape Town, South Africa, reported that over 60% of the pathogenic *E. coli* isolates obtained from the samples exhibited multiresistance against ampicillin, cefuroxime, cephalexin, ceftazidime, and tetracycline with multidrug resistance indices ranging between 4.2 and 5.6 (Doughari et al. 2011).

1.3.1.4 Detection Methods

No specific African guidelines or standards exist regarding the detection of pathogenic *E. coli* in water; thus, most of these countries depend on existing international standards. From previously published works, the most used detection method has been the culture method, followed by confirmation using different biochemical, serological, or immunological tests and PCR for the detection of virulence genes of pathogenic *E. coli*. Other molecular methods like restriction fragments length polymorphism (RFLP)

Table 1.2 Some virulence factors associated with human-related *E. coli* pathotypes

Infection site	Pathotype	Abbreviation	Virulence factors	Gene regulation	References
Intestinal	Enterohemorrhagic *E. coli*	EHEC	Shiga toxins (Stx1 and Stx2); Attaching and effacing adherence (*eaeA*; intimin); fimbriae and fimbriae-related adhesins; hemolysins (bacteriophage- and plasmid-encoded)	Bacterial chromosome (pathogenicity islands and phage) and plasmid-mediated	Viazis and Diez-Gonzalez (2011)
	Enteropathogenic *E. coli*	EPEC	EAF (EPEC adherence factor; the bundle forming pilus (BFP) only present in typical EPEC and absent in atypical strains)	Plasmid-mediated	Trabulsi et al. (2002a)
	Enterotoxigenic *E. coli*	ETEC	Toxins (heat-stable—ST, and heat-labile—LT); fimbrial-associated colonizing factors (CFs); autotransporters (serine—TibA and serine protease—EatA); suggested iron acquisition systems, i.e., *irp2, fyuA*	Plasmid	Sahl et al. (2017)
	Enteroaggregative *E. coli*	EAEC	Adhesins; toxins	Chromosomal (pathogenicity islands); plasmids	Okeke and Nataro (2001)
	Enteroinvasive *E. coli*	EIEC	Cell invasion mechanism	Plasmid	Gomes et al. (2016)
	Diffusely adherent *E. coli*	DAEC	Adhesins	Not determined	Robins-Browne et al. (2016)
Extraintestinal	Uropathogenic *E. coli*	UPEC	Adhesins; siderophores	Chromosomal; plasmid	Mainil (2013)
	Neonatal meningitis *E. coli*	NMEC	Adhesins; invasins; iron uptake	Chromosomal; plasmid	Sarowska et al. (2019)

PCR, pulsed-field gel electrophoresis (PFGE), and enterobacterial repetitive intergenic consensus (ERIC) PCR have been used to discriminate or establish the clonal relationship between isolates. Although culture suffers from some limitations such as not being able to detect viable but not culturable bacteria (VBNC) and interference from some nonpathogenic flora that naturally occur in such environments, the culture method is still the preferred choice in developing countries, probably backed by its simplicity and cost for routine analysis.

Culture-Based Methods

The culture-based method may involve plating on selective media or the multiple tube technique (Prats et al. 2007). While some of the media used may not be specific to pathogenic *E. coli*, some like Sorbitol MacConkey (SMAC) agar are most widely used and designed for *E. coli* O157:H7 (Effler et al. 2001; El-Shenawy and El-Shenawy 2005) because of its significant involvement in numerous diseases outbreaks worldwide. Thus, the following paragraphs in this section will focus on methods for the isolation of this strain.

Pre-enrichment in broth followed by streaking on selective agar media has been proposed for analysis of water samples to ensure higher recovery of pathogenic *E. coli* while inhibiting the competing background microflora. For examples, tryptic soy broth (TSB), modified *E. coli* (mEC) broth, and peptone water (Effler et al. 2001) supplemented with antibiotics have been used (Meyer-Broseta et al. 2001). SMAC agar media containing cefixime and potassium tellurite has also been used (Meyer-Broseta et al. 2001). The cefixime and potassium tellurite in this agar-based medium delay sorbitol fermentation and the absence of ß-glucuronidase (GUD) activity of *E. coli* O157, leading to characteristic clear colonies while the other nonpathogenic strains would appear bright pink (Manafi and Kremsmaier 2001). The primary role of cefixime is to target *Proteus* spp. (Chapman et al. 1991), while potassium tellurite aids in inhibiting or eliminating the growth of other *E. coli* and other bacterial species like *Aeromonas* spp. (Zadik et al. 1993). In South Africa, the immuno-magnetic separation (IMS) enrichment procedure has been used to investigate the presence of disease-causing *E. coli* in wastewater samples (Grabow et al. 2003).

Many fluorescence-based and color-based selective differential agars have been reported to perform better than SMAC in environmental samples (Manafi and Kremsmaier 2001). Some of these include Rainbow Agar O157, CHROM agar O157, HiCrome EC O157:H7 selective HiVeg agar base, and HiCrome SMAC agar base supplemented with tellurite-cefixime (Zadik et al. 1993). In Egypt, HiCrome EC O157:H7 selective agar base which contains sorbitol and a proprietary chromogenic mixture instead of lactose and indicator dyes and supplemented with cefixime and potassium tellurite was used. *E. coli* O157:H7 selectively and precisely breaks down the chromogenic substrate in this media producing a dark purple to a magenta-colored moiety. *E. coli* O157:H7 exhibit light pink to mauve-colored colonies (El-Leithy et al. 2012; El-Shatoury et al. 2015) (Fig. 1.3).

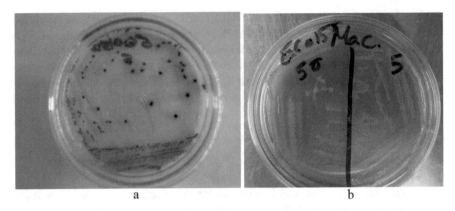

Fig. 1.3 *E. coli* O157:H7 on chromogenic selective agar media. (**a**) HiCrome EC O157:H7 selective agar, (**b**) HiCrome MacConky sorbitol agar

Moreover, HiCrome MacConky sorbitol agar base supplemented with tellurite and cefixime has been used to confirm *E. coli* O157:H7. The bacterium *E. coli* O157:H7 showed colorless colony confirming non-sorbitol fermentation (El-Liethy 2013). In South Africa, Müller et al. (2001) used Rainbow agar O157 to isolate *E. coli* O157 and other EHEC strains. Rainbow agar O157 contains a chromogenic substrate which is specific for two *E. coli*-associated enzymes: β-galactosidase (blue-black substrate) and β-glucuronidase (red substrate). The strain characteristically lacks glucuronidase activity, and thus produces unique charcoal gray or black colonies on this media.

Molecular-Based Methods

In Africa, PCR-based detection protocols targeting virulence genes of the different pathogenic *E. coli* have been used after an initial culture step. These reactions may target a single gene (El-Jakee et al. 2009; Grabow et al. 2003) or multiple gene pairs (El-Leithy et al. 2012; El-Shatoury et al. 2015). Genes used to identify the different pathotypes have been reported in numerous studies in Africa (Omar and Barnard 2010, 2014; Abia et al. 2016c, 2017)

1.3.1.5 Occurrence in Different African Water Sources

Pathogenic *E. coli* have been detected in many studies in Africa from humans, animals, food, and water. In this chapter, we focus on the occurrence of pathogenic *E. coli* in different water sources.

Occurrence in Drinking Water

Abong'o et al. (2008) examined 30 water samples collected from drinking water and boreholes in South Africa. The samples were examined using culture methods followed by PCR. The PCR results showed 8.6% of the examined water samples harbored *E. coli* O157:H7 bearing homologous *fliCH7, rfbEO157,* and *eaeA* genes (Abong'o and Momba 2008). Also, Momba et al. (2008) investigated the prevalence of *E. coli* O157:H7 in 180 drinking water samples collected from selected distribution systems within the Amathole District of the Eastern Cape using enrichment culture and confirmed isolates using molecular methods. The authors detected *E. coli* O157 in 46 (25.56%) of the samples. In another study, seven virulence genes screened for (*Ial, Stx1, Stx2, EaeA, Eagg, ST,* and *LT*) were isolated from different drinking water sources (Jagals et al. 2013). In Southern Africa, mainly in South Africa and Swaziland, heavy rainfall occurred several days before the occurrence of one of the most significant outbreaks involving *E. coli* O157:H7 in 1992 that contaminate surface and drinking water (Table 1.3) (Effler et al. 2001).

Occurrence in Surface Water

Various *E. coli* pathotypes have been detected in different surface water sources, e.g., rivers, lakes, sea, and agricultural drainage water. *E. coli* O157:H7 can survive in river water at room temperature up to 3 months (Ibrahim et al. 2019). In South Africa, Müller et al. (2001) investigated the occurrence of *E. coli* O157:H7 in various river waters in South Africa and reported that 4% of the isolates carried virulence genes. In a separate study in Cape Town, South Africa, the authors used culture, followed by biochemical and serological tests and reported several pathogenic *E. coli* serotypes

Table 1.3 The occurrence of pathogenic *E. coli* in drinking water

Country	Source	Organisms	Year	Detection methods	Counts	References
South Africa	Drinking water and their containers	EIEC and EHEC	2012	PCR	46% virulence genes	Jagals et al. (2013)
South Africa	Drinking water	*E. coli* O157:H7	NM	Culture and PCR methods	46 out of 180 positive	Momba et al. (2008)
South Africa	Drinking water and boreholes	*E. coli* O157:H7	NM	Culture and PCR methods	25.5%	Abong'o and Momba (2008)
Swaziland and South Africa	Drinking water	*E. coli* O157:H7	1992	Culture and PFGE	42% (327/778)	Effler et al. (2001)

NM Not mentioned

including *E. coli* O26, *E. coli* O55, *E. coli* O111:NM, *E. coli* O126, *E. coli* O44, *E. coli* O124, *E. coli* O96:H9, *E. coli* O103:H2, *E. coli* O145:NM, and *E. coli* O145:H2 (Doughari et al. 2011).

Saline water samples collected from the seashore of the Mediterranean Sea at Alexandria City, Egypt, were examined for the presence of *E. coli* O157 on Sorbitol MacConky agar followed by antiserum. EHEC was recovered from 23 out of 48 seawater samples (48%) (El-Shenawy and El-Shenawy 2005). Furthermore, in 2009, El-Jakee and his group investigated water samples from different sources, including the River Nile, and found the presence of numerous virulence genes in the isolates (El-Jakee et al. 2009). El-Leithy et al. (2012) examined 175 water samples collected from River Nile, groundwater, agriculture drainage water, and the Mediterranean Sea. *E. coli* O157:H7 was detected using the culture method on HiCrome EC O157:H7 agar and six virulence genes including Shiga toxin 1 and 2 genes (*stx1* and *stx2*), intimin gene (*eae*), hemolysin gene (*hly*), somatic O157 antigen gene (*rfbE*), and flagellar antigen gene (*fliC*) of the detected *E. coli* O157:H7 isolates were investigated using multiplex PCR. It was found that the *stx2, eae,* and *rfbE* were the most frequent genes in *E. coli* strains (98%) followed by *stx1* (84%) and *fliC* (66%) while the *hly* gene was not detected. Also, in the same study, *E. coli* O157 was detected in 64% River Nile (Rossita Branch) samples using multiplex PCR and 100% in El-Rahawy Drain water samples (Table 1.4).

Table 1.4 Some African studies on the presence of pathogenic *E. coli* in surface water

Country	Water matrix	Organisms	Studying year	Detection methods	Counts	References
Egypt	Mediterranean Sea	EHEC	2002	Sorbitol MacConky agar, Antisera	48%	El-Shenawy and El-Shenawy (2005)
Egypt	River Nile, Drainage water	*E. coli* O157:H7	2009–2011	HiCrome EC O157:H7 agar, multiplex PCR	32%	El-Leithy et al. (2012)
Egypt	River Nile	*E. coli* O157	2009–2011	Multiplex PCR	32 out of 50 (64%)	El-Leithy et al. (2012)
Egypt	El-Rahawy Drain	*E. coli* O157	2009–2011	Multiplex PCR	100%	El-Leithy et al. (2012)
Egypt	Agriculture drains	Fecal coliforms	NM	MPN method	>90%	El-Jakee et al. (2009)
Egypt	Surface water	Pathogenic *E. coli*	NM	Culture followed by PCR	57.1% (*stx1*); 28.6% (stx2); 21.4% (*eae*) 28.6% (*fliCh7*)	El-Jakee et al. (2009)
South Africa	Rivers	*E. coli* O157:H7	1998	Culture followed by Antisera	4% containing virulence genes	Müller et al. (2001)

NM Not mentioned

The occurrence of pathogenic *E. coli* in El-Rahawy Drain originated from huge amounts of agricultural drainage water, and untreated wastewater from surrounding areas of El-Rahawy villages. The dam also received a large amount of partially treated wastewater from Zenin and Abou-Rawash wastewater treatment plants from the Greater Cairo that discharged directly into Rossita Branch by passing through El-Rahawy Drain (El-Leithy et al. 2012).

Occurrence in Wastewater

Wastewater treatment plants (WWTPs) have been identified as significant contributors to poor surface water quality due to the constant discharge of poorly treated or untreated wastewater into these water bodies. Thus, the presence of pathogenic bacteria, especially in the final effluent of these plants would indicate the quality of the water downstream from the WWTP discharge point. Thus, numerous studies have been conducted in some African countries to detect the presence of pathogenic organisms, including *E. coli* pathotypes.

For example, Grabow and his colleagues (2003) successfully isolated these pathotypes from sewage samples in South Africa using the enrichment Immuno-Magnetic Separation (IMS) selective agar. The samples included raw sewage and primary, secondary, and tertiary effluents from six wastewater treatment plants around Johannesburg, Gauteng province, South Africa. The authors reported the presence of all the pathotypes tested in the raw influent while ETEC and EAEC were detected in all treated wastewater samples after primary and secondary treatments. Also, Omar and Barnard (2010) used a multiplex PCR targeting eight virulence genes belonging to different *E. coli* pathotypes and reported the presence of EAEC (90%), ETEC (80%), and EIEC (10%) in the samples.

Other researchers have reported the presence of *E.coli* pathotypes in wastewater in Tunisia (Salem et al. 2011), South Africa (Adefisoye and Okoh 2016), and Senegal (Alpha et al. 2017).

1.4 *Salmonella* Spp.

1.4.1 *Brief History*

Theobald Smith first successfully isolated the organisms from pigs and the bacterium was named *Salmonella* after Theobald's partner, Daniel Elmer Salmon, who was a veterinary pathologist (Salmon and Smith 1886). In Africa, invasive *Salmonella enterica* infections appear to have been widespread in 1955 (Watson 1955). A further study on the distribution of the typhoid bacteria was carried out by Crocker (1957). In many countries in sub-Saharan Africa, *Salmonella* was sporadic and a correlation between malaria, HIV, and typhoid was suggested in 1929 and confirmed in 1987 (Giglioli 1929; Mabey et al. 1987; Reddy et al. 2010). However, the first cases of invasive non-typhoidal salmonella together with the report of AIDS occurred in Africa in 1984 (Feasey et al. 2012). Today, *Salmonella* is a primary etiologic agent of invasive bacterial febrile illness throughout sub-Saharan Africa with a high prevalence mostly observed in children younger than 15 years (Marks et al. 2017).

1.4.2 Microbiological Characteristics

The genus *Salmonella* is in the family *Enterobacteriaceae*, which consists of rod-shaped, Gram-negative bacteria measuring approximately 2–5 μm in length and 0.7–1.5 μm in diameter (Costa et al. 2008). Salmonella is nonencapsulated, non-spore-forming, and mostly motile (Ryan and Ray 2004). Salmonellae, *S. arizonae*, are lactose, sucrose, and malonate non-fermenters (Andrews et al. 1995). The organisms can grow in wide pH (4.05–9.0) and temperature (5.3–5.3 °C) depending on the surrounding conditions (Chung and Goepfert 1970; Jay 1996). These bacteria are facultative anaerobes that have the potential of producing hydrogen sulfide (Fabrega and Vila 2013).

Salmonella species have been divided into different groups based on two antigenic features, the somatic (O) and flagellar (H) antigens, and these together include more than 2500 serotypes (Agbaje et al. 2011). *Salmonella* is divided into two species according to the Center for Disease Control and Prevention (CDC), that is, *S. enterica* and *S. bongori* (Popoff and Minor 1997). The nomenclature of *Salmonella* serotypes is not italicized, and the first letter is capitalized such as *Salmonella enterica* subspecies *enterica* I serotypes Enteritidis, Typhimurium, Typhi, or Choleraesuis (Brenner et al. 2000).

1.4.3 Virulence Factors and Antibiotic Resistance

Salmonella Typhi is considered as a human pathogen and the etiologic agent of typhoid fever that is a life-threatening disease. The WHO (World Health Organization) (2018) estimated that between 11,000,000 and 21,000,000 cases of typhoid fever are recorded worldwide every year, with an associated 128,000 to 161,000 deaths.

It has been reported that the virulence factors contributing to pathogenicity in most enteric bacterial pathogens like *E. coli*, *Yersinia*, and *Shigella* spp. are mostly plasmid encoded. On the other hand, virulence plasmids contribution to pathogenesis is less significant in *Salmonella* than in the other bacterial pathogens. However, some *Salmonella* serovars contain virulence plasmids (Bäumler et al. 1998). Salmonellae carry some virulence genes such as *viz, sef, pef, spv,* or *inv* that are involved in adhesion to and invasion of the host cells (Clouthier et al. 1993; Bäumler et al. 1996). Others genes are linked to the survival of the pathogens in the host system, e.g., *mgtC*

(Blanc-Potard and Groisman 1997), while some genes such as the *sop, stn,* and/or *pip A, B,* and *D* play a critical role in pathogenicity processes (Wallis and Galyov 2000). *Salmonella* Typhi is the typical model for *Salmonella* pathogenesis, and causes typhoid fever through secretion of the typhoid toxin; secretion only occurs when the bacteria are within mammalian cells (Chang et al. 2016). The typhoid toxin belongs to a family called AB5 subunits and is a remarkable exotoxin produced by all *S.* Typhi, Paratyphi A, and B, but is uncommon in *S.* Paratyphi C (Chang et al. 2016).

In several studies that have been carried out in African countries, *Salmonella* species showed resistance to different groups of antibiotics. In a study conducted in Nigeria by Oluyege et al. (2009), 13.3% of *Salmonella* strains isolated from wells, streams, and borehole water were multiresistant with resistance recorded against antibiotics like tetracycline, amoxicillin, cotrimoxazole, nitrofurantoin, gentamicin, nalidixic acid, and ofloxacin. In another study in Nigeria, 24 out of the 54 *Salmonella* spp. that were isolated from different water sources were susceptible to ciprofloxacin while all species were resistant to penicillin (Stella et al. 2018). Similarly, *Salmonella* isolates recovered from well water samples in Ghana showed high resistance to many antibiotics still commonly used within the country. The authors, however, showed that the majority of the isolates were susceptible to antibiotics investigated (Dekker et al. 2015). Antibiotic-resistant *Salmonella* spp. have been reported in many other African countries like Cameroon (Moctar et al. 2019), Egypt (Elkenany et al. 2019), South Africa (Ekwanzala et al. 2018), and Gabon (Ehrhardt et al. 2017).

1.4.4 Detection Methods

1.4.4.1 Culture-Based Methods

Most developing countries, especially in Africa, depended on culture methods for detecting *Salmonella*. The most frequently used culture media for *Salmonella* detection include Salmonella-Shigella (SS) agar, bismuth sulfite agar, xylose lysine desoxycholate (XLD) agar, and/or chromogenic selective salmonella agar followed by morphological and biochemical confirmation and serotyping (Akinyemi et al. 2006; Momba et al. 2006; El-Taweel et al. 2010; Traoré et al. 2015; Maysa and Abd-Elall 2015) (Fig. 1.3). For example, Momba et al. (2006) isolated *Salmonella* from drinking water samples using XLD agar followed by confirmation of suspected isolates using biochemical tests. In another study, the organism was isolated from water using a chromogenic selective *Salmonella* agar following a pre-enrichment in Selenite F; confirmation was done using a latex agglutination test and the API 20E (Dekker et al. 2015). Also, a recent study used a pre-enrichment step in buffered peptone water and subcultured into both Tetrathionate broth and Rappaport-Vassiliadis Soy Peptone (RVS) broth followed by streaking onto two selective agar, namely, Chromogenic Salmonella agar and Miller Mallinson agar, and confirmed using biochemical and serological tests using polyvalent antiserum (Stella et al. 2018).

1.4.4.2 Molecular-Based Methods

Although molecular methods like PCR have been used to detect Salmonella in water samples in Africa, this has almost always been used for the confirmation of presumptive isolates obtained from culture media. For example, *Salmonella* species were detected in ground and surface water samples in South Africa, using culture method on XLD media, and all suspected *S.* Typhimurium isolates were negative using PCR (Momba et al. 2006). However, another study conducted in South Africa investigated the presence of Salmonella in river water and riverbed sediments by targeting the *invA* gene directly from the samples using real-time PCR, following pre-enrichment in peptone water, but without prior isolation on solid media (Abia et al. 2016b). Pulsed-field gel electrophoresis (PFGE) has also been used to establish the clonal relatedness of *Salmonella* isolated from well water (Dekker et al. 2015). PCR has also been used to characterize *Salmonellae* isolated from African waters. For example, a study conducted in Egypt used molecular methods for the identification of *Salmonella* from irrigation drain water samples and further identification of specific virulence genes in the isolates (Maysa and Abd-Elall 2015). Other studies that used molecular methods for the identification and characterization of *Salmonella* have been conducted in many other countries. Some of these were conducted in Moroccan coastal waters (Setti et al. 2009), South African river water and riverbed sediments (Ekwanzala et al. 2017), Tanzanian water sources (Kweka 2013), and irrigation water in Nigeria (Abakpa et al. 2015).

1.4.5 Occurrence in Different African Water Sources

Typhoid fever and salmonellosis might be caused via direct consumption of untreated contaminated water or poorly cooked or raw food (Eng et al. 2015). Intentional or accidental exposure to such contaminated water sources could lead to the risk of infection from salmonellae (Abia et al. 2016a). This is particularly crucial for most African countries where access to safe water is limited or absent and contaminated water sources serve as alternatives, and at times the only available water source for personal and household use. Despite this, the prevalence of *Salmonella* in surface and drinking water has not been widely investigated in many African countries. The presence of this pathogen in fresh surface water environments is limited although surveys involving drinking water are more common, a reflection of poor water quality in these countries. The lack of safe drinking water coupled with poor sanitation and hygiene have therefore led to the increased incidence and spread of typhoid fever in many of these countries (Akinyemi et al. 2006; Momba et al. 2006; Oluyege et al. 2009).

1.4.5.1 Occurrence in Drinking Water

In some countries in Africa, borehole, rain, and groundwater sources are often used as a drinking water source and mostly consumed without any treatment. For example, numerous reports of the presence of *Salmonella* in drinking water in South Africa and Nigeria have been published. In Nigeria, Akinyemi et al. (2006) investigated 180 water samples from wells and taps sampled from 18 different points and found *S.* Typhi in three of the well water samples. In a separate study conducted by Oguntoke et al. (2009) who collected water samples from rain, wells, and boreholes in Ibadan City, the authors found that 23.6% of well water samples were positive for *S.* Typhi (Table 1.5). In a recent study, *Salmonella* species were the most prevalent in reservoir (35.0%) and borehole (7.45%) water samples collected from Nnokwa, Alor, Nnobi, Abatete, and Oraukwu areas in Nigeria (Stella et al. 2018). The authors reported the isolation of *S.* Typhi, *S.* Paratyphi A, *S.* Paratyphi C, and *S.* Enterica, with *S.* Typhi being the most isolated species.

Table 1.5 The detection of *Salmonella* spp. in drinking water

Country	Water matrix	Organisms	Year	Detection methods	Counts	References
Burkina Faso	Tap water	*Salmonella enterica*	2008–2010	Culture and serotype	Negative	Traoré et al. (2015)
Burkina Faso	Well water	*Salmonella enterica*	2008–2010	Culture and serotype	2%	Traoré et al. (2015)
Burkina Faso	Reservoir water	*Salmonella enterica*	2008–2010	Culture and serotype	15–20%	Traoré et al. (2015)
Burkina Faso	Channels water	*Salmonella enterica*	2008–2010	Culture and serotype	20–31%	Traoré et al. (2015)
Ghana	Dug well water	*Salmonella* spp.	2009–2010	Culture and API	6.5% positive	Dekker et al. (2015)
Nigeria	Well water	*Salmonella* Typhi	NM	SS Agar	3 out of 18 samples	Akinyemi et al. (2006)
Nigeria	Well water	*Salmonella* Typhi	2005	NM	23% positive	Oguntoke et al. (2009)
Nigeria	Reservoir water	*Salmonella* spp.	2013	Culture and biochemical tests	35.0% Positive	Stella et al. (2018)
Nigeria	Boreholes	*Salmonella* spp.	2013	Culture and biochemical tests	7.45% positive	Stella et al. (2018)
South Africa	Boreholes and groundwater	*Salmonella* spp.	2001–2002	XLD agar, biochemical tests	100%	Momba et al. (2006)
South Africa	Boreholes and groundwater	*Salmonella* Typhimurium	2001–2002	XLD agar and PCR	Negative	Momba et al. (2006)

NM Not mentioned

In a study in South Africa, ground and surface water sources were tested for 4 months, and *Salmonella* was present in two of the borehole water samples used as primary drinking water sources (Momba et al. 2006) (Table 1.5). In Ghana, Dekker et al. (2015) collected 398 well water samples, and their results demonstrated that 26 samples (6.5%) were positive for *Salmonella* spp. with serological testing revealing the presence of different serovars including serovars *S.* Colindale, *S.* Stanleyville, *S.* Duisburg, *S.* Rubislaw, *S.* Saabruecken, *S.* Nima, and *S.* Give.

Similarly, Traoré et al. (2015) investigated 218 water samples for *Salmonella* presence from different sources in Ouagadougou, Burkina Faso, including taps, wells, channels, and reservoirs. Their results indicated *Salmonella* was absent in all tap water and was noticed in only one well sample. However, *Salmonella* was commonly detected in the samples from the reservoir and channel.

1.4.5.2 Occurrence in Surface Water

Salmonella species reach to surface water via surface runoff such as during heavy rains and when these water bodies overflow their banks. Wild animals also play an important role in water contamination as the organism has been isolated in such animals (Skov et al. 2008) and the relationship between the wildlife and human isolates has also been established (Smith et al. 2014). In Africa, aquatic environments are considered natural reservoirs and play a crucial role in transmitting salmonellae when used for drinking (Momba et al. 2006; Maysa and Abd-Elall 2015; Stella et al. 2018). It has also been found that *S.* Typhimurium is able to survive in surface water (River Nile) for more than 3 months (Ibrahim et al. 2019). Similarly, Abia et al. (2016d) demonstrated using laboratory experiments that the organism could survive in water and sediments of the Apies River in South Africa, for over 30 days.

In a study conducted on drains from farm settings in Egypt, the authors reported the presence of *Salmonella* in 18.3% of the agricultural drains, reporting the isolation of *S.* Typhimurium, *S.* Enteritidis, and *S.* Virchow (Osman et al. 2011). Also, in the same study *Salmonella* was detected in 7.5% of irrigation canals in which *S.* Typhimurium and *S.* Senftenberg were confirmed using PCR (Maysa and Abd-Elall 2015). Moreover, El-Taweel et al. (2010) analyzed 60 samples collected from the River Nile at Cairo and 14 water samples collected from the Rossita Branch. In this study, *Salmonella* was detected using two conventional methods, and presumptive isolates were confirmed using PCR. The results showed that 83.3% and 98.3% of the water samples were positive for *Salmonella* using the MPN and membrane filtration methods, respectively. Similarly, in a recent study in Nigeria, *Salmonella* species were the most prevalent in surface water by 42.9%, and 16 of these were confirmed as being *S.* Typhi, *S.* Paratyphi A, *S.* Paratyphi C, or *S.* Enterica (Stella et al. 2018). These studies, the methods used, and the results obtained, are summarized in Table 1.6.

Table 1.6 The occurrence of *Salmonella* spp. in surface water

Country	Water matrix	Organisms	Year	Detection methods	Counts	References
Egypt	Agriculture drains	*Salmonella* spp.	2014	Culture, biochemical and serological tests	18.3% positive	Maysa and Abd-Elall (2015)
Egypt	Irrigation drains	*Salmonella* spp.	2014	Culture, biochemical and serological tests	7.5% positive	Maysa and Abd-Elall (2015)
Egypt	Wastewater effluent and River Nile	*Salmonella* spp.	NM	Culture and biochemical tests	10.6% positive	Osman et al. (2011)
Egypt	River Nile	*Salmonella* spp.	2005–2008	MPN and MF methods	83.3–98.3%	El-Taweel et al. (2010)
Egypt	River Nile (tributary)	*Salmonella* spp.	2005–2008	MPN and MF methods	100%	El-Taweel et al. (2010)
Nigeria	Surface water	*Salmonella* spp.	2013	Culture and biochemical tests	42.9% positive	Stella et al. (2018)
South Africa	Surface water	*Salmonella* Typhimurium	2001–2002	Culture method	100% positive	Momba et al. (2006)
South Africa	Surface water	*Salmonella* spp.	2016	Real-time PCR	23.5% positive	Abia et al. (2016b)
Zimbabwe	Surface water	*Salmonella* spp.	2007	Culture and biochemical tests	NM	Zvidzai et al. (2007)

NM Not mentioned

1.4.5.3   Occurrence in Wastewater

Typhoid patients are carriers of *Salmonella* and can shed these bacterial pathogens in their feces over long periods. Thus, considerable counts of *Salmonella* species are frequently reported in raw wastewater. It has been reported that the organisms may even be resistant to conventional wastewater treatment processes and find their way into the final effluent of these plants (Toze et al. 2004). For example, the effluents of two wastewater treatment plants in Durban, South Africa, were investigated for *Salmonella*. The authors reported counts ranging between 0 and 4.14 logs CFU/mL in both the wastewater treatment plants. Also, an overall high percentage of the 200 confirmed *Salmonella* isolates possessed the *spiC, misL, orfL,* and *pipD* genes (Odjadjare and Olaniran 2015). In another study, the prevalence of *Salmonella* species in four sewage treatment plants in Gauteng Province, South Africa, was investigated, and the findings revealed significantly high *Salmonella* Typhimurium counts in all the treatment plants (Dungeni et al. 2010). Similarly, a study in

Al-Sharqiya Governorate, Egypt, reported the presence of *Salmonella* in wastewater after analyzing samples from untreated, aeration, oxidation, and anaerobically treated wastewater for the presence of *Salmonella* (Mahgoub et al. 2016). Other examples of studies that have investigated the presence of *Salmonella* in Egyptian wastewater samples include Zaki et al. (2009) and El-Lathy et al. (2009)

The occurrences of *Salmonella* in the final effluent and all wastewater treatment steps of the Agadir wastewater treatment plant, Morocco, were examined and the researchers reported the presence of *Salmonella* in samples from all the treatment steps and further reported a high diversity in the identified serotypes (El Boulani et al. 2016). Table 1.7 summarizes the findings of some of the studies on the presence of *Salmonella* in African wastewater.

Table 1.7 The occurrence of *Salmonella* spp. in wastewater

Country	Water matrix	Organisms	Year	Detection methods	Counts	References
Egypt	Inlet of oxidation pond of El-Sadat City	*Salmonella* spp.	2005–2008	MPN and Surface plate methods	2.8×10^4 MPN/100 mL and 9.6×10^4 CFU/100 mL	El-Lathy et al. (2009)
Egypt	The outlet of oxidation pond of El-Sadat City	*Salmonella* spp.	2005–2008	MPN and Surface plate methods	1.1×10^2 MPN/100 mL and 3.3×10^2 CFU/100 mL	El-Lathy et al. (2009)
Egypt	Inlet of Zenin wastewater treatment plant	*Salmonella* spp.	2005–2008	MPN and Surface plate methods	4.0×10^4 MPN/100 mL and 1.5×10^5 CFU/100 mL	El-Lathy et al. (2009)
Egypt	The outlet of the Zenin wastewater treatment plant	*Salmonella* spp.	2005–2008	MPN and Surface plate methods	1.0×10^2 MPN/100 mL and 3.3×10^2 CFU/100 mL	El-Lathy et al. (2009)
Morocco	Agadir wastewater treatment plant	*Salmonella* spp.	NM	Culture, API and molecular methods	19 Salmonella serotype	El Boulani et al. (2016)
South Africa	Wastewater treatment plant effluent	*Salmonella* spp.	2012–2013	Culture followed biochemical and PCR	$0–10^4$ CFU/mL	Odjadjare and Olaniran (2015)
South Africa	Wastewater	*S.* Typhimurium	2007–2008	Culture and API confirmation	0–88.2%	Dungeni et al. (2010)

NM Not mentioned

1.5 Conclusion

Pathogenic *E. coli* and *Salmonella* species are present in water sources in many African countries. The presence of these organisms in wastewater represent a menace for the quality of the receiving water bodies. However, advanced scientific equipment are strongly needed for the effective detection and characterization of these pathogens in the different African water sources. Also, many African countries, especially in the sub-Saharan region, lack centralized sewage systems. Thus, reports on these pathogens in wastewater in such countries will be limited or completely absent. This could have adverse effects on effective monitoring and reporting of the actual prevalence of these pathogens in these countries, with possible adverse consequences to public health.

References

Abakpa GO, Umoh VJ, Ameh JB et al (2015) Diversity and antimicrobial resistance of Salmonella enterica isolated from fresh produce and environmental samples. Environ Nanotechnol Monit Manag 3:38–46

Abia ALK, Schaefer L, Ubomba-Jaswa E, Le Roux W (2017) Abundance of pathogenic escherichia coli virulence-associated genes in well and borehole water used for domestic purposes in a peri-urban community of South Africa. Int J Environ Res Public Health doi: https://doi.org/10.3390/ijerph14030320

Abia ALK, Ubomba-Jaswa E, du Preez M, Momba MNB (2015) Riverbed sediments in the Apies River, South Africa: recommending the use of both Clostridium perfringens and Escherichia coli as indicators of faecal pollution. J Soils Sediments 15:2412–2424

Abia ALK, Ubomba-Jaswa E, Genthe B, Momba MNB (2016a) Quantitative microbial risk assessment (QMRA) shows increased public health risk associated with exposure to river water under conditions of riverbed sediment resuspension. Sci Total Environ. https://doi.org/10.1016/j.scitotenv.2016.05.155

Abia ALK, Ubomba-Jaswa E, Momba MNB (2016b) Prevalence of pathogenic microorganisms and their correlation with the abundance of indicator organisms in riverbed sediments. Int J Environ Sci Technol 13:2905–2916

Abia ALK, Ubomba-Jaswa E, Momba MNB (2016c) Occurrence of diarrhoeagenic Escherichia coli virulence genes in water and bed sediments of a river used by communities in Gauteng, South Africa. Environ Sci Pollut Res 23:15665–15674

Abia ALK, Ubomba-Jaswa E, Momba MNB (2016d) Competitive Survival of Escherichia coli, Vibrio cholerae, Salmonella typhimurium and Shigella dysenteriae in Riverbed Sediments. Microb Ecol 72:881–889

Abong'o BO, Momba MNB (2008) Prevalence and potential link between E. coli O157:H7 isolated from drinking water, meat and vegetables and stools of diarrhoeic confirmed and non-confirmed HIV/AIDS patients in the Amathole District – South Africa, J Appl Microbiol. 105:424–431

Abong'o BO, Momba MNB, Mwambakane JN, Mwambakana JN (2008) Prevalence of Escherichia coli O157:H7 among diarrhoeic HIV/AIDS patients in Eastern Cape Province-South Africa. Pak J Biol Sci 11(8):1066–1075

Adefisoye MA, Okoh AI (2016) Identification and antimicrobial resistance prevalence of pathogenic Escherichia coli strains from treated wastewater effluents in Eastern Cape, South Africa. Microbiologyopen 5:143–151African Union (2017) Hamburg Model United Nations. www. hammun.de. Nov. 30 - Dec. 3, 2017

Agbaje M, Begum RH, Oyekunle MA et al (2011) Evolution of *Salmonella* nomenclature: a critical note. Folia Microbiol 56:497–503

Akinyemi KO, Oyefolu AOB, Adewale SOB et al (2006) Bacterial pathogens associated with tap and well waters in Lagos, Nigeria. East Cent Afr J Surg 11(1):110–117

Alpha AD, Delphine B, Fatou TL et al (2017) Prevalence of pathogenic and antibiotics resistant Escherichia coli from effluents of a slaughterhouse and a municipal wastewater treatment plant in Dakar. Afr J Microbiol Res 11:1035–1042

Andrews WH, June GA, Sherrod PS (1995) *Salmonella.* In: FDA bacteriological analytical manual, 8th ed, 5.01–5.20APHA (American Public health Association) (2012) Standard methods for the examination of water and wastewater, 22nd edn. APHA (American Public health Association), Washington, D.C.

Bartram J, Cotruvo J, Exner M et al (2003) Heterotrophic plate counts and drinking-water Safety. Published on behalf of (WHO) by *IWA* Publishing, ISBN: 92 4 156226 9, UK.

Bäumler AJ, Tsolis RM, Bowe FA et al (1996) The peffimbrial operon of *Salmonella typhimurium* mediates adhesion to murine small intestine and is necessary for fluid accumulation in the infant mouse. Infect Immun 64(1):61–68

Bäumler AJ, Tsolis RM, Ficht TA, Adams LG (1998) Evolution of Host Adaptation in *Salmonella enterica.* Infect Immun 66(10):4579–4587

Blanc-Potard AB, Groisman EA (1997) The Salmonella *selC* locus contains a pathogenicity island mediating intramacrophage survival. EMBO J 16(17):5376–5385

Bray J, Beavan TED (1948) Slide agglutination of Bacterium coli var. neapolitanum in summer diarrhæa. J Pathol Bacteriol 60:395–401

Brenner FW, Villar RG, Angulo FJ et al (2000) Salmonella Nomenclature. Guest Commentary. J Clin Microbiol 38(7):2465–2467

Chang S-J, Song J, Galán JE (2016) Receptor-Mediated Sorting of Typhoid Toxin during Its Export from Salmonella Typhi-Infected Cells. Cell Host Microbe 20:682–689

Chapman PA, Siddons CA, Zadik PM, Jewes L (1991) An improved selective medium for the isolation of Escherichia coli O157. J Med Microbiol 35:107–110

Chigor VN, Umoh VJ, Smith SI et al (2010) Multidrug resistance and plasmid patterns of *Escherichia coli* O157 and other *E. coli* isolated from diarrhoeal stools and surface waters from some selected sources in Zaria, Nigeria. Int J Environ Res Public Health 7(10):3831–3841

Chung KC, Goepfert JM (1970) Growth of Salmonella at low pH. J Food Sci 35:326–328

CIWA (Cooperation in International Water in Africa) (2016). www.worldbank.org. Retrieved 13-11-2016

Clouthier SC, Müller KH, Doran JL et al (1993) Characterization of three fimbrial genes, sefABC, of *Salmonella enteritidis.* J Bacteriol 175(9):2523–2533

Costa RA, Cristiane F, De Carvalho T et al (2008) Antibiotic Resistance in Salmonella: A Risk for Tropical Aquaculture. In: Kumar Y (ed) Salmonella - A Diversified Superbug. Intech Open, London

Crocker CG (1957) Distribution of types of typhoid bacteria over Africa. S Afr Med J 31:169–172

Cross AS (2008) What is a virulence factor? Crit Care 12:196. https://doi.org/10.1186/cc7127

De SN, Bhattacharya K, Sarkar JK (1956) A study of the pathogenicity of strains ofbacterium coli from acute and chronic enteritis. J Pathol Bacteriol 71:201–209

Deborah Chen H, Frankel G, Chen DH et al (2005) Enteropathogenic Escherichia coli: unravelling pathogenesis. FEMS Microbiol Rev 29:83–98

DebRoy C, Fratamico PM, Yan X et al (2016) Comparison of O-antigen gene clusters of all O-serogroups of Escherichia coli and proposal for adopting a new nomenclature for O-typing. PLoS One 11:1–13. https://doi.org/10.1371/journal.pone.0147434

Dekker DM, Krumkamp R, Sarpong N et al (2015) Drinking water from dug wells in rural Ghana-*Salmonella* contamination, environmental factors, and genotypes. Int J Environ Res Public Health 12:3535–3546

Don J, Brenner J, Farmer J (1986) Family I. Enterobacteriaceae. Rahn, Nom. Fam. Comm. Ewing, Farmer, and Brenner. Bergey's Manual of Systemic Bacteriology, 2nd edn. Williams & Wilkins, Baltimore, MD, p 587

Doughari HJ, Ndakidemi PA, Human IS, Benade S (2011) Virulence factors and antibiotic susceptibility among verotoxic non-O157: H7 Escherichia coli isolates obtained from water and wastewater samples in Cape Town. South Africa. Afr J Biotechnol 10(64):14160–14168

Dungeni M, van der Merwe RR, Momba MM (2010) Abundance of pathogenic bacteria and viral indicators in chlorinated effluents produced by four wastewater treatment plants in the Gauteng Province, South Africa. Water SA 36(5):607–614

DuPont HL, Formal SB, Hornick RB et al (1971) Pathogenesis of Escherichia coli Diarrhea. N Engl J Med 285:1–9. https://doi.org/10.1056/NEJM197107012850101

Edmunds WM (2008) Groundwater in Africa – palaeowater, climate change and modern recharge. In: SMA A, AM MD (eds) Applied groundwater studies in Africa (IAH selected papers on hydrogeology vol 13). CRC Press, Balkema Leiden, pp 305–336

Effler E, Isaacson M, Arntzen L et al (2001) Factors contributing to the emergence of *Escherichia coli* 0157 in Africa. Emerg Infect Dis 7:812–819

Egyptian Standard (2007) Ministry of Health: Minister's Office, Egyptian Standards for potable water, Dissection No. (458)

Ehrhardt J, Alabi AS, Kremsner PG et al (2017) Bacterial contamination of water samples in Gabon, 2013. J Microbiol Immunol Infect 50:718–722

Ekwanzala MD, Abia ALK, Keshri J, Momba MNB (2017) Genetic characterization of Salmonella and Shigella spp. Isolates recovered from water and riverbed sediment of the Apies River, South Africa. Water SA 43:387–397

Ekwanzala MD, Dewar JB, Kamika I, Momba MNB (2018) Systematic review in South Africa reveals antibiotic resistance genes shared between clinical and environmental settings. Infect Drug Resist 11:1907–1920

El Boulani A, Mimouni R, Chaouqy NE et al (2016) Characterization and antibiotic susceptibility of *Salmonella* strains isolated from wastewater treated by infiltration percolation process. Moroccan J Biol 13:44–51

EL-Alfy SM, Ahmed SF, Selim SA et al (2013) Prevalence and characterization of Shiga toxin O157 and non-O157 enterohemorrhagic Escherichia coli isolated from different sources in Ismailia, Egypt. Afr J Microbiol Res 7(21):2637–2645

El-Jakee J, El-moussa K, Mohamed F, Mohamed G (2009) Using molecular techniques for characterization of E. coli isolated from water sources in Egypt. Glob. Veterinaria 3:354–362

Elkenany R, Elsayed MM, Zakaria AI et al (2019) Antimicrobial resistance profiles and virulence genotyping of Salmonella enterica serovars recovered from broiler chickens and chicken carcasses in Egypt. BMC Vet Res 15:1–9. https://doi.org/10.1186/s12917-019-1867-z

El-Lathy MA, El-Taweel GE, El-Sonosy WM et al (2009) Determination of pathogenic bacteria in wastewater using conventional and PCR techniques. Environ Biotechnol 5(2):73–80

El-Leithy MA, El-Shatoury EH, El-Senousy WM et al (2012) Detection of six E. coli O157 virulence genes in water samples using multiplex PCR. Egypt. J Microbiol 47:171–188

El-Leithy MA, Hemdan BA, El-Shatoury EH et al (2016) Prevalence of *Legionella* spp. and *Helicobacter pylori* in different water resources in Egypt. Egypt. J Environ Res 4:1–12

El-Leithy MAR (2013) Molecular detection and characterization of some pathogenic bacteria in water. PhD thesis. Microbiology Department, Ain Shams University, Cairo, Egypt

El-Shatoury EH, El-Leithy MA, Abou-Zeid MA et al (2015) Antibiotic susceptibility of Shiga Toxin-Producing E. *coli* O157:H7 isolated from different water sources. Open Conf Proc J 6:30–34

El-Shenawy MA, El-Shenawy M (2005) Enterohaemorrhagic *Escherichia coli* O157 in coastal environment of Alexandria, Egypt. Microb Ecol Health Dis 17:103–106

El-Taweel GE, Moussa TAA, Samhan FA et al (2010) Nested PCR and Conventional Techniques for Detection of *Salmonella* spp. in River Nile Water, Egypt. Egypt J Microbiol 45:63–76

Eng SK, Pusparajah P, Ab Mutalib NS et al (2015) Salmonella: A review on pathogenesis, epidemiology and antibiotic resistance. Front Life Sci 8:284–293

EPA (2003) US Environmental Protection Agency Safe Drinking Water Act. EPA 816 – F – 03 –016

Eseherich T (1885) Die Darmbakterien des Neugeboreiien und Säuglings. Dtsch Medizinische Wochenschrift 11:740–741

Fabrega A, Vila J (2013) Salmonella enterica Serovar Typhimurium skills to succeed in the host: virulence and regulation. Clin Microbiol Rev 26:308–341

Feasey NA, Dougan G, Kingsley RA et al (2012) Invasive non-typhoidal salmonella disease: An emerging and neglected tropical disease in Africa. Lancet 379:2489–2499

Fleisher JM (1990) The effects of measurement error on previously reported mathematical relationships between indicator organism density and swimming-associated illness: a quantitative estimate of the resulting bias. Int J Epidemiol 19:1100–1106

Giglioli G (1929) Paratyphoid C, an endemic disease of British Guiana: a clinical and pathological outline, B. paratyphosum C as a pyogenic organism: case reports. Proc R Soc Med 23:165–177

Gomes TAT, Elias WP, Scaletsky ICA et al (2016) Diarrheagenic Escherichia coli. Braz J Microbiol 47:3–30

Grabow WOK, Müller EE, Ehlers MM, et al (2003) Occurrence of E. coli O157:H7 and other Pathogenic E. coli strains in Water sources intended for direct and Indirect human consumption. WRC Report No 1068/1/03. ISBN No: 1–86845–988-8

Grabow WOK, Neubrech TE, Holtzhausen CS, Jofre J (1995) Bacteroides fragilis and Escherichia coli bacteriophages: excretion by humans and animals. Water Sci Technol 31(5–6):223–230

Grey D, Sadoff CW (2007) Sink or swim? Water security for growth and development. Water Policy 9:545–571 Gruber JS, Ercumen A, Colford JM Jr (2014) Coliform bacteria as indicators of diarrheal risk in household drinking water: systematic review and metanalysis. PLoS One 9(9):e107429

Gyles CL (2007) Shiga toxin-producing Escherichia coli: An overview. J Anim Sci 85:45–62

Haller L, Pote´ J, Loizeau J-L, Wildi W (2009) Distribution and survival of faecal indicator bacteria in the sediments of the Bay of Vidy, Lake Geneva, Switzerland. Biol Indic 9:540–547

Hassard F, Andrews A, Jones DL et al (2017) Physicochemical factors influence the abundance and culturability of human enteric pathogens and fecal indicator organisms in estuarine water and sediment. Front Microbiol 8:1–18. https://doi.org/10.3389/fmicb.2017.01996

Hazen TH, Michalski J, Luo Q et al (2017) Comparative genomics and transcriptomics of Escherichia coli isolates carrying virulence factors of both enteropathogenic and enterotoxigenic E. coli. Sci Rep 7:1–17. https://doi.org/10.1038/s41598-017-03489-z

Hendricks CW (1978) Exceptions to the coliforms and the fecal coliforms tests. In: Berg G (ed) Indicators of viruses in water and food. Ann Arbor Sci, Michigan, p 99

Horman A, Rimhanen-Finne R, Maunula L et al (2004) Campylobacter spp., Giardia spp., Cryptosporidium spp., Noroviruses, and Indicator Organisms in Surface Water in Southwestern Finland, 2000-2001. Appl Environ Microbiol 70:87–95

Hunter PR, MacDonald AM, Carter RC (2010) Water supply and health. PLoS Med 7:e1000361I

brahim EME, El-Liethy MA, Abia ALK et al (2019) Survival of E. coli O157:H7, Salmonella Typhimurium, HAdV2 and MNV-1 in river water under dark conditions and varying storage temperatures. Sci Total Environ 648:1297–1304

Iweriebor BC, Iwu CJ, Obi LC et al (2015) Multiple antibiotic resistances among Shiga toxin producing Escherichia coli O157 in feces of dairy cattle farms in Eastern Cape of South Africa. BMC Microbiol 15(1):213

Jagals P, Barnard TG, Mokoena MM et al (2013) Pathogenic Escherichia coli in rural household container waters. Water Sci Technol 67(6):1230–1237

Jay JM (1996) Foodborne gastroenteritis caused by Salmonella and Shigella. In: Modern food microbiology, 5th edn. Chapman & Hall, New York, pp 507–526

Kaper JB, Nataro JP, Mobley HLT (2004) Pathogenic Escherichia coli. Nat Rev Microbiol 2:123–140

Kweka E (2013) Characterization of Salmonella species from water bodies in Dar-Es-Salaam city, Tanzania. J Health Biol Sci 1:16–20

Lebughe M, Phaku P, Niemann S et al (2017) The impact of the Staphylococcus aureus virulome on infection in a developing country: A cohort study. Front Microbiol 8:1662. https://doi.org/10.3389/fmicb.2017.01662

Leiter WL (1929) Water bacteriology: The Eijkman fermentation test as an aid in the detection of fecal organisms in water. J Lab Clin Med 15(2):194–195

Levine M, Nalin D, Hornick R et al (1978) Escherichia coli strains that cause diarrhœa but do not produce heat-labile or heat-stable enterotoxins and are non-invasive. Lancet 311:1119–1122

Logue CM, Doetkott C, Mangiamele P et al (2012) Genotypic and phenotypic traits that distinguish neonatal meningitis-associated Escherichia coli from fecal E. coli isolates of healthy human hosts. Appl Environ Microbiol 78:5824–5830

Luna GM, Vignaroli C, Rinaldi C et al (2010) Extraintestinal Escherichia coli carrying virulence genes in coastal marine sediments. Appl Environ Microbiol 76:5659–5668

Mabey DCW, Brown A, Greenwood BM (1987) Plasmodium falciparum malaria and Salmonella infections in Gambian children. J Infect Dis 155:1319–1321

MacDonald AM, Bonsor HC, Dochartaigh BÉÓ, Taylor RG (2012) Quantitative maps of groundwater resources in Africa. Environ Res Lett. https://doi.org/10.1088/1748-9326/7/2/024009

Mahgoub S, Samaras P, Abdelbasit H, Abdelfattah H (2016) Seasonal variation in microbiological and physicochemical characteristics of municipal wastewater in Al-Sharqiya province, Egypt (case study). Desalination Water Treat 57(5):2355–2364

Mainil J (2013) Escherichia coli virulence factors. Vet Immunol Immunopathol 152:2–12

Manafi M, Kremsmaier B (2001) Comparative evaluation of different chromogenic/flurogenic media for detecting Escherichia coli O157:H7 in food. Int Food Microbiol 71:257–262

Mansan-almeida R, Pereira AL, Giugliano LG (2013) Diffusely adherent Escherichia coli strains isolated from children and adults constitute two different populations. BMC Microbiol 13(1). https://doi.org/10.1186/1471-2180-13-22

Marks F, von Kalckreuth V, Aaby P et al (2017) Incidence of invasive salmonella disease in sub-Saharan Africa: a multicentre population-based surveillance study. Lancet Glob Health 5:e310–e323

Maysa AIA, Abd-Elall AMM (2015) Diversity and virulence-associated genes of Salmonella enterica serovars isolated from wastewater agricultural drains, leafy green producing farms, cattle and human along their courses. Revue Méd Vét 166(3–4):96–106

Meyer-Broseta S, Bastian SN, Arne PD et al (2001) Review of epidemiological surveys on the prevalence of contamination of healthy cattle with Escherichia coli serogroup O157:H7. Int J Hyg Environ Health 203:347–361

Moctar M, Mouiche M, Moffo F et al (2019) Antimicrobial resistance from a one health perspective in Cameroon: a systematic review and meta-analysis. BMC Public Health 19:1–20

Momba MN, Malakate VK, Theron J (2006) Abundance of pathogenic Escherichia coli, Salmonella Typhimurium and Vibrio cholerae in Nkonkobe drinking water sources. J Water Health 4(3):289–296

Momba MNB, Abong'o BO, Mwambakana JN (2008) Prevalence of enterohaemorrhagic Escherichia coli O157:H7 in drinking water and its predicted impact on diarrhoeic HIV/AIDS patients in the Amathole District, Eastern Cape Province, South Africa. Water SA 34(3):365–372

Mudrak B, Kuehn MJ (2010) Heat-labile enterotoxin: Beyond GM1 binding. Toxins (Basel) 2:1445–1470

Müller EE, Ehlers MM, Grabow WOK (2001) The occurrence of E. coli O157:H7 in South African water sources intended for direct and indirect human consumption. Water Res 35(13):3085–3088

Nataro JP, Kaper JB (1998) Diarrheagenic Escherichia coli. Clin Microbiol Rev 11:142–201

Nataro JP, Kaper JB, Robins-Browne R et al (1987) Patterns of adherence of diarrheagenic escherichia coli to HEp-2 cells. Pediatr Infect Dis J 6:829–831

Neter E, Wstphal O, Luderritz O et al (1955) Demonstration of antibodies against enteropathogenic Escherichia coli in sera of children of various ages. Pediatrics 16:801–808

Nontongana N, Sibanda T, Ngwenya E, Okoh AI (2014) Prevalence and antibiogram profiling of Escherichia coli pathotypes isolated from the kat river and the fort beaufort abstraction water. Int J Environ Res Public Health 11:8213–8227

Odjadjare EC, Olaniran AO (2015) Prevalence of antimicrobial resistant and virulent *Salmonella* spp. in treated effluent and receiving aquatic Milieu of wastewater treatment Plants in Durban, South Africa. Int J Environ Res Public Health 12:9692–9713

Odonkor ST, Joseph K, Ampofo JK (2013) Escherichia coli as an indicator of bacteriological quality of water: an overview. Microbiol Res 4(1). https://doi.org/10.4081/mr.2013.e2Oguntoke O, Aboderin OJ, Bankole AM (2009) Association of water-borne diseases morbidity pattern and water quality in parts of Ibadan City, Nigeria. Tanzania J Health Res 11(4):189–195

Okeke IN, Nataro JP (2001) Enteroaggregative Escherichia coli. Lancet Infect Dis 1:304–313

Oliveira M, Freirea D, Pedroso NM (2018) *Escherichia coli* is not a suitable fecal indicator to assess water fecal contamination by otters. Braz J Biol 78(1):155–159

Oluyege JO, Dada AC, Odeyemi AT (2009) Incidence of multiple antibiotic resistant Gram-negative bacteria isolated from surface and underground water sources in southwestern region of Nigeria. Water Sci Technol 59(10):1929–1936

Omar KB, Barnard TG (2010) The occurrence of pathogenic Escherichia coli in South African wastewater treatment plants as detected by multiplex PCR. Water SA 36:172–176

Omar KB, Barnard TG (2014) Detection of diarrhoeagenic Escherichia coli in clinical and environmental water sources in South Africa using single-step 11-gene m-PCR. World J Microbiol Biotechnol 30:2663–2671

Osman GA, Hassan HM, Kamel MM (2011) Resistance and sensitivity of some bacterial strains isolated from hospital wastewater and Nile water using chlorination and some antibiotics in Cairo (Egypt). J Am Sci 7:1033–1041

Pasqua M, Michelacci V, Di Martino ML et al (2017) The Intriguing Evolutionary Journey of Enteroinvasive E. coli (EIEC) toward Pathogenicity. Front Microbiol 8:2390. https://doi.org/10.3389/fmicb.2017.02390

Popoff MY, Le Minor L (1997) Antigenic formulas of the *Salmonella* serovars, 7th revision. World Health Organization Collaborating Centre for Reference and Research on *Salmonella*. Pasteur Institute, Paris, France

Prats J, Garcia-Armisen T, Larrea J, Servais P (2007) Comparison of culture-based methods to enumerate Escherichia coli in tropical and temperate freshwaters. Lett Appl Microbiol 46:243–248

Ranjbar R, Safarpoor Dehkordi F, Sakhaei Shahreza MH, Rahimi E (2018) Prevalence, identification of virulence factors, O-serogroups and antibiotic resistance properties of Shiga-toxin producing Escherichia coli strains isolated from raw milk and traditional dairy products. Antimicrob Resist Infect Control 7:1–11. https://doi.org/10.1186/s13756-018-0345-x

Reddy, EA, Shaw AV, Crump JA (2010) Community-acquired bloodstream infections in Africa: a systematic review and meta-analysis. Lancet infect dis 10:417–432

Riley LW, Remis RS, Helgerson SD et al (1983) Hemorrhagic Colitis associated with a rare Escherichia coli serotype. N Engl J Med 308:681–685

Robins-Browne RM, Holt KE, Ingle DJ et al (2016) Are Escherichia coli pathotypes still relevant in the era of whole-genome sequencing? Front Cell Infect Microbiol 6:1–9. https://doi.org/10.3389/fcimb.2016.00141

Ryan KJ, Ray CG (eds) (2004) Sherris medical microbiology, 4th edn, McGraw Hill. ISBN 0-8385-8529-9

SABS (2001) Specification: drinking water (SABS 241: 2001). South African Bureau of Standards. South Africa, Pretoria

Sahl JW, Sistrunk JR, Baby NI et al (2017) Insights into enterotoxigenic Escherichia coli diversity in Bangladesh utilizing genomic epidemiology. Sci Rep 7:1–12. https://doi.org/10.1038/s41598-017-03631-x

Salem IB, Ouardani I, Hassine M, Aouni M (2011) Bacteriological and physico-chemical assessment of wastewater in different region of Tunisia: impact on human health. BMC Res Notes 4:144. https://doi.org/10.1186/1756-0500-4-144

Salmon DE, Smith T (1886) Discovery of *Salmonella* Choleraesuis. The bacterium of swine-plague. Amer Monthly Microscopic J 7:204–205Sarowska J, Futoma-Koloch B, Jama-Kmiecik A et al (2019) Virulence factors, prevalence and potential transmission of extraintestinal pathogenic Escherichia coli isolated from different sources: recent reports. Gut Pathog 11:1–16. https://doi.org/10.1186/s13099-019-0290-0

Scanlon BR, Keese KE, Flint AF et al (2006) Lobal synthesis of groundwater recharge in semiarid and arid regions. Hydrol Process 20:3335–3370

Setti I, Rodriguez-Castro A, Pata MP et al (2009) Characteristics and dynamics of Salmonella contamination along the coast of agadir, Morocco. Appl Environ Microbiol 75:7700–7709

Skov MN, Madsen JJ, Rahbek C et al (2008) Transmission of Salmonella between wildlife and meat-production animals in Denmark. J Appl Microbiol 105:1558–1568

Smith AM, Ismail H, Henton MM, Keddy KH (2014) Similarities between salmonella enteritidis isolated from humans and captive wild animals in South Africa. J Infect Dev Ctries 8:1615–1619

SSMO (Sudanese Standards and Metrology Organization) (2002) Sudan online. Water guideline in Sudan. SSMO Publication, Khartoum, Sudan

Stella EI, Ifeoma EM, Ochiabuto OMTB et al (2018) Evaluation of *Salmonella* Species in water sources in two local government areas of Anambra State. Cohesive J Microbiol Infect Dis 1:1–9

Sunny CJ, Weiping C, Betty HO et al (2007) Microbial source tracking in a small southern California urban watershed indicates wild animals and growth as the source of fecal bacteria. Appl Microbiol Biotechnol 76:927–934

Thomas MS, Wigneshweraraj S (2014) Regulation of virulence gene expression. Virulence 5:832–834

Toze S, Hanna J, Smith T et al (2004) Determination of water quality improvements due to the artificial recharge of treated effluent. In: Steenvoorden J, Endreny T (eds) Wastewater re-use and groundwater quality. Wallingford, Int Assoc Hydrological Sciences

Trabulsi LR, Keller R, Gomes TAT (2002a) Typical and Atypical Enteropathogenic Escherichia coli. Emerg Infect Dis 8:508–513

Trabulsi LR, Keller R, Tardelli Gomes TA (2002b) Typical and atypical enteropathogenic Escherichia coli. Emerg Infect Dis 8:508–513

Traoré O, Nyholm O, Siitonen A et al (2015) Prevalence and diversity of Salmonella enterica in water, fish and lettuce in Ouagadougou, Burkina Faso. BMC Microbiol 15:151–158UNESCO (2016). The United Nations World Water Development Report 2016: Water and Jobs. UNESCO, Paris, p 2016. ISBN 978-92-3-100146-8

United Nations (2018) United Nations Sustainable Development Goal 6 Synthesis Report 2018 on Water and Sanitation. United Nations, New YorkViazis S, Diez-Gonzalez F (2011) Enterohemorrhagic Escherichia coli. The twentieth century's emerging foodborne pathogen: a review, 1st edn. Adv Agron. https://doi.org/10.1016/B978-0-12-387689-8.00006-0

Wallis TS, Galyov EE (2000) Molecular basis of Salmonella-induced enteritis. Mol Microbiol 36(5):997–1005

Watson KC (1955) Isolation of *Salmonella* Typhi from the bloodstream. J Lab Clin Med 46:128–134

Welch RA, Burland V, Plunkett G et al (2002) Extensive mosaic structure revealed by the complete genome sequence of uropathogenic Escherichia coli. Proc Natl Acad Sci 99:17020–17024

WHO (2008) Guidelines for drinking-water quality: incorporating the first and second addenda volume 1: recommendations, 3rd edn. World Health Organization, Geneva, Switzerland., ISBN-13: 9789241547611, p 688

WHO (World Health Organization) (2018) Typhoid vaccines: WHO position paper-March 2018. Wkly Epidemiol Rec 93(13):153–172

Zadik PM, Chapman PA, Siddons CA (1993) Use of tellurite for the selection of verocytotoxigenic Escherichia coli O157. J Med Microbiol 39:155–158

Zaki S, Abd-El-Haleem D, El-Helow E, Mustafa M (2009) Molecular and biochemical diagnosis of *Salmonella* in wastewater. J Appl Sci Environ Manage 13(2):83–92

Zvidzai C, Mukutirwa T, Mundembe R (2007) Microbial community analysis of drinking water sources from rural areas of Zimbabwe. Afr J Microbiol Res 1:100–103

Chapter 2
Emerging and Reemerging Bacterial Pathogens of Humans in Environmental and Hospital Settings

Ubani Esther K. Fono-Tamo, Martina Oyedi Chukwu, Eunice Ubomba-Jaswa, C. L. Obi, John Barr Dewar, and Akebe Luther King Abia

2.1 Introduction

A survey of human pathogens from the World Health Organization (WHO), the Centres of Disease Control and Prevention (CDC), and PubMed lists identified 1407 human pathogens of which 177 (13%) were recognized as emerging pathogens. Prevalence data, more than a decade ago, for the various groups of pathogens for total and emerging pathogens was already, respectively, 208 and 77 (44%) for viruses/prions, 538 and 54 (30%) for bacteria or rickettsia, 317 and 22 for fungi, and 57 and 14 for protozoa (Woolhouse and Gowtage-Sequeria 2005). Emerging pathogens can be zoonotic or non-zoonotic. However, about 60% of emerging pathogens are reported as being zoonotic (Jones et al. 2008; Cutler et al. 2010). Zoonotic pathogens are pathogens for which humans or other vertebrates are either the main hosts or occasional hosts. As such, these pathogens are easily transmitted between

Ubani Esther K. Fono-Tamo and Martina Oyedi Chukwu contributed equally to this work.

U. E. K. Fono-Tamo (✉) · M. O. Chukwu (✉) · J. B. Dewar
Department of Life and Consumer Science, College of Agriculture and Environmental Sciences, University of South Africa, Roodepoort, Gauteng, South Africa

E. Ubomba-Jaswa
Department of Biotechnology, University of Johannesburg, Doornfontein, Gauteng, South Africa

Water Research Commission, Lynnwood Bridge Office Park, Bloukrans Building, Lynnwood Manor, Pretoria, South Africa

C. L. Obi
Division of Academic Affairs, Sefako Makgatho Health Science University, Pretoria North, South Africa

A. L. K. Abia
Antimicrobial Research Unit, College of Health Sciences, University of KwaZulu-Natal, Durban, South Africa

© Springer Nature Switzerland AG 2020
A. L. K. Abia, G. R. Lanza (eds.), *Current Microbiological Research in Africa*,
https://doi.org/10.1007/978-3-030-35296-7_2

humans and vertebrate animals (Woolhouse and Gowtage-Sequeria 2005). These pathogens can be transmitted to humans via direct contact, scratches or bites, arthropod vectors, exposure to contaminated feces, feces-contaminated environmental sources and carcasses, or consumption of contaminated food and water (Chikeka and Dimler 2015). Water and food are notable environmental reservoirs and vehicles of transmission of zoonotic and non-zoonotic bacterial pathogens (Vouga and Greub 2016). Also, significant environmental and sociodemographic changes have resulted in increased human exposure to environmental pathogenic species and transmission between persons (Vouga and Greub 2016).

2.2   Definition of Emerging and Reemerging Pathogens

Emerging pathogens can be defined as organisms that were previously not recognized to cause any human disease or that are newly appearing. They also include pathogens that have existed but are rapidly increasing in prevalence in a population due to increased human exposure or pathogens that are expanding to new geographical regions (Hoogenboezem 2007). Reemerging pathogens are pathogens that were previously easily controlled by antibiotics and treatable but have now developed resistance and appear in epidemics (United Nations Environment Programme Global Environment Monitoring System/Water Programme (UNEP/GEMS) 2008; Vouga and Greub 2016; Nii-Trebi 2017).

2.3   Reasons for the Emergence/Reemergence of Pathogens

Many variables interplay in the emergence and reemergence of pathogens. These variables can be grouped into human/demographic, pathogen, socioeconomic, environmental, and scientific/technological factors (Table 2.1) (Woolhouse and Gowtage-Sequeria 2005; Louten 2016).

Emerging bacterial pathogens may have long been present in the environment but remained undetected due to inadequate diagnostic techniques or humans only recently became exposed to them. Thus, pathogen emergence can mainly be attributed to increased human exposure to natural reservoirs of bacterial pathogens, the emergence of more virulent bacterial strains with the potential to cause opportunistic infections, and development of improved and novel diagnostic techniques (Vouga and Greub 2016). Other factors that facilitate pathogen's emergence include improved disease surveillance, climate change, agricultural intensification and development, microbial adaptation, immigration, and travel, increase in natural reservoirs and vectors, development of biological weapons, breakdown of public health measures and mutation (Ohimian 2017). Also, of major concern is the emergence of resistance of infectious pathogens to many first-line antibiotics worldwide (WHO 2017).

Table 2.1 Factors that contribute to the emergence of pathogens

Factors	Variables	Mechanism	References
Human factors	Human demographics	Increase in high-risk populations, rapid population growth, population density, overcrowding	Jones et al. (2008); Louten (2016)
	Migration/transportation	Continuous mass travel allows infectious pathogens in isolated places to quickly become global threats, dispersal of pathogenic species facilitated by the transportation of food, and animals	Tulchinsky and Varavikova (2015)
	Urbanization	Urbanization increases the population of persons in cities and increases transmission of pathogens by bringing people in closer contact with one another. Unplanned and inadequately planned urbanization or development leads to the growth of peri-urban slums in cities	Sharma et al. (2003); Hoogenboezem (2007); La Rosa et al. (2012); Knobler et al. (2006)
	Globalization	Brings people together in trade and culture and fosters the transfer of pathogens	Louten (2016); Sellman and Pederson (2007); Morens and Fauci (2013)
	Human behavior, vulnerability, and intent to harm (bioterrorism)	Individual's level of fitness, nutrition, stress, deliberate or accidental release of pathogens into the environment, pregnancy, age, immune status, chronic sickness, increase and unprescribed use of antibiotics, drug abuse/addiction, deliberate human action directed at the development of biological weapons of destruction	Morens and Fauci (2013); Nii-Trebi (2017)
Socioeconomic factors	Poor sanitation	Littering of the environment, improper waste disposal	Woolhouse and Gowtage-Sequeria (2005)
	Poor population health	Access to health infrastructure, inadequate healthcare, availability of health services, level of healthcare providers' expertise, insufficient number of health personnel, lack of public health programs, malnutrition	Woolhouse and Gowtage-Sequeria (2005)
	Basic amenities	Lack of proper sewage disposal systems, access to safe or treated water, lack of or breakdown of public health systems, the decline in infrastructure	Sharma et al. (2003); Hoogenboezem (2007); La Rosa et al. (2007)

(continued)

Table 2.1 (continued)

Factors	Variables	Mechanism	References
Environmental factors	Climatic and weather change	Increased rainfall and temperature often correlate with the spread of diseases to areas where they were not previously found	Knobler et al. (2006); Louten (2016)
	Vectors/wildlife	Increased vector movement leads to increased exposure to vectors and reservoirs of pathogens, wildlife host species richness	Knobler et al. (2006)
	Pollution	Contamination of food sources and water supply	Woolhouse and Gowtage-Sequeria (2005)
Pathogen factors	Genetic composition	Type of nucleic acid (DNA or RNA)	Louten (2016)
	Genetic evolution	Antimicrobial resistance arising from exposure to sublethal doses of antibiotics or acquisition of resistance genes, increased virulence, mutation, recombination, selection, and deliberate manipulation result in nucleotide substitution rates which affect adaptation and rise of new strains/variants of known pathogens with improved pathogenic potential, persistence and spread to the human host	Sellman and Pederson (2007); Tulchinsky and Varavikova (2015); Woolhouse and Gowtage-Sequeria (2005); Cutler et al. (2010)
Scientific and Technological factors	Modernization in agricultural practices	Excessive use of antibiotics and antiparasitic medicine in agriculture and livestock farming, increase in insecticides use	Woolhouse and Gowtage-Sequeria (2005)
	Development and industry	Changes in land use and industrialization lead to modification of the environment by deforestation or reforestation, development of dams which change the ecosystems and their relations to humans, e.g., dams increase arthropod populations	Woolhouse and Gowtage-Sequeria (2005); Vouga and Greub (2016); Morens and Fauci (2013)
	Improved diagnostic tools	Recent advances in microbiological diagnosis have increased the number of identifiable bacteria	Vouga and Greub (2016)

2.4 The Global Significance of Emerging Pathogen

Despite the tremendous progress that has been recorded in the prevention and control of infectious diseases, these diseases remain among the top causes of deaths globally. The WHO estimates that about one-third of annual deaths in the world results from infectious diseases, of which 70% are caused by emerging pathogens. This represents a significant burden on human health and the global economy. Measures that calculate mortality rates, such as public health burden, can be estimated by disability-adjusted life years (DALY). This was developed in the 1990s as a means of weighing the differences in general health and life expectancy of different countries. DALY is denoted as the aggregate of years lost ascribed to illness, disability, or premature death, and one DALY can be considered as one lost year of a healthy life. The sum of all DALYs across a population represents the gap between the present health condition and a perfectly healthy population (Devleesschauwer et al. 2014). For instance, the burden of foodborne diseases including campylobacteriosis is substantial: annually one in every ten persons fall ill resulting to an average of 500 million cases and 33 million of healthy life years are lost (WHO 2018e). The burden of *Campylobacter* is estimated at 7.5 million DALY or 8.4% of the total burden of diarrheal disease in the 2010 Global Burden of Disease Study (World Health Organisation 2013).

The impact of pathogen emergence extends beyond increased health risk and has broader implications on economic development. The socioeconomic development and prospect for a better future are crippled by the burden of infectious diseases (World Health Organisation 2015). There is a massive loss of income to combating epidemics in many countries. Hence, most countries have lost focus on development, improved health, and alleviating poverty by fighting diseases. The economic impact of epidemics is depicted by rising treatment cost and reduced productivity caused by sickness. Emerging infectious diseases are estimated to have caused hundreds of billions of dollars worth of damage (McCloskey et al. 2014).

A summary of some reported outbreaks caused by some of these emerging and reemerging pathogens is presented in Table 2.2.

Table 2.2 Most recent common outbreaks in countries in Africa

Country	Number of districts affected	Pathogen/serotype	Year	Suspected cases (lab-confirmed cases)	Deaths
Angola	14 Provinces	*V. cholerae*	2006	46,758	1893 (4.0%)
Burkina Faso	34 Districts	*N. meningitidis* (serogroup A)	2007	22,255	1490 (7%)
Cameroon	6 Regions	*V. cholerae*	2010	7869	515 (6.5%)
	5 Districts	*N. meningitidis* A	2011	923	57 (6.2%)
Côte d'Ivoire	1 District: Divot	*V. cholerae*	2003	70	15
D.R. Congo	1 Province: Kinshasa	*V. cholerae*	2018	1065 (83)	43 (4%)
Ethiopia	5 Regions	*V. cholerae* O1 *Inaba*	2006	22,101	219 (1.0%)
Guinea-Bissau	Nationwide	*V. cholerae*	2008	7166	133 (1.9%)
Kenya	7 Counties	*V. cholerae*		3967 (596)	76 (1.9%)
Liberia	1 District	*N. meningitidis* C	2017	31 (14)	13
Mozambique	2 Provinces	*V. cholerae*	2018	1799	1 (0.06%)
Niger	4 Regions	*V. cholerae* O1 *Inaba*	2018	3692	68 (1.8%)
	13 Districts	*N. meningitidis* C and W	2015	8500	573

(continued)

Table 2.2 (continued)

Country	Number of districts affected	Pathogen/serotype	Year	Suspected cases (lab-confirmed cases)	Deaths
Nigeria	1 State	*V. cholerae*	2017	1558 (13)	11 (0.7%)
	5 States	*N. meningitidis* C	2017	1407	211 (15%)
Northern Sudan	9 Provinces	*V. cholerae* O1 Inaba	2006	2007	77 (3.8%)
Senegal		*V. cholerae*	2005	23,325	303 (1.2%)
Sierra Leone	13 Districts	*V. cholerae*	2012	20,736	280 (1.35%)
Somalia	4 Districts	*V. cholerae* O1 Ogawa	2018	1613	9 (0.6%)
South Africa	3 Provinces	*L. monocyte-genes* ST6, and 19 other strains	2017–2018	(1024)	200
	1 Province: Mpumalanga	*V. cholerae*	2003	179	5 (2.79%)
	1 District: Juba	*V. cholerae*	2014	586	22 (3.8%)
	Nationwide	*N. meningitidis* serogroup A	2007	6946	430 (6.19%)
Tanzania	23 Regions	*V. cholerae*	2016	24,108	378
Togo	19 Health districts	*N. meningitis* W	2017	201	17
Zambia		*V. cholerae* O1 Ogawa		547	15 (1.8%)
Zimbabwe		*V. cholerae*		8535 (163)	50 (0.6%)
West Africa	8 Countries	*V. cholerae*	2005	31, 259	517
Central Africa	Cameroon, Chad, Niger	*V. cholerae*	2010	40,468	1879
African Meningitis Belt	18 Countries	*N. meningitidis* serogroup W135 and A	2013	9249	857 (9.3%)

Prepared from data extracted WHO reports from http://www.who.int/en/news-room/fact-sheets/detail/cholera, http://www.who.int/csr/don/2013_06_06_menin/en/, https://www.who.int/media-centre/factsheets/listeriosis/en/ and http://www.who.int/csr/don/02-may-2018-listeriosis-south-africa/en/

2.5 Some Notable Emerging/Reemerging Bacterial Pathogens in Africa

2.5.1 Campylobacter *Species*

Campylobacter species are now recognized as significant emerging pathogens due to increasing prevalence in clinical diseases. Campylobacters are also known to cause a wide range of gastrointestinal and extra-gastrointestinal diseases in human and animals. Globally, *Campylobacter* is one of four key etiological agents of diarrheal diseases and is now considered as the most common bacterial cause of human gastroenteritis in the world. As a result, *Campylobacter* species have emerged as prominent etiological agents of gastroenteritis and diarrhea in humans with infectious doses of 300 and 800 colony forming units (CFU), respectively (Coker et al. 2002; Sandberg et al. 2006; Gallo et al. 2016). The gastroenteritis caused by *Campylobacter* species is termed campylobacteriosis, caused primarily by *C. jejuni* and *C. coli*, although other species such as *C. upsaliensis* and *C. lari* are gaining recognition (MacDonald et al. 2015). The average incubation period preceding the onset of *Campylobacter* infection is usually 2–5 days with a range from 1 to 10 days, depending on the infective dose. The most clinical manifestation of *Campylobacter* infections includes diarrhea, abdominal pain, fever, headache, nausea, and vomiting. Sequelae such as bacteremia, pancreatitis, miscarriage, hepatitis, reactive arthritis, and neurological disorders (e.g., Guillain-Barre syndrome) may arise from *Campylobacter* infections (WHO 2018e). Electrolyte replacement and rehydration is the usual treatment option for *Campylobacter* infection, unless for invasive cases where antimicrobial treatment is required (WHO 2018e).

2.5.1.1 History, Description, and Identification

Theodor Escherich first observed spiral-shaped organisms that resembled *Campylobacter* in diarrheic stool samples from children in 1886. MacFaydean and Stockman identified *Campylobacter* in the tissue of fetus of aborted sheep in the early twentieth century. Moreover, in 1919, Smith studied virulent abortions of bovines and isolated a spiral-shaped bacterium and suggested the name *Vibrio fetus*. The genus *Campylobacter* was established in 1963 by the renaming of *Vibrio fetus* to *Campylobacter fetus* (Sebald and Veron 1963; Kist 1986; Butzler 2004). *Campylobacter* belongs to the family *Campylobacteraceae*, and of the phylum *Proteobacteria* (Kaakoush et al. 2015). *Campylobacter* species are non-spore-forming, Gram-negative bacteria and are approximately 0.2–0.8 μm by 0.5–5 μm. Campylobacters are spiral, rod-shaped, or curved bacteria, having either a single polar flagellum, bipolar flagella, or no flagellum, depending on the species (Man 2011). Most *Campylobacter* species grow under microaerobic conditions, but certain species prefer anaerobic conditions (Vandamme and De Ley 1991; Kassa et al. 2005). Many species of *Campylobacter* have been identified including *Campylobacter concisus*, *C. curvus*, *C. rectus*, *C. mucosalis*, *C. showae*, *C. gracilis*, *C. lari*, *C. upsaliensis*, *C. jejuni*, *C. coli*, and *C. hyointestinalis*. However, *C. jejuni* and *C. coli* are considered the major gastrointestinal disease-causing pathogens in this genus (Kaakoush et al. 2015).

Traditional culture media for the isolation of campylobacters include blood agar, *Campylobacter* blood-free selective agar, charcoal cefoperazone deoxycholate agar (CDDA), Preston broth, modified Exeter broth (MEB), *Campylobacter* enrichment broth, cefoperazone, vancomycin, and amphotericin B agar (CVA) and Bolton broth (Diergaardt et al. 2003; Horman et al. 2004; Bull et al. 2006; Kwan et al. 2008; Kim et al. 2016). Antibiotics are often added to these media for selective recovery of *Campylobacter* species from specimens with a high background of competing bacteria. Successful molecular detection of *Campylobacter* species is based on the amplification of the *rpo*, *23S rRNA*, *16S rRNA,* and *rpoB* (the subunit of the RNA polymerase) sequences (Diergaardt et al. 2003; Korczak et al. 2009; Lang et al. 2010; Kim et al. 2016). In addition, amplification of various housekeeping genes such as *aspA* (aspartase A), *glnA* (glutamine synthetase), *gyrA* (gyrase), *gltA* (citrate synthase), *glyA* (serine hydroxymethyltransferase), *pgm* (phosphoglucomutase), *tkt* (transketolase), and *uncA* (ATP synthase alpha subunit) has been adopted for the molecular identification of *Campylobacter* species (Dingle et al. 2001; Korczak et al. 2009).

2.5.1.2 Epidemiology, Risk Factors, and Transmission Modes

Campylobacter infections are among the most common and widespread infectious diseases of the last century. Both developed and developing countries have experienced an increased incidence of these diseases over the last decades (Coker et al. 2002). Data from Australia, North America, and Europe have revealed dramatic increases in campylobacteriosis while data from parts of Africa suggest that this infection is endemic in Africa, especially among children (Kaakoush et al. 2015). Globally, about 5–14% of total diarrheal cases is believed to be caused by *Campylobacter* spp., particularly *C. jejuni* (Lang et al. 2010; Kaakoush et al. 2015). Among diagnosed diarrhea cases, *Campylobacter* spp. have been observed to cause diarrhea either singly or as coinfection.

Campylobacter infections are predominant among persons that are <5 years and ≥65 years (Sopwith et al. 2010). Geographical location and seasonal variation are other factors that appear to influence the epidemiology of *Campylobacter* infections and could contribute to an increase in the concentration of a single strain of *Campylobacter* species (Levesque et al. 2013). Higher incidences of *Campylobacter* infections have been reported in rural areas than in cities and is associated with warmer summer months (Sopwith et al. 2008; Lengerh et al. 2013).

The development of *Campylobacter* infection after exposure to a reservoir or vehicle depends on certain risk factors at various times for sporadic *Campylobacter* infections (Sandberg et al. 2006). Besides contact with animals or infected persons, identified risk factors associated with the incidence and spread of *Campylobacter* disease include consumption of animal products, milk, and raw vegetables, fishing, patient age and gender, dwelling in rural areas with no or inadequately treated water supply, rainfall, international travel, and exposure to polluted water (Kemp et al. 2005; Sandberg et al. 2006; Kwan et al. 2008; Sopwith et al. 2008; Sopwith et al. 2010; Lengerh et al. 2013; Levesque et al. 2013; Pham-Duc et al. 2014; Kaakoush et al. 2015; MacDonald et al. 2015). *Campylobacter* infections were formerly considered mainly as foodborne, but it is now recognized that transmission via contaminated water is possible (Clark et al. 2003; Jakopanec et al. 2008; Silva et al. 2011). While the role of water as a reservoir of *Campylobacter* remains debatable owing to their low numbers, survival rates, and persistence in water, the role of water as a vehicle of *Campylobacter* transmission is highly documented (Kwan et al. 2008).

2.5.1.3 Distribution and Reservoirs of *Campylobacter* Spp.

Campylobacter spp. are distributed in a wide variety of environmental matrices with a high recovery rate of ≥50% from various environmental sources. Poultry is considered the major animal reservoir and source of transmission of *Campylobacter* species to humans (Axelsson-Olsson et al. 2005; Bull et al. 2006; MacDonald et al. 2015). However, other reservoirs including environmental waters (such as rivers, lakes, freshwater and recreational beaches, wastewater effluent), bird droppings, irrigation water, slaughterhouses and food-processing plants, animal and bird carcasses, farm litter and fecal matter, river sediment, soils, cattle, wildlife, pets, humans, and pigs can harbor and transmit campylobacters (Horman et al. 2004; Kemp et al. 2005; Bull et al. 2006; Sopwith et al. 2008; Lang et al. 2010; Berghaus et al. 2013; Lengerh et al. 2013; Khan et al. 2013a; Khan et al. 2013b; Banting et al. 2016). Each of these hosts harbors a unique strain or group of strains of these organisms. For instance, *Campylobacter* strains, ST-21, and ST-61 are highly prevalent among ruminants, while ST-45 are especially prevalent among wildlife and the environment (soil and water).

2.5.2 Arcobacter *Species*

Arcobacter is a leading cause of traveler's diarrhea and infections in immunosuppressed people. Comparatively, infections by *Arcobacter* appear to be more severe than those by *Campylobacter* species, because most patients with *Arcobacter* infections are treated as inpatients. Infections by *Arcobacter* spp. include acute diarrhea lasting 3–15 days, and persistent or recurrent diarrhea lasting >2 weeks to 2 months (Vandenberg et al. 2004; Prouzet-Mauléon et al. 2006). Clinical symptoms of *Arcobacter* spp. infections are manifested differently in humans. Sometimes the diarrhea is characterized by abdominal pain or the patient may experience abdominal pain alone, bloody and watery stools, vomiting, acute renal failure, rectal bleeding, fever, anorexia, weight loss, and asthenia (Vandenberg et al. 2004; Wybo et al. 2004; Prouzet-Mauléon et al. 2006; Douidah et al. 2012; Figueras et al. 2014). Most diarrhea caused by *Arcobacter* spp. can be attributed to *A. butzleri*, which appears to be the major clinically significant organism in this genus although other species like *A. cryaerophilus* and *A. skirrowii* are also potential causal agents of bacteremia, pyelonephritis, septicemia, and peritonitis (Lau et al. 2002). However, the extent to which *Arcobacter* spp. are involved in these conditions is unknown because of the high rate of coinfection with other organisms such as *Salmonella*, toxigenic *C. difficile,* and rotavirus which has been reported (Van den Abeele et al. 2014). *Arcobacter* can be recovered from several clinical specimens, including stool, peritonitis pus, and blood (Prouzet-Mauléon et al. 2006).

Arcobacter is a common contaminant of surface water (Collado et al. 2008; Webb et al. 2017; Moreno et al. 2003). *Arcobacter* species, notably *A. butzleri*, often occur in nature together with *Campylobacter* species and appear to be more prevalent in most cases (Banting et al. 2017). Several factors, including misidentification, are responsible for their delayed detection and recognition as waterborne pathogens. Following the development and adoption of molecular methods such as PCR in the detection of pathogens in humans, water, animals, and other environmental samples, *Arcobacter* species have become recognized as potential emerging pathogens. Their regular presence in feces allows them to be grouped as potential enteric pathogens (Khan et al. 2013b; Ugarte-Ruiz et al. 2015).

2.5.2.1 History, Description, and Identification

Arcobacter was first isolated in 1977 by Ellis from aborted bovine fetal tissues and previously named *Campylobacter butzleri* until the genus *Arcobacter* was created in the early 1990s. *Arcobacter* spp. are aerotolerant, Gram-negative, motile, curved-rod, non-spore-forming bacteria capable of growing at temperatures <30 °C; this differentiates them from *Campylobacter* spp. (Houf et al. 2005; Prouzet-Mauléon et al. 2006; Van den Abeele et al. 2014). They are positive for catalase, oxidase, and nitrate reduction but negative for hippurate hydrolysis and urease activity (Ertas et al. 2010; Fisher et al. 2015).

Arcobacter growth media include *Arcobacter* enrichment broth, blood agar, MacConkey agar, CCDA agar, JM broth, CVA, and *Arcobacter* selective agar incubated aerobically at 30 °C for 48–72 h (Collado et al. 2010; Glacometti et al. 2013). Selective supplements comprising cefoperazone, amphotericin B, and teicoplanin are used for the selective isolation of *Arcobacter* species. Other supplements used with *Arcobacter* broth are 5-fluorouracil, novobiocin, and trimethoprim (Chinivasagam et al. 2007). Molecular identification of *Arcobacter* spp. may involve amplification of the *23S rRNA* and *16S rRNA* genes; however, identification based on *23S rRNA* appears to be less accurate than *16S rRNA* (Houf et al. 2002; Collado et al. 2011; Douidah et al. 2012). Various housekeeping genes, namely *gyr*A, *gyr*B, *rpo*B-*rpo*C, and *hsp*-60, can be used to differentiate between species and show their phylogenetic relationship. In addition, putative *Arcobacter* virulence genes (*cia*B, *cad*F, *cj*1349, *hec*B, *hec*A, *mvi*N, *pld*A, and *irg*A) that confer capacity to cause infection can be adopted for their identification (Collado et al. 2010; Fera et al. 2010; Collado et al. 2011; Zacharow et al. 2015).

2.5.2.2 Epidemiology, Risk Factors, and Transmission Modes

Arcobacter species cause sporadic clinical infections and are of increasing incidence, particularly among diarrheal patients (Douidah et al. 2012; Jiang et al. 2010). Several factors, including misidentification, are responsible for their delayed detection and recognition as emerging pathogens (Khan et al. 2013b; Ugarte-Ruiz et al. 2015). Improved diagnostic techniques have enabled the efficient isolation of this organism and revealed that *Arcobacter butzleri* and, occasionally, *A. skirrowii and A. cryaerophilus* constitute a significant proportion of organisms that were formerly identified as *Campylobacter*-related organisms (Prouzet-Mauléon et al. 2006; Van den Abeele et al. 2014). Vandenberg et al. (2004) reported that *Arcobacter* species made up 3.5% (97 isolates) of *Campylobacter* and related organisms that were recovered from 77 patients in Brussels, Belgium. Although *Arcobacter* spp. have been identified in temperate regions, they are endemic in tropical and developing countries, especially among travelers (Jiang et al. 2010). *Arcobacter* spp. infect patients within the age bracket of 30 days to 90 years, but infection with *Arcobacter* spp. is more common among members of older age groups (Vandenberg et al. 2004; Van den Abeele et al. 2014). They are present in asymptomatic persons (healthy carriers), and comparative infection rates have been observed in males and females (Vandenberg et al. 2004). *Arcobacter*-induced diarrhea is common among international travelers and immunocompromised patients and is considered the fourth most frequently isolated *Campylobacteraceae* in human clinical samples; in some cases, *Arcobacter* has been detected as the only enteropathogen in patients with diarrhea. This observation of high incidence and being the sole pathogen in human gastrointestinal disease emphasizes their role in the disease (Prouzet-Mauléon et al. 2006; Figueras et al. 2014; Wybo et al. 2004).

Arcobacter infections appear to share similar risk factors and exposure routes as *Campylobacter* infections as well as display related microbiological and clinical features as *C. jejuni* (Vandenberg et al. 2004). Frequent detection of *Arcobacter* species in water, food, and feces allows them to be grouped as potential water/food-borne and enteric pathogens (Khan et al. 2013b; Ugarte-Ruiz et al. 2015). Thus, the transmission of *Arcobacter* could be via person-person, consumption of under-cooked meat and contaminated water, or contact with pets (such as dogs) or animal feces (Wesley 1997; Figueras et al. 2014; Houf et al. 2008). There are scanty reports on possible risk factors that predispose humans to *Arcobacter* infections. Figueras et al. (2014) reported a persistent case in a healthy 26-year-old Spanish male of bloody diarrhea caused by *A. cryaerophilus,* and despite not being able to find the contagious source of *Arcobacter* in the environment of the patient, they speculated that the infection could have been acquired through the consumption of poorly cooked poultry meat or fish. Living in a place with inadequate water supplies as well as residing in a household which has dogs as pets have also been implicated as possible risk factors for *Arcobacter* infections (Wesley 1997; Houf et al. 2008).

2.5.2.3　Distribution and Reservoirs of *Arcobacter* Spp.

Arcobacters are widely distributed in nature and can be recovered from several environmental sources that could serve as vehicles of transmission. Also, this organism tends to show spatial clustering in the environment, suggesting that its incidence varies with location. Their presence in environmental waters of varying salinity gradients has also been established, suggesting that *Arcobacter* is predominantly found in environmental waters (Collado et al. 2008; Lee et al. 2012). Other reported reservoirs of *Arcobacter* spp. include agricultural produce, raw milk, poultry carcasses, pork, beef, sewage, livestock feces, domestic animals, and stormwater (Houf et al. 2005; Andersen et al. 2007; Collado et al. 2008; Ertas et al. 2010; Houf et al. 2008; Douidah et al. 2012; Mottola et al. 2016; Zacharow et al. 2015). These reservoirs should be regarded as a hazard for human health as they are potential vehicles for transmission to humans and dissemination in the environment.

2.5.3 Plesiomonas shigelloides

Plesiomonas shigelloides has been identified as a cause of both locally acquired and travelers' diarrhea. *P. shigelloides* can cause diarrhea in humans and animals that is usually mild, self-limiting, and may not require treatment (Wong et al. 2000). However, the organism can be invasive, has been associated with several outbreaks, travelers' diarrhea, and local diarrhea episodes, and some patients have experienced severe and prolonged symptoms (Theodoropoulos et al. 2001; Nwokocha and Onyemelukwe 2014). The diarrhea is occasionally cholera-like, with numerous bowel movements (up to 30) occurring during the peak of the illness (Arai et al. 1980; Kain and Kelly 1989; Matsuyama et al. 2015). The diarrhea, which often lasts between 2 to several weeks (7 weeks) may be presented as watery and/or bloody stools (Xia et al. 2015; Kain and Kelly 1989). Other associated symptoms characterizing *P. shigelloides* infection include fever, abdominal pain and discomfort, fecal leukocytes, dehydration, mucus in stools, colitis, hypogastric pain, peritonitis, and vomiting, which are generally more common among female patients (Shigematsu et al. 2000; Kain and Kelly 1989; Wong et al. 2000). Generally, three major types of *P. shigelloides* diarrheal infections have been noted, viz. a watery secretory type, an invasive cholera-like type, and a subacute or chronic form lasting between 2 weeks and 3 months (Salerno et al. 2007).

Although infection by *P. shigelloides* can be self-limiting, chronic cases and cases with more severe symptoms including death have been reported, particularly in patients that are immunocompromised by underlying medical conditions (Rolston and Hopfer 1984). Wong et al. (2000) reported two deaths and 25 patients presented with prolonged symptoms from *P. shigelloides* infection. Also, the authors presented evidence of an increasing prevalence of *P. shigelloides* among patients with diarrhea, suggesting that the organism probably occurs more commonly than is currently recognized. They reported severe gastrointestinal symptoms, with prominent cramping, dehydration, and febrility characterizing infections by *P. shigelloides* in cancer patients whose immunity was further suppressed by chemotherapy.

Coinfection with other enteric bacteria, namely *Campylobacter*, *Shigella*, *Salmonella*, and *Yersinia*, has been observed among patients with diarrhea, which has been estimated to increase the risk for disease sixfold compared with single infections by *P. shigelloides* (Escobar et al. 2012). Moreover, other extraintestinal diseases have been associated with *P. shigelloides* infection, namely proctitis, meningitis/meningoencephalitis, bacteremia, fatal septicemia, and biliary tract and liver diseases (Paul et al. 1990; Chen et al. 2013). Infection with *P. shigelloides* in infants can cause meningitis/meningoencephalitis with consequent effects on the cardiovascular, hematologic, gastrointestinal, respiratory systems, and coagulation (Xia et al. 2015). The clinical manifestations of sepsis and meningoencephalitis caused by this bacterium in the early stage are like those of common infections. However, *P. shigelloides*-induced sepsis and meningoencephalitis progress fast and destroy multiple organs, and so the mortality rate may increase with delayed diagnosis or incorrect antibiotics treatment (Xia et al. 2015).

P. shigelloides strains capacity for pathogenicity is unclear, but it is likely, as this pathogen has demonstrated the ability to produce putative virulence factors related to enteropathogenicity (Falcon et al. 2003). Furthermore, *P. shigelloides* possesses virulence-associated properties of *Shigella* spp. such as harboring plasmids and production of toxins that are hydrophobic and cytotoxic to HeLa cells and are Hep-invasive (Olsvik et al. 1990). Isolates of *P. shigelloides* culture can cause intracellular vacuolation in mammalian cells and produce hemolytic and enterotoxic activities (Falcon et al. 2003). Furthermore, *P. shigelloides* can adhere and multiply in host intestinal cells (Theodoropoulos et al. 2001; Falcon et al. 2003).

2.5.3.1 History, Description, and Identification

Plesiomonas shigelloides garnered attention from researchers in the mid-twentieth century. It was initially considered to be related to *Vibrio cholerae* and *Aeromonas,* and thus the genus *Plesiomonas* was placed within the family *Vibrionaceae.* However, phylogenetic analyses and antigenic profiling revealed a closer relationship with members of the family *Enterobacteriaceae* (Theodoropoulos et al. 2001). These organisms are motile (with lophotrichous polar flagella), short, Gram-negative, rod-shaped (generally 0.3–1.0 μm in width, 0.6–6.0 μm in length), non-spore-forming, and facultatively anaerobic bacteria, recently classified in the family *Enterobacteriaceae* (Salerno et al. 2007).

Plesiomonas shigelloides can be isolated with a variety of media such as blood agar, MacConkey agar, salmonella-shigella agar, alkaline peptone water, and inositol-brilliant green-bile salts agar, where they produce non-lactose-, non-sucrose-fermenting colonies (Fisher et al. 2015). However, alkaline peptone-water and inositol-brilliant green-bile salts agar are optimal for its growth (Von Graevenitz and Bucher 1983). On inositol Brilliant Green Bile Agar, it appears as pink colonies, while on blood agar, large, gray, opaque, and convex β-hemolytic colonies usually about 2–3 mm in diameter are obtained after incubation at 35–37 °C for 16–24 h (Public Health England 2015). This organism can be typed and grouped into different serovars, namely the O and H antigen group (Aldova 2000). The O antigen includes some somatic antigen groups that are also found in *Shigella* spp. (Shepherd et al. 2000; Farmer et al. 2006). Presumptive identification of *P. shigelloides* rests on their motility with multiple flagella, ability to produce cytochrome, oxidase, and ferment inositol and mannitol (Shigematsu et al. 2000). They can grow at salt concentrations of 0–5%, at a pH of 4.0–8.0, and at temperatures of 8–44 °C (Tseng et al. 2002; Murray et al. 2003). The molecular identification of *P. shigelloides* can be achieved by the amplification of the *23S rRNA*, *hug*A, and various housekeeping (*fusA*, *leuS*, *pyrG*, *recG*, and *rpoB*) genes (Gonzalez-Rey et al. 2003; Salerno et al. 2007).

2.5.3.2 Epidemiology, Risk Factors, and Transmission Modes

Plesiomonas shigelloides was initially considered a foodborne or zoonotic organism, but it is now recognized as an emerging and significant enteric and extraintestinal pathogen that is widespread in water (Salerno et al. 2007; Gonzalez-Rey et al. 2003). The bacterium has been identified in healthy symptomless adults and children, but carrier rates in children are higher than in adults. An estimated incidence rate of about 0.0078% among healthy carriers has been reported (Arai et al. 1980; Matsuyama et al. 2015). However, contrasting reports were presented by Chen et al. (2013) and Nwokocha and Onyemelukwe (2014). These authors did not recover *P. shigelloides* from healthy humans. This implies that *P. shigelloides* may not be an indigenous microflora of the human intestine, but a high prevalence in an area may result in its presence in healthy humans.

The distribution of *P. shigelloides* infection differs significantly between regions and periods. Characteristically, *P. shigelloides* is predominant in regions with tropical and subtropical climates, but this bacterium has also been detected in environments of extreme cold climate such as the Arctic regions (Gonzalez-Rey et al. 2003; Chen et al. 2013; Paul et al. 1990). Studies in Asia (Shigematsu et al. 2000), Europe (Paul et al. 1990), North America (Kain and Kelly 1989), and Africa (Nwokocha and Onyemelukwe 2014) have revealed a substantial prevalence of *P. shigelloides* in diarrheic stools collected from travelers as well as out- and inpatients who have complained of diarrhea and other symptoms. *Plesiomonas shigelloides* has been ranked as the third to fourth bacterial pathogen occurring in these stools, and this bacterium is often in association with other enteric bacteria and viruses (Chen et al. 2013; Nwokocha and Onyemelukwe 2014). The prevalence of *P. shigelloides* among patients with diarrhea has been reported to range from 7.2% to 8.6% and has been linked to patients of all age groups (with an age distribution of 3 months to 86 years) (Escobar et al. 2012; Nwokocha and Onyemelukwe 2014). Nevertheless, its incidence is higher in patients from the age of 19 to 44 years, and the organism infects both males and females equally (Kain and Kelly 1989; Wong et al. 2000; Chen et al. 2013). Also, seasonal effects have been observed with the occurrence of this organism among healthy humans, animal carriers, and diarrhea outbreaks. Most of the locally acquired infectious gastroenteritis outbreaks in humans caused by *P. shigelloides* occurred during the summer season, a time which corresponds to the period of major environmental contamination (Arai et al. 1980). However, *P. shigelloides*-induced travelers' diarrhea usually occurs throughout the year (Wong et al. 2000; Shigematsu et al. 2000).

The pathogen is transmitted through the fecal-oral route, consumption of seafood and unsafe water, exposure to animals that are carriers of this pathogen, and traveling (particularly to developing countries). Chen et al. (2013) reported that 15.4% of patients with diarrhea had a history of consuming seafood and uncooked food in the week before the onset of their illness. In Africa, the sources of *P. shigelloides*-induced diarrhea are not fully known, but water sources seem to be the most likely cause. Obi et al. (2002) and Nwokocha and Onyemelukwe (2014) reported that most of the patients from whom *P. shigelloides* was isolated depended largely on streams, ponds, rivers, and hand-dug wells of doubtful water quality as their water source in South Africa and Nigeria, respectively. These water sources were observed to be liable to contamination by filthy litter in many areas, indiscriminate defecation, and animals freely moving around the vicinities. Living in an overcrowded condition also fosters the transmission of this pathogens.

Risk factors such as age, gender, length of trip, and immunity status most often are better associated with additional symptoms (e.g., vomiting, fever, abdominal pain) characterizing an infection by *P. shigelloides* rather than with the predisposition to infection (Shigematsu et al. 2000).

2.5.3.3 Distribution and Reservoirs of *Plesiomonas shigelloides*

The natural reservoirs of *P. shigelloides* in temperate and tropical environments are water and aquatic life such as fish, shellfish, seafood, amphibians, reptiles, and waterfowls. Some animals, such as dogs and cats, are frequent intestinal carriers of this pathogen (Chen et al. 2013). Also, reports indicate that birds are also potential hosts of *P. shigelloides* (Matsuyama et al. 2015). These environmental reservoirs help in the dissemination and widespread occurrence of *P. shigelloides*.

2.5.4   Aeromonas hydrophilia

Aeromonas hydrophila is now considered as one of the prominent enteric pathogens of public health risk, and this pathogen is listed on the Contaminant Candidate list by the US Environmental Protection Agency (Igbinosa et al. 2012). Aeromonas is a free-living bacterium with a vast host range that includes humans, animals, birds, fish, and reptiles. *Aeromonas hydrophila* is known to primarily cause gastrointestinal disease. However, it has been implicated in the etiology of non-gastrointestinal health complications such as cellulitis, bacteremia, septicemia, pneumonia, osteomyelitis, neonatal urinary tract infection, meningitis, kidney, ocular and respiratory tract infections with infectious doses and incubation periods $\geq 10^{10}$ organisms and 1–38 days, respectively, depending on the clinical case (Public Health Agency of Canada 2011).

2.5.4.1   History, Description, and Identification

Historically, *Aeromonas* species were allocated to the family *Enterobacteriaceae* based on phenotypic characteristics (Janda and Abbott 2010). However, recent advances in molecular phylogenetic studies indicated that *Enterobacteriaceae* evolutionary history is different from Aeromonads and this observation warranted a proposal for a distinct family *Aeromonadaceae* and placement of Aeromonads in their own family (Janda and Abbott 2010; Stainer 1943). The genus *Aeromonas* is divided into two groups: the nonmotile (psychrophilic) spp. represented by

Aeromonas salmonicida, and the motile spp. known as the mesophilic species, which includes *Aeromonas hydrophilia* (*A. hydrophilia*).

A. hydrophilia was previously referred to as *Proteus hydrophilic* (Stainer 1943). It is the type species of the motile group of *Aeromonas* (Janda and Abbott 2010). Aeromonads are mesophilic, facultative anaerobe, and grow in a temperature range of 35–37 °C (Janda and Abbott 2010). They are indole and oxidase-positive, non-sporulating, polar-flagellated, Gram-negative rods ranging from 0.3 to 1.0 μm in width by 1.0 to 3.5 μm in length (Public Health Agency of Canada 2011; Khor et al. 2015). They are frequently isolated with media that contain cefsulodin-irgasan-novobiocin (CIN) and starch ampicillin agar (SAA) (Hoel et al. 2017), Aeromonas selective media (ASM) (Chen et al. 2015) or with Tryptic Soy Agar (TSA) (Zhang et al. 2014). Accurate identification and confirmation of *A. hydrophilia* based on biochemical tests is a challenge due to the heterogeneous nature of the species (Persson et al. 2015). However, molecular detection of some highly conserved genes, such as the *16S RNA-23S RNA* intergenic spacer regions (Musa and Ahmed 2017), *gyrB* gene which codes for the B-subunit of DNA gyrase (Yáñez et al. 2003), *rpoD* gene, encoding the σ70 factor conferring promoter specific transition initiation on RNA polymerase (Soler et al. 2004); the RNA polymerase D subunit (*rpoD*) has offered easy and precise identification of *A. hydrophilia* (Hoel et al. 2017).

A. hydrophila is a highly pathogenic organism; detection of virulence genes is very important in determining the potential virulence capacity of strains (Janda and Abbott 2010). Pathogenic strains of *A. hydrophila* have been observed to possess several virulence factors, genes or secreted enzymes, that may have supported pathogenicity and influenced infection (Janda and Abbott 2010; Soler et al. 2002). Such factors include lipase, hemolysins genes (*hlyA, ahhl,* and *asa1*), enterotoxins (heat labile and heat stable; *act, alt ast,* respectively) (Hoel et al. 2017), protease, nuclease, aerolysin (*aerA),* quorum sensing, and secretion mechanisms (Varoutas et al. 2002; Aguilera-Arreola et al. 2007; Janda and Abbott 2010; Rasmussen-Ivey et al. 2016). Involvement of some genes coding for aerolysin and hemolysin has been extensively investigated; 91% and 52% were reported among clinical samples and chicken isolates, respectively (Aguilera-Arreola et al. 2007; Musa and Ahmed 2017). These genes are highly conserved in *A. hydrophila* and are associated with the activation of serine protease and glycerophospholipid-cholesterol acyl-transferase (GCAT) (Chacón et al. 2003).

2.5.4.2 Epidemiology, Risk Factors, and Transmission Modes

A. hydrophila is an aquatic organism occurring naturally in freshwater bodies such as rivers and lakes (Chacon et al. 2003), estuaries, seashores, and sewage (Dahdouh et al. 2016). It rarely occurs in water with high salinity and extremely polluted rivers (Dahdouh et al. 2016). It is the primary agent of necrotizing fasciitis or sepsis in patients with diabetics mellitus (Abuhammour et al. 2006; Janda and Abbott 2010) and immunocompromised patients especially those with burns or aquatic trauma (Hiransuthikul et al. 2005). Until recently, *A. hydrophilia* was believed to be a marine and amphibian pathogenic organism (Yang et al. 2018; Teka et al. 1999) and an opportunistic pathogen in immunocompromised patients (Chao et al. 2013). However, increasing number of cases of intestinal and extraintestinal diseases documented globally has revealed that *A. hydrophilia* is a highly virulent emerging pathogen, capable of causing infection in both immunocompetent and immunocompromised host (Bogdanović et al. 1991; Janda and Abbott 2010; Hiransuthikul et al. 2005; Chao et al. 2013; Chen et al. 2015; Li et al. 2015). *A. hydrophilia* was the only agent found in a blood culture of a preterm neonate with meningitis (Kali et al. 2016). In China, 8.2% was reported among diarrhea patients, and it was also involved with sepsis in a bone graft donor (Hasan et al. 2018). Among the aftermates of the 2004 natural disaster (Tsunami), *A. hydrophilia* was the most isolated agent of bone and soft tissue infection (Hiransuthikul et al. 2005). Its association with hemolytic uremic syndrome (Bogdanović et al. 1991) and necrotizing fasciitis has also been reported (Abuhammour et al. 2006). The mortality rate due to bacteremia caused by *A. hydrophila* ranges from 28% to 46% (Okumura et al. 2011; Chao et al. 2013).

The organism is also the principal pathogenic agent responsible for motile *Aeromonas* septicemia (MAS) epidemic outbreak in fishing industries (Pang et al. 2015; Yang et al. 2018), especially the recent event that occurred in China, which led to a massive economic loss as a result of MAS in catfish and other sea vertebrates (Pang et al. 2015). Most commonly associated symptoms observed in fishes are skin hemorrhage, red sores, body ulceration, lethargy, pop-eye, and fin-rot (Yang et al. 2018). It is transmitted to man through the consumption of contaminated seafood or water (Hoel et al. 2017; Vivekanandhan et al. 2002) and via recreational activities such as boating, skiing, and diving (Zhang et al. 2014). Whole genome sequencing and comparative genomic report studies show that *A. hydrophilia* isolated from different countries processes similar virulence factors (Hossain et al. 2014).

2.5.4.3 Distribution and Reservoirs of *A. hydrophila*

Aeromonas hydrophila is ubiquitous in aquatic systems, evident by its consistent presence in a wide range of water bodies. *Aeromonas hydrophila* presence has been established in flood water, dams, coastal and marine waters, springs, lake, private wells, and drinking water. Prevalences of up to 60% or more have been documented in some of these waters, and the organism was found to occur at concentrations over 1000 CFU/100 mL (Biamon and Hazen 1983; Razzolini et al. 2010; Miyagi et al. 2016). As such, water appears to be a major reservoir of this pathogen in the environment. Wastewater systems are also known to harbor *A. hydrophila* as well as contribute to its spread in the environment.

Furthermore, *Aeromonas hydrophila* can be found in fresh produce, mud, beef, poultry, pork, shellfish, shrimp, dairy products, and soil (Public Health Agency of Canada 2011).

2.5.5 Listeria monocytogenes

Listeria monocytogenes is a foodborne bacterial pathogen that causes listeriosis. It is responsible for foodborne infections in humans, which is often severe and causes between 20% and 50% mortality in most susceptible populations. Although the number of cases of *L. monocytogenes* infections is small, the high rate of death associated with this pathogen makes it a significant public health concern (WHO 2018c). Recently, this pathogen resulted in an outbreak in South Africa of which 1024 laboratory-confirmed cases were reported to the National Institute of Communicable Diseases (NICD) between 1 January 2017 and April 2018. A fatality rate of 28.6% (200 deaths) was recorded within this period, and the spread of this outbreak was observed in Namibia (WHO 2018b).

Listeria monocytogenes can cause noninvasive and invasive listeriosis. The non-invasive form often occurs as gastroenteritis and is a mild form of the disease affecting mainly healthy people, while the severe and invasive form affects specific high-risk populations and is characterized by severe symptoms and high mortality rates (WHO 2018c). Infection caused by *L. monocytogenes* is characterized by diarrhea, fever, body pains, weakness, vomiting, skin rash, sepsis, and a flu-like illness; severe infection can also result to health complications like convulsions, bacteremia or septicemia, endocarditis, meningoencephalitis, brain abscess, and meningitis (Huang et al. 2006; Muriana and Kushwaha 2017; New York State Department of Health 2017). *Listeria monocytogenes* is considered the fourth most common cause of meningeal infection and accounts for 10% of community-acquired bacterial meningitis. However, in certain high-risk groups such as pregnant women, patients with or undertaking treatment for cancer, AIDS and organ transplants, older adults and infants, *L. monocytogenes* is the most common cause of bacterial meningitis. In pregnant women infected by the organism, miscarriage or stillbirth due to septicemia or meningitis form of the disease, premature delivery, decreased fetal movement, and fetal tachycardia have been reported (Farber et al. 1996; Huang et al. 2006; New York State Department of Health 2017; Shange 2017). The organism can infect many parts of the body, including the brain, spinal cord membranes, bloodstream, and gastrointestinal tract (New York State Department of Health 2017).

The average incubation period of *Listeria monocytogenes* before the onset of an infection is 8 days (range: 1–67 days) but differs significantly by the type of clinical disease. The average/(total span of) incubation period of 27.5 days (17–67 days), 9 days (1–14 days), 2 days (1–12 days) and 24 h (6–240 h) have been reported for the onset of the *Listeria monocytogenes* induced pregnancy-associated, central nervous system (CNS), bacteremia and gastrointestinal infections, respectively (Goulet et al. 2013). The estimated infectious dose of *L. monocytogenes* is approximately 10^7–10^9 CFU in healthy humans and 10^5–10^7 in high-risk persons (Farber et al. 1996). The organism can be recovered from several human specimens, including blood, stool, placenta, and cerebrospinal fluid (CSF) (Huang et al. 2006; Goulet et al. 2013). *Listeria monocytogenes* is susceptible to ampicillin, gentamicin, and trimethoprim-sulfamethoxazole (Huang et al. 2006).

2.5.5.1 History, Description, and Identification

The first isolation and description of *Listeria monocytogenes* was documented in 1926 by Everitt Murray, and he named the organism *Bacterium monocytogenes*. The organism was renamed as *Listeria monocytogenes* to honor Dr. Joseph Lister. *Listeria monocytogenes* is taxonomically classified into the family *Listeriaceae* and phylum Firmicutes. *Listeria monocytogenes* is a Gram-positive, motile, facultatively anaerobic, non-sporulating, facultatively intracellular short rod with round ends measuring 0.4–0.5 μm in diameter by 0.5–2 μm in length (Low and Donachie 1997; Muriana and Kushwaha 2017). The organism is motile with the aid of peritrichous flagella and may occur singly or in short chains. The bacterium can persist within a temperature range of 4–50 °C but grows optimally at a temperature range of 30–37 °C. Unlike many other foodborne bacterial pathogens, *L. monocytogenes* can multiply at refrigeration temperature (4 °C). The organism can grow well in the pH range of 4.4–9.6 and at high salt concentrations of up to 10–12% NaCl (Vallim et al. 2015; Muriana and Kushwaha 2017).

Several growth media can be adopted for the enriched and no-enriched isolation of *L. monocytogenes,* namely, Buffered Listeria Enrichment Broth, Brain Heart Infusion (BHI) broth, Brilliance Listeria agar, BHI agar, TSA, Oxford medium base, and Columbia agar, usually at 37 °C for 24 h (Vallim et al. 2015; Ajayeoba et al. 2016). Selective isolation of *Listeria monocytogenes* requires the use of supplement formulations containing cycloheximide, colistin sulfate, acriflavine, cefotetan, or moxalactam and colistin methanesulfonate. The bacterium grows best in anaerobic to microaerobic conditions, preferring a 10% CO_2 environment (Muriana and Kushwaha 2017). Presumptive identification of *L. monocytogenes* can be done with API Coryne system, API Listeria kit, standard biochemical methods, and Christie-Atkins-Munch-Petersen (CAMP) test (Vallim et al. 2015; Braga et al. 2017). Standard biochemical tests for presumptive confirmation of this pathogen include β-hemolysis on blood agar, catalase, motility, and ability to ferment rhamnose and methyl mannoside but not xylose (Zhou and Jiao, 2005). Serological characterization has enabled the identification of different serotypes of *L. monocytogenes,* including serotypes 1/2a, 1/2b, 1/2c, 3a, 3b, 3c, 4b, 4c, 4d, and 4e. Classical serotyping of the bacterium can be carried out using commercially available Listeria Antisera (Vallim et al. 2015; Braga et al. 2017). Also, molecular typing by whole genome sequencing, pulsed-field gel electrophoresis, and multilocus sequence typing have demarcated clonal complexes and several sequence types of *L. monocytogenes* belonging to different serogroups (Wang et al. 2012; Jensen et al. 2016)

Successful molecular identification of *L. monocytogenes* can be achieved via the amplification of various gene fragments including *16S rRNA*, virulence (*hly*A, *iap*), *act*A, *in*IA, *in*I, *in*IAB, and *dth*-18) genes using PCR, RAPD, and PFGE (Almeida and Almeida 2000; Pangallo et al. 2001; Aznar and Alarcon 2003; Zhou and Jiao, 2005; Huang et al. 2006; Braga et al. 2017).

2.5.5.2 Epidemiology, Risk Factors, and Transmission Modes

Listeria monocytogenes infection is reported majorly in industrialized nations. Several strains of *L. monocytogenes* could cause listeriosis; however, according to epidemiological investigations, only serotypes 1/2a, 1/2b, and 4b play a predominant role in human disease processes. Also, serotypes 4d and 4e are rarely found in human samples. *L. monocytogenes* sequence types ST1, ST2, ST6, ST8, and ST9 tend to be most predominant during outbreaks. ST8 appears to be globally distributed, while ST6 is predominantly associated with cases in Africa (Wang et al. 2012; Jensen et al. 2016). Most vulnerable groups to infection by *L. monocytogenes* include neonates, pregnant women, aged persons, and immunocompromised persons (WHO 2018b). Case fatality rate (CFR) associated with *L. monocytogenes* infection is age related and higher in patients ≥70 years. *L. monocytogenes* infection rate and the fatality rate are similar between male and female patients (Jensen et al. 2016).

The bacterium is mainly spread through the consumption of contaminated food and sometimes fluid. Primary vehicles for transmission of *L. monocytogenes* during most outbreaks include cheeses, vegetables, processed foods and meats, raw meats (beef, pork, chicken), milk, and fish products (Vallim et al. 2015; Ajayeoba et al. 2016; Shange 2017). In general, nosocomial and person-to-person transfer of *L. monocytogenes* is known but rare; person-to-person transmission is primarily through vertical transfer from mother to fetus (Huang et al. 2006). Among all *L. monocytogenes* strains, the serotype 1/2a, 1/2b, and 4b are highly predominant in food, storage, and distribution environments. In terms of survival, the serotype 1/2b is capable of prolonged viability than other strains (Vallim et al. 2015; Braga et al. 2017; Ranjbar and Halaji 2018).

Although healthy individuals may acquire *L. monocytogenes* infections, risk factors such as chronic diseases, alcohol, chemotherapy, organ or stem cell transplant, pregnancy, and age dramatically increase the chances of getting infected by *L. monocytogenes* (Lorber 1997; Fernàndez-Sabé et al. 2009).

2.5.5.3 Distribution and Reservoirs of *Listeria monocytogenes*

In nature, *Listeria monocytogenes* is widely distributed and soil, water, and animals are known reservoirs of this bacterium. Most animals are asymptomatic carriers of this pathogen in the environment, and it has been isolated from a variety of animals including goats, pigs, cattle, dogs, mice, horses, and sheep (Muriana and Kushwaha 2017; New York State Department of Health 2017). Silage, sewage, fecal matter, decaying vegetation, and vegetables are also environmental sources of this bacterium. *Listeria monocytogenes* naturally thrives in decomposing plant matter, where they live as saprophytes (Vázquez-Boland et al. 2001). Also, the organism is commonly recovered with ready-to-eat vegetables such as cabbage, carrot, cucumber, lettuce, and tomato (Ajayeoba et al. 2016). Listeria monocytogenes serotypes often display variations in occurrence by geographical regions, and this has been linked to location and concentration of industries that process food products of animal origin (Vallim et al. 2015; Ranjbar and Halaji 2018).

2.5.6   Vibrio cholerae *O1*

Vibrio cholerae O1 is a serogroup of the species *Vibrio cholerae*, which is endemic in Africa and has caused almost all recent endemic and epidemic cholera outbreaks. Globally, an annual estimate of 1.3–4.0 million cases and 21,000 to 143,000 deaths are attributed to cholera (WHO 2018a). Most people infected with the organism remain asymptomatic. However, the bacterium can be present in their feces for up to 10 days and is shed back to the environment, potentially infecting other people. Among symptomatic people, the majority show mild or moderate symptoms characterized by acute watery diarrhea, but severe cholera can lead to dehydration which may result in death if not treated (Handa 2017; WHO 2018a). Virulence traits such as the toxin-coregulated adhesion pili (TCP) and cholera toxin (CT) are certain key pathogenicity factors in *V. cholerae* that are essential for the progress of cholera disease (Rashed et al. 2012). During these outbreaks, *V. cholerae* was treated with antibiotics such as azithromycin but was found to be resistant to tetracycline and ampicillin.

2.5.6.1   History, Description, and Identification

The bacterium, *Vibrio cholerae,* was first discovered and described with a microscope by the Italian scientist Filippo Pacini in 1854 during the 1846–1863 Asiatic cholera pandemic, but most of the scientific community was unaware of his work and as such his work was completely ignored. Then, 30 years later, in 1884, unaware of Pacini's work, the German scientist Robert Koch again discovered and successfully isolated the organism during the 1883 epidemics in Egypt and India. Although *V. cholerae* was first discovered by the Italian Filippo Pacini, the German Robert Koch became the acknowledged discoverer of this bacterium because of his fame in the scientific community, ability to obtain a pure culture of the bacterium, and publication in a popular press (Howard-Jones 1984; Bentivoglio and Pacini 1995; Lippi and Gotuzzo 2014).

Vibrio cholerae O1 is a Gram-negative comma-shaped, aerobic, or facultatively anaerobic bacillus that varies in size from 1 to 3 μm in length by 0.5–0.8 μm in diameter (Handa 2017). The bacterium belongs to the family *Vibrionaceae* and the phylum Proteobacterium. *Vibrio cholerae* O1 has two biotypes, namely the classical and EI Tor, and each biotype has three serotypes, viz., Ogawa, Inaba, and Hikojima (Rashed et al. 2012). Traditional culture media for the isolation of *V. cholerae* include Luria Bertani broth, Luria Bertani agar, MacConkey agar, gelatin agar, and meat extract agar. Alkaline peptone water and estuarine peptone water are used for the enrichment of this pathogen before isolation on agar media. Thiosulfate-citrate-bile salt-sucrose agar, taurocholate tellurite gelatin agar, CHROMagar *Vibrio,* and cellobiose-polymyxin B-colistin agar are media that are presumed selective for the isolation of *V. cholerae* (Raquel et al. 2010; Huq et al. 2013; Morris and Acheson 2013). Serological identification of *V. cholerae* serotypes is performed by the agglutination test using polyvalent antisera raised against the O1 antigen (Keddy et al. 2013). *V. cholerae* can be identified by the amplification of the *hly*A, *ctx*A, *ctx*B, *tcp*A, *omp*W, and *rfb*O1 genes using molecular techniques (Lyon 2001; Huang et al. 2012; Yadava et al. 2013).

2.5.6.2 Epidemiology, Transmission Modes, and Risk Factors

Approximately 210 serogroups of *V. cholerae* are known, but the O1 and O139 serogroups are the primary etiological agents of cholera with the potential to lead to endemic, epidemic, or pandemic outbreaks. However, epidemiological investigations have revealed that most outbreaks in Africa are caused by the O1 serogroup (De et al. 2013; Oladokun and Okoh 2016). *Vibrio cholerae* O1 seems to be endemic in Africa and has led to sporadic cholera outbreaks in several African countries since its spread from Asia in the 1970s (De et al. 2013). In recent decades, the WHO has received reports of cholera outbreaks with cases that range from 547 to 98,424 and fatality rates of 0.6% to 4.6% from Zimbabwe, Niger, Sudan, Tanzania, Nigeria, and Cameroon. Characterization of stool and water samples collected during different outbreaks have revealed Ogawa and Inaba strains as the serotypes responsible for most epidemics in Africa (Rashed et al. 2012). Epidemiological reports of past outbreaks indicate that the bacterium can infect both males and females equally and all age groups.

The bacterium can be transmitted via the fecal-oral route by the consumption of contaminated food or water. The infectious dose of the organism varies with the mode of acquisition. An infectious dose of 10^3–10^6 or 10^2–10^4 is required when ingested with water and food, respectively (Handa 2017). The author also identifies that seasonal factors, overcrowding in peri-urban slums and camps, malnutrition, inadequate access to clean water and sanitation facilities, and lack of proper hygiene practices are among several factors that facilitate the spread of the bacterium in a community.

2.5.6.3 Distribution and Reservoirs of *Vibrio cholerae* O1

In the environment, the organism is a primary inhabitant of the marine ecosystem and appears to be more prevalent in warmer waters (with temperatures >17–20 °C). The major reservoirs of the bacterium are shellfish, human, and water. The bacterium is rarely isolated from animals, and there has been no evidence that animals play a role in its transmission (Handa 2017; Morris and Acheson 2013).

2.5.7 Neisseria meningitidis

N. meningitidis is the most common etiological agent of meningococcal disease, a severe infection of the meninges that affects the brain membrane with the potential to result in brain damage. The disease is fatal in 5–15% with treatment, and 50% of cases if not treated (World Health Organisation 2011). *N. meningitidis* is endemic in Africa and has caused several outbreaks with high fatality rates over the past 100 years (World Health Organisation 2011).

Clinical manifestations of meningitis include a stiff neck, high fever, light sensitivity, confusion, headaches, vomiting, and in severe cases brain damage, deafness, or a learning disability in 10–20% of survivors; a rare but severe form of *N. menin-*

gitidis infection is meningococcal septicemia, which causes rapid circulatory collapse (WHO 2018d). Infection by *N. meningitidis* is potentially fatal due to the rapid onset of the disease in susceptible hosts (Hill et al. 2010); hence, admission to hospital and emergency treatment is expedient. The average incubation period of the bacterium is 4 days, but this may range between 2 and 10 days. *N. meningitidis* has developed several mechanisms to evade killing by host immune system. Strategies adopted for successful evasion of host defense mechanism include mimicking host structures, horizontal genetic exchange, and frequent antigenic variation. *N. meningitidis* expresses several virulence factors such as capsular polysaccharide, lipopolysaccharide (endotoxin, LOS), iron sequestration, and surface-adhesive proteins (such as pili, porins PorA and PorB, adhesion molecules Opa and Opc). *N. meningitidis* survival in a host is contingent on its possession of a capsule that provides resistance to antibodies or complement-mediated killing and inhibits phagocytosis, and the ability to adhere to the host cells and tissues (Stephen 2009). A range of antibiotics is effective against the bacterium, including ampicillin, penicillin, ciprofloxacin, and ceftriaxone; ceftriaxone is the drug of choice during epidemics in Africa, particularly in areas with inadequate health resources (WHO 2018d).

2.5.7.1 History, Description, and Identification

The Italian pathologists Marchiafava and Celli were the first to describe an intracellular oval micrococcus observed in a sample of CSF in 1884 (Marchiafava and Celli 1884). However, *N. meningitidis* was first isolated by Anton Weichselbaum in 1887 from cerebrospinal fluids of patients with clinical presentations of meningococcal disease. It was initially named *Neisseria intracellularis* because of its relatedness to *Neisseria gonorrhoeae* and the intracellular oval micrococci of the bacterium (Weichselbaum 1887; Manchanda et al. 2006). *N. meningitidis* is an encapsulated, Gram-negative, nonmotile, aerobic diplococcus bacterium with a kidney shape, belonging to the phylum β-proteobacterium and a member of the family *Neisseriaceae* and genus *Neisseria* (Rouphael and Stephens 2015).

The bacterium is a fastidious organism which can be isolated from spinal fluid or blood from infected patients on several growth media including blood agar, Brain Heart Infusion agar, Muller-Hinton agar, supplemented chocolate agar, and trypticase agar. The addition of an antibiotic mix such as vancomycin, colistin, and nystatin to chocolate agar increases selectivity for *N. meningitidis*. Optimal incubation conditions for the isolation of *N. meningitidis* is 35–37 °C in humid 5% CO_2 (Olcen and Fredlund 2001; Hill et al. 2010). Confirmation of this bacterium can be successfully performed by biochemical tests, agglutination tests, and PCR while characterization of serogroups and clonal groups can be achieved by slide agglutination tests and DNA sequencing or monoclonal antibodies typing, respectively (Hill et al. 2010). Thirteen serogroups of *N. meningitidis* are recognized based on the polysaccharide capsule expressed and immunological reactivity. These serogroups include A, B, C, E-29, H, I, K, L, W-135, X, Y, Z, and Z. Further classification into serotype, serosubtype, and immunotype is based on the PorB, PorA, and LOS structure, respectively (Hill et al. 2010).

2.5.7.2 Incidence, Epidemiology, Risk Factors, and Transmission Modes

N. meningitidis remains a global concern but with uneven impact, triggering sporadic cases, case clusters, epidemics, and pandemics. *N. meningitidis* infection patterns vary extensively with geographical range, age groups, and bacterial serogroups. In most developed countries, the peak of *N. meningitidis* infections coincides with respiratory viral illnesses during winter months; however, in Africa, the infection peak is usually during the dry season and has been associated with mycoplasma infections (Young et al. 1972; Greenwood et al. 1985). Most outbreaks in Africa occur regularly in the African meningitis belt which spans sub-Saharan Africa from Senegal in the west to Ethiopia in the east (26 countries). Epidemics in this belt began in 1905 and occurred every 8–10 years, the largest epidemics being recorded in 1996–1997, causing over 250,000 cases, 25,000 deaths, and disability in 50,000 persons (World Health Organisation 2011). Another major outbreak of *Neisseria meningitis* was experienced in 2013, affecting several African countries, including Guinea, South Sudan, Benin, Burkina Faso, and Nigeria. About 9249 suspected cases and 857 (9.3%) deaths were documented during this epidemy (WHO 2013). Although 13 serogroups of *N. meningitidis* have been identified, meningitis is mostly caused by five serogroups, namely, A, B, C, W135, X, and Y (Hill et al. 2010; WHO 2018d). Reported outbreaks in Africa were predominantly caused by the serogroups A and W135 (WHO 2013). Globally, particularly in sporadic cases of meningitis, the bacterium infects mostly children that are less than 2 years of age attributable to the waning protective maternal antibody. However, in epidemic situations, older children and adolescents are reportedly highly affected as well.

According to the Centre for Disease Prevention and Control (2015), human-to-human transmission of *N. meningitidis* is possible through inhalation of respiratory droplets and throat secretions. As such, close and prolonged contact or living with an infected person predisposes someone to the bacterium. Estimated risk of secondary transmission is generally 2–4 cases per 1000 household members, although the risk is 500–800 times higher in the general population. Dusty winds, humidity, cold nights, underlying chronic disease, antecedent upper respiratory tract viral infections, and overcrowding increase the risk of contracting the bacterium (WHO 2018d).

2.5.7.3 Distribution and Reservoirs of *N. meningitidis*

Humans are natural hosts of *N. meningitidis* mainly because the organism exists commensally in the nasopharynx of 1% to 10% of healthy persons at any given time and no established animal reservoirs have been identified till date (Manchanda et al. 2006; Hill et al. 2010; Centre for Disease Prevention and Control 2015).

2.5.8 Mycobacterium tuberculosis

Mycobacterium tuberculosis is the etiological agent of tuberculosis and has emerged as one of the world's leading curable infectious agent (Knobler et al. 2006). Tuberculosis is associated with a high fatality rate and has been estimated to cause 1.4 million deaths (WHO 2018a). Its emergence as a priority infectious bacterium is mainly due to increased acquisition of drug resistance (Knobler et al. 2006), fostered by delay and gaps in treatment, and inefficient drug susceptibility testing (van der Werf et al. 2012). Infection by *M. tuberculosis* is characterized by severe coughing that last ≥3 weeks, chest pain, coughing blood or mucus, fatigue, weight loss, lack of appetite, fever, and profuse sweating (Centres for Disease Control and Prevention 2018a, b). Rifampicin, Isoniazid, Ethambutol, and Pyrazinamide are drugs of choice in the treatment of *M. tuberculosis*.

2.5.8.1 History, Description, and Identification

The disease tuberculosis has been hypothesized to date back to approximately 3 million years ago (Knobler et al. 2006); however, the first published scientific discovery of the tubercle bacillus, *Mycobacterium tuberculosis* was done by Robert Koch in 1882 (Barberis et al. 2017).

Mycobacterium tuberculosis belongs to the family *Mycobacteriaceae* and is an aerobic, nonmotile, non-spore forming, non-capsulated bacillus, and can vary in shape from coccobacilli to long rods measuring 1–10 μm in length and 0.2–0.6 μm in width (Ali and Parissa 2012). The bacterium can be recovered from numerous human specimens, including sputum, urine, blood, tissue biopsy, gastric lavage, and other body fluids on special nutrient media (JotScroll 2017; Knobler et al. 2006). This bacterium is fastidious, grows slowly, and takes between 2 and 8 weeks for colonies to appear after culturing on nutrient-rich media such as Lowenstein-Jensen, Petragnini, Dorset, Tarshis, Loeffler serum slope, Pawlowsky agar as well as Dubos, Middlebrooks, Proskauer and Becks, and Sauton media (JotScroll 2017).

2.5.8.2 Epidemiology, Risk Factors, and Transmission Modes

An estimated one-third of the world's population is infected with *Mycobacterium tuberculosis* latently, with an annual 3% rise in new cases, of which 10% of this culminates into active disease (WHO 2018b). Approximately ten million cases of tuberculosis are speculated to occur around the world annually. About 4–5 million cases are highly infectious, and 2–3 million cases result in death (WHO 2018a). As a result, a single *Mycobacterium tuberculosis*-infected person can infect up to 10–15 persons in a year (WHO 2018b). About 95% of Mycobacterium tuberculosis infections occur in developing nations, and in Africa, an increase of 10% attributed to coinfection with HIV has been reported (WHO 2018a). Persons infected with HIV are 800 times more vulnerable to the development of active tuberculosis and death in 15% of cases, especially in West and sub-Saharan Africa (Knobler et al. 2006).

Mycobacterium tuberculosis is highly infectious and spreads from person to person, through inhalation of airborne respiratory droplets or contamination of skin abrasion from an infected person (Centres for Disease Control and Prevention 2018a, b). Migration from poorer high prevalence countries has led to reemergence of this bacterium in countries where it had previously been controlled such as Russia, the USA, Australia, and Canada (Knobler et al. 2006). Poorly ventilated, unsanitary, and overcrowded living conditions, malnutrition, excessive alcohol use, chronic diseases, and smoking are associated with *M. tuberculosis* infection (Barberis et al. 2017).

2.5.8.3   Distribution and Reservoirs of *Mycobacterium tuberculosis*

Mycobacterium tuberculosis is an obligate pathogen; hence, the human host is the major reservoir of this pathogen in the environment (Knobler et al. 2006). Although the organism is known to affect a wide range of livestock and wildlife including lion, buffalo, wild boars, meerkats, and cattle (Patterson et al. 2017), genomic data suggest that tuberculosis did not commence as a zoonosis (Wirth et al. 2008). However, *M. tuberculosis* has been detected in soil, sediment, and water, but the role of these environments as reservoirs or vehicles of transmission of *M. tuberculosis* is yet to be proven (Santos et al. 2015; Barberis et al. 2017).

2.6   Resistance to Antibiotics: A Critical Factor in Bacterial Pathogen Emergence

Antibiotic resistance develops when bacteria adapt and grow in the presence of antibiotics, thereby resulting in the emergence of resistance of infectious agents to many primary treatments (Centre for Disease Prevention and Control 2015; World Health Organisation 2015, van der Werf et al. 2012). The development of resistance is accelerated by the frequency of antibiotic usage, abuse or overuse, the use of drugs in agriculture and livestock farming and spread through the exchange of genetic material between different bacteria (Centre for Disease Prevention and Control 2015; World Health Organisation 2015). Furthermore, incomplete antibiotic treatment exposes pathogens to low doses of antibiotics and consequently results in the formation of a bacterial population that is resistant to the antibiotics irrespective of subsequent dosage (Centre for Disease Prevention and Control 2015; van der Werf et al. 2012). Antibiotic resistance is of great global concern for the emergence or reemergence of pathogens (WHO 2017). Novel antibiotic resistance mechanisms are emerging and spreading globally, and this jeopardizes the underlying principle of controlling infectious diseases incidence and emergence of new pathogenic strains (Knobler et al. 2006). The increasing rate of antimicrobial resistance among emerging bacterial pathogens has obvious drawbacks on the clinical management of infections and compromises treatment (van der Werf et al. 2012). The direct consequences of infection by resistant microbes can be severe, viz. prolonged illnesses, prolonged hospitalization, disability, increased mortality, and increased treatment costs (World Health Organisation 2015).

Antimicrobial resistance is highest in developing countries where the use of antibiotic drugs in humans and animals is greatly unrestricted (World Health Organisation 2013). Also, developing countries, including Africa, are particularly vulnerable to antibiotic resistance because of conditions that enable the spread of infectious pathogens such as poor sanitation, lack of guidance and control of antibiotic use, inadequate infection control measures, and inadequate healthcare and public infrastructure (WHO 2017). The surveillance of antibiotic resistance in *Campylobacter* has identified levels of resistance to tetracyclines and fluoroquinolones in many parts of the world (World Health Organisation 2013). *Aeromonas hydrophila* has developed resistance to most broad-spectrum antibiotics, including ampicillin, methicillin, tetracycline, streptomycin, and nalidixic acid (Dahdouh et al. 2016; Yang et al. 2018). *Mycobacterium tuberculosis* exhibits resistance to different drugs used to treat tuberculosis and shows high resistance to rifampin, streptomycin, ethambutol, and isoniazid (Chang et al. 2015). The WHO Global TB report indicated an average of 580,000 new cases of multidrug-resistant (MDR) *M. tuberculosis* infections in 2015, with an estimated 43% fatality rate. Most of these cases were concentrated in African countries and often linked with HIV (WHO 2017).

Circulation of drug-resistant bacteria in any population can be via food, water, and other environmental components (Chang et al. 2015; WHO 2017). Also, the transmission is influenced by trade, globalization, and migration (WHO 2017). Different antibiotics may belong to the same class; as such, resistance to a specific antibiotic may trigger resistance to a whole class in an organism (WHO 2017, 2018f).

Several probable mechanisms of drug resistance have been highlighted but could be grouped as genetic (origin) and epidemiological (spread/dissemination) mechanisms (Chang et al. 2015). However, these mechanisms are not mutually exclusive; as such, the simultaneous contribution of several mechanisms to MDR is highly plausible (Hooper 2001). Possible origins behind the development of drug resistance include single biochemical mechanism responsible for phenotypic expression of MDR strains (e.g., efflux pumps and thickening of cell wall), genetic linkage, multidrug therapy with accelerated treatment failure, and mutable and recombinant bacterial lineages that confer resistance to multiple drugs (Hooper 2001; Munita and Arias 2016). For instance, the bacterial efflux pumps extrude antibiotics out of cells in a way that the intracellular antibiotic concentration reduces and resistance to the antibiotics results (Hooper 2001; Chang et al. 2015). Also, MDR strains may arise because resistance to multiple drug classes are genetically linked and are either closely located on the bacterial chromosomes or the same horizontally transmitted element such as a plasmid or transposon (Chang et al. 2015). Mutation and recombination processes generate drug resistance genes and are highly variable in bacterial lineages and as such may contribute to the disproportionate frequency of drug resistance within a population (Chang et al. 2015; Munita and Arias 2016). Therefore, understanding the genetic and epidemiological basis of antibiotic resistance is critical to design strategies to curb the emergence and dissemination of antibiotic resistance. Also, it is useful in developing innovative therapeutic approaches against multidrug-resistant organisms (Munita and Arias 2016).

Mechanisms for the dissemination of MDR strains include bystander selection, positive epistasis, niche differentiation and importation of MDR from a high-use population, followed by a spread in a "recipient population" (Hooper 2001; Chang et al. 2015). Commensal organisms in host cells such as microflora acquire resistance by bystander mechanism, which occurs when a drug is used to treat infections caused by infectious pathogens carried by the same host. This mechanism can also occur when an infection is misdiagnosed or undiagnosed, and treatment is directed at a nonexistent infection (Chang et al. 2015).

2.7 Measures Adopted for the Control of Emerging Pathogens During Outbreaks

Control mechanisms for managing outbreaks are anything but simple and encompass multitask to combat disease spread. These tasks are directed towards emergency response activities, treatment, surveillance, and prevention strategies, which may include legislation review and reform.

- An emergency response plan is developed to curtail and end outbreak situations. A comprehensive response is very multifaceted, comprising many components that should be coordinated harmoniously. The response plan should include coordinating responders, acquiring health information, communicating associated risk, and implementing health interventions (WHO 2018g).
- Treatment: Antibiotics can serve as preventive (chemoprophylaxis) or curative measures steered to either prevent or treat infections by emerging bacterial pathogens. Antibiotic prophylaxis reduces the risk of transmission. Consequently, treatment has a survival benefit for infected persons and reduces the incidence of emerging pathogens in a community. For instance, the use of chemoprophylaxis with sulfonamides reduced the incidence of *N. meningitis* infections among United States army recruit during World War II. Also, the development of penicillin G significantly improved the treatment of meningococcal meningitis with resultant reduction in mortality rates (World Health Organisation 2015; Mathema et al. 2017).
- Surveillance, from case detection to investigation and laboratory confirmation, are imperative control measures for managing epidemics by emerging bacterial pathogens. Surveillance is useful in detecting and confirming outbreaks or sporadic cases, monitoring incidence trends, distribution and evolution of causal agents, estimating disease burden, monitoring antibiotic resistance profile and evaluating the effectiveness of control or preventive strategies (WHO 2018d). It provides isolates that can be used for attribution models based on serotyping or genotyping, provides information to inform national decision-making by determining the relative importance of disease agents in relation to others, showing reservoirs for infection and transmission. In developing countries, surveillance provides baseline data on incidence, seasonal patterns, and risk factors (World Health Organisation 2013).

- Prevention Strategies: Deterrence of future outbreaks is only feasible when prevention measures are implemented. Effectiveness of preventive measures can be seen when such measures are continuous and consistent. Universal prevention strategies adopted in control with the overall goal of eliminating future outbreaks include vaccination, inspection activities to facilities that are likely outbreak sources, and strengthening health systems (WHO 2018b). To illustrate this, the launch of vaccination campaigns in many African countries is concomitant with a reported decrease in recorded cases during epidemic seasons. For instance, Burkina Faso, Mali, and Niger during 2011 epidemic season reported the lowest number of meningitis A cases ever recorded after a new meningococcal conjugate A vaccine was introduced, and a total of 20 million persons were vaccinated in these regions (World Health Organisation 2011). Similarly, prevention is performed by the *vaccination of children at birth* using the BCG vaccine, which protects against tuberculosis, and even when infected, prevents a severe form of the disease (JotScroll 2017).

2.8 Conclusion

Africa continues to be the most disease-stricken continent in the world. Thus, emerging bacterial pathogens are a major cause of concern because of the significant public health burden posed and the impact on development seen in divergence of focus and funds from capital projects to treatments. Interaction between environmental, host, and pathogen factors contribute significantly to the emergence of pathogens. Mounting evidence of the high prevalence and increased multidrug resistance of emerging pathogens as seen in sporadic cases or epidemics points to the significance of these pathogens as potential threats to public health and the economy in Africa. Transmission and risk factors are critical components that facilitate the sustenance and circulation of emerging pathogens in a population. These factors can be socioeconomic (e.g., living conditions), environmental (such as climatic or seasonal factors), or host related (health status and lifestyle).

Heterogeneity and complexity of factors that contribute to pathogen emergence and spread of pathogens significantly hamper progress in controlling and predicting the dynamics of epidemics caused by emerging pathogens and result in disproportionate burden, particularly in Africa. In Africa, prevention and control strategies implemented for the control of emerging pathogens are hampered by inequalities in necessary infrastructure, healthcare provision, environmental conditions, hygiene practices, sanitary conditions, inefficient sewage disposal and water treatment facilities and lack of epidemiological data due to inadequate surveillance and investigation. To improve the effectiveness of control measures targeted at reducing the impact of emerging pathogens prevalence in Africa, the focus should be given to community amplifiers of emerging pathogens, molecular identification, surveillance, treatment, identification of high-risk groups, and environmental control measures in places where these organisms are endemic. In summary, understanding the individual effect of these contributors to pathogen emergence will have a strong implication for the control and prevention of pathogen emergence.

References

Abuhammour W, Hasan RA, Rogers D (2006) Necrotizing fasciitis caused by Aeromonas hydrophilia in an immunocompetent child. Pediatr Emerg Care 22:48–51. https://doi.org/10.1097/01. pec.0000195755.66705.f8

Aguilera-Arreola MG, Hernández-Rodríguez C, Zúñiga G et al (2007) Virulence potential and genetic diversity of Aeromonas caviae, Aeromonas veronii, and Aeromonas hydrophila clinical isolates from Mexico and Spain: a comparative study. Can J Microbiol 53:877–887

Ajayeoba TA, Atanda OO, ObadinaAdewale O et al (2016) The incidence and distribution of Listeria monocytogenes in ready-to-eat vegetables in South-Western Nigeria. Food Sci Nutr 4(1):59–66

Aldova E (2000) New serovars of Plesiomonas shigelloides: 1992-1998. Cent Eur J Public Health 8:150–151

Ali AV, Parissa F (2012) Morphological characterisation of Mycobacterium tuberculosis. In: Cardona P-J (ed) Understanding tuberculosis-deciphering the secret life of the bacilli, p 334. InTech Open Science. http://cdn.intechopen.com/pdfs/28419/InTech-Morphological_characterization_of_mycobacterium_tuberculosis.pdf

Almeida PF, Almeida RC (2000) A PCR protocol using inl gene as a target for specific detection of Listeria monocytogenes. Food Control 11:97–101

Andersen MME, Wesley IV, Nestor E, Trampe DW (2007) Prevalence of Arcobacter species in market-weight commercial turkeys. Antonie Van Leeuwenhoek 92:309–317

Arai T, Ikejima N, Shimada T, Sakazaki R (1980) A survey of Plesiomonas shigelloides from aquatic environments, domestic animals, pets and humans. J Hyg 84:203–211

Axelsson-Olsson D, Waldenstrom J, Broman T et al (2005) Protozoan Acanthamoeba polyphaga as a potential reservoir for Campylobacter jejuni. Appl Environ Microbiol 71(2):987–992

Aznar R, Alarcon B (2003) PCR detection of Listeria monocytogenes: a study of multiple factors affecting sensitivity. J Appl Microbiol 95(5):958–966

Banting GS, Braithwaite S, Scott C et al (2016) Evaluation of various Campylobacter-specific quantitative PCR (qPCR) assays for detection and enumeration of Campylobacteraceae in irrigation water and wastewater via a miniaturized most-probable-number–qPCR assay. Appl Environ Microbiol 82(15):4743–4756

Banting G, Figueras Salvat, MJ (2017) Arcobacter. In: J.B. Rose and B. Jiménez-Cisneros, (eds) Global water pathogens project. http://www.waterpathogens.org (J.S Meschke, and R. Girones (eds) Part 3 Viruses) http://www.waterpathogens.org/book/arcobacter Michigan State University, E. Lansing, MI, UNESCO

Barberis I, Bragazzi NL, Galluzo L, Martini M (2017) The history of tuberculosis: from the first historical records to the isolation of Koch's bacillus. J Prevent Med Hyg 58(1):E9–E12

Bentivoglio M, Pacini P (1995) Filippo Pacini: a determined observer. Brain Res Bull 38(2):161–165

Berghaus RD, Thayer SG, Law BF et al (2013) Enumeration of Salmonella and Campylobacter spp. in environmental farm samples and processing plant carcass rinses from commercial broiler chicken flocks. Appl Environ Microbiol 7:4106–4114

Biamon EJ, Hazen TC (1983) Survival and distribution of Aeromonas hydrophila in near-shore coastal waters of Puerto Rico receiving rum distillery effluent. Water Res 17(3):319–326

Bogdanović R, Čobeljić M, Marković M et al (1991) Haemolytic-uraemic syndrome associated with Aeromonas hydrophila enterocolitis. Pediatr Nephrol 5(3):293–295

Braga V, Vazquez S, Vico V et al (2017) Prevalence and serotype distribution of Listeria monocytogenes isolated from foods in Montevideo-Uruguay. Braz J Microbiol 48(4):689–694

Bull SA, Allen VM, Domingue G, Jorgensen F, Frost JA, Ure R, Whyte R, Tinker D, Corry JEL, Gillard-King J, Humphrey TJ (2006) Sources of Campylobacter spp. colonising housed broiler flocks during rearing. Appl Environ Microbiol 72(1):645–652

Butzler JP (2004) Campylobacter, from obscurity to celebrity. Clin Microbiol Infect 10(10):868–876

Centres for Disease Control and Prevention (2018a) CDC features: tuberculosis (TB) disease: symptoms and risk factors. Retrieved October 29, 2018, from https://www.cdc.gov/features/tbsymptoms/index.html

Centres for Disease Control and Prevention (2018b) *World TB Day: History of World TB Day*. Retrieved October 29, 2018, from Centres for Disease Control and Prevention: https://www.cdc.gov/tb/worldtbday/history.htm

Chacón MR, Figueras MJ, Castro-Escarpulli G et al (2003) Distribution of virulence genes in clinical and environmental isolates of *Aeromonas* spp. Antonie Van Leeuwenhoek 84(4):269–278. https://doi.org/10.1023/A:1026042125243

Chang H-H, Cohen T, Grad YH et al (2015) Origin and Proliferation of Multiple-Drug Resistance in Bacterial Pathogens. Microbiol Mol Biol Rev 79(1):101–116

Chao C-M, Lai C-C Gau S-J, Hsueh P-R (2013) Skin and soft tissue infection caused by *Aeromonas* species in cancer patients. J Microbiol Immunol Infect 46(2):144–146

Chen PL, Tsai PJ, Chen CS et al (2015) *Aeromonas* stool isolates from individuals with or without diarrhea in southern Taiwan: Predominance of *Aeromonas veronii*. J Microbiol Immunol Infect 48(6):618–624

Chen X, Chen Y, Yang Q et al (2013) *Plesiomonas shigelloides* infection in Southeast China. PLoS One 8(11):e77877. https://doi.org/10.1371/journal.pone.007787

Chikeka I, Dimler JS (2015) Neglected bacterial zoonoses. Clin Microbiol Infect 21:404–415

Chinivasagam HN, Corney BG, Wright LL et al (2007) Detection of *Arcobacter* spp. in piggery effluent and effluented irrigated soils in Southeast Queensland. J Appl Microbiol 103:418–426

Chacón, M. R., Figueras, M. J., Castro-Escarpulli, G., Soler, L., & Guarro, J. (2003). Distribution of virulence genes in clinical and environmental isolates of Aeromonas spp. Antonie van Leeuwenhoek, International Journal of General and Molecular Microbiology, 84(4), 269–278. https://doi.org/10.1023/A:1026042125243

Clark CG, Price L, Ahmed R et al (2003) Characterisation of waterborne outbreak–associated *Campylobacter jejuni*, Walkerton, Ontario. Emerg Infect Dis 9(10):1232–1241

Coker AO, Isokpehi RD, Thomas BN et al (2002) Human campylobacteriosis in developing countries. Emerg Infect Dis 8(3):237–243

Collado L, Inza I, Guarro J, Figueras MJ (2008) Presence of *Arcobacter* spp. in environmental waters correlates with high levels of faecal pollution. Environ Microbiol 10(6):1635–1640

Collado L, Kasimir G, Perez U et al (2010) Occurrence and diversity of *Arcobacter* spp. along the Llobregat River catchment, at sewage effluents and in a drinking water treatment plant. Water Res 44:3696–3702

Collado L, Figueras, MJ (2011) Taxonomy, Epidemiology, and Clinical Relevance of the Genus Arcobacter. Clin Microbiol Rev 24(1):174–192

Cutler SJ, Fooks AR, van der Poel WHM (2010) Public health threat of new, reemerging and neglected zoonoses in industrialised world. Emerg Infect Dis 16(1):1–7

Dahdouh B, Basha O, Khalil S, Tanekhy M (2016) Molecular characterization, antimicrobial susceptibility and salt tolerance of *Aeromonas hydrophila* from fresh, brackish and marine fishes. Alex J Vet Sci 48(2):46

De R, Ghosh JB, Gupta SS et al (2013) The role of *Vibrio cholerae* genotyping in Africa. J Infect Dis 208(Suppl_1):S32–S38

Devleesschauwer B, Havelaar AH, Maertens de Noordhout C et al (2014) Calculating disability-adjusted life years to quantify burden. Int J Public Health. https://doi.org/10.1007/s00038-014-0552-z

Diergaardt SM, Venter SN, Chalmers M et al (2003) Evaluation of the Cape Town Protocol for the isolation of *Campylobacter* spp. from environmental waters. Water SA 29(2):225–229

Dingle KE, Colles FM, Wareing DRA, Ure R, Fox AJ, Bolton FE, Bootsma HJ, Willems RJL, Urwin R, Maiden MCJ (2001) Multilocus Sequence Typing System for Campylobacter jejuni. J Clin Microbiol 39(1):14–23

Douidah L, de Zutter L, Baré J et al (2012) Occurrence of putative virulence genes in *Arcobacter* species isolated from humans and animals. J Clin Microbiol 50(3):735–741

Ertas N, Dogruer Y, Gonulalan Z et al (2010) Prevalence of *Arcobacter* species in drinking water, spring water, and raw milk as determined by multiplex PCR. J Food Prot 73(11):2099–2102

Escobar JC, Bhavnani D, Trueba G et al (2012) *Plesiomonas shigelloides* infection, Ecuador, 2004–2008. Emerg Infect Dis 18(2):322–324

Falcon R, Carbonell GV, Figueredo PMS et al (2003) Intracellular vacuolation induced by culture filtrates of *Plesiomonas shigelloides* isolated environmental sources. J Appl Microbiol 95(2):273–278

Farber JM, Ross WH, Harwig J (1996) Health risk assessment of Listeria *monocytogenes* in Canada. Int J Food Microbiol 30(1–2):145–156

Farmer JJ, Arduino MJ, Hickman-Brenner FW (2006) The genera *Aeromonas* and *Plesiomonas*. In: Dworkin M, Falkow S, Rosenberg E, Schleifer K-H, Stackebrandt E (eds) The prokaryotes, a handbook on the biology of bacteria. *Proteobacteria*: gamma subclass, vol 6, 3rd edn. Springer, New York, NY, pp 564–596

Fera MT, Gugliandolo C, Lentini V et al (2010) Specific detection of *Arcobacter* spp. Lett Appl Microbiol 50:65–70

Fernàndez-Sabé N, Cervera C, López-Medrano F et al (2009) Risk factors, clinical features, and outcomes of listeriosis in solid-organ transplant recipients: a matched case-control study. Clin Infect Dis 49(8):1153–1159

Figueras MJ, Levican A, Pujol I et al (2014) A severe case of persistent diarrhoea associated with *Arcobacter cryaerophilus* but attributed to *Campylobacter* sp. and a review of the clinical incidence of *Arcobacter* spp. New Microbes New Infect 2(2):31–37

Fisher JC, Newton RJ, Dila DK, McLellan SL (2015) Urban microbial ecology of a freshwater estuary of Lake Michigan. Elementa 3:1–14. https://doi.org/10.12952/journal.elementa.000064

Gallo MT, Di Domenico EG, Toma L et al (2016) *Campylobacter jejuni* fatal sepsis in a patient with Non-Hodgkin's lymphoma: Case report and literature review of a difficult diagnosis. Int J Mol Sci 17:544

Glacometti F, Lucchi A, Manfreda G et al (2013) Occurrence and genetic diversity of *Arcobacter butzleri* in an artisanal dairy plant in Italy. Appl Environ Microbiol 79(21):6665–6669

Gonzalez-Rey C, Svenson SB, Eriksson LM, Ciznar I, Krovacek K (2003) Unexpected finding of the "tropical" bacterial pathogen *Plesiomonas shigelloides* from lake water north of the Polar Circle. Polar Biol 26:495–499

Goulet V, King LA Vaillant V, de Valk H (2013) What is the incubation period for listeriosis. BMC Infect Dis 13:11–17

Greenwood B, Bradley AK, Wall RA (1985) Meningococcal disease and season in sub-Saharan Africa. Lancet 2:829–830

Handa S (2017) Pediatrics: general medicine-cholera. In: Steele RW (eds) Retrieved October 12, 2018, from Medscape: https://emedicine.medscape.com/article/962643-overview#a2

Hasan O, Khan W, Jessar M et al (2018) Bone graft donor site infection with a rare organism, *Aeromonas Hydrophila*. A typical location, presentation and organism with 2 years follow-up. Case report. Int J Surg Case Rep 51:154–157

Hill DJ, Griffiths NJ, Borodina E, Virji M (2010) Cellular and molecular biology of *N. meningitidis* colonization and invasive disease. Clin Sci 118(9):547–564

Hiransuthikul N, Tantisiriwat W, Lertutsahakul K et al (2005) Skin and soft-tissue infections among tsunami survivors in southern Thailand. Clin Infect Dis 41(10):e93–e96. https://doi.org/10.1086/497372

Hoel S, Vadstein O, Jakobsen AN (2017) Species distribution and prevalence of putative virulence factors in Mesophilic *Aeromonas* spp. isolated from fresh retail sushi. Front Microbiol 8:931. https://doi.org/10.3389/fmicb.2017.00931

Houf K, De Zutter L, Van Hoof J, Vandamme P, (2002) Occurrence and Distribution of Arcobacter Species in Poultry Processing. J Food Prot 65(8):1233–1239

Hoogenboezem W (2007) Influences of sewage treatment plant effluents on the occurrence of emerging waterborne pathogens in surface water, July 2007. http://www.riwa-rijn.org/wp-content/uploads/2015/05/146_WWTP_pathogens.pdf

Hooper DC (2001) Emerging mechanisms of fluoroquinolone resistance. Emerg Infect Dis 7(2):337–341

Horman A, Rimhanen-Finne R, Maunula L et al (2004) *Campylobacter* spp., *Giardia* spp., *Cryptosporidium* spp., noroviruses, and indicator organisms in surface water in Southwestern Finland, 2000-2001. Appl Environ Microbiol 70(1):87–95

Hossain MJ, Sun D, McGarey DJ et al (2014) An asian origin of virulent *Aeromonas hydrophila* responsible for disease epidemics in united states-farmed catfish. MBio 5(3):e00848–e00814

Houf K, De Smet S, Bare J, Daminet S (2008) Dogs as carriers of the emerging pathogen *Arcobacter*. Vet Microbiol 130(1/2):208–213

Houf K, On SLW, Coenye T et al (2005) *Arcobacter cibarius* sp. nov., isolated from broiler carcasses. Int J Syst Evol Microbiol 55:713–717

Howard-Jones N (1984) Robert Koch and the cholera vibrio: a centenary. Br Med J 288:379–381

Huang J, Zhu Y, Wen H et al (2012) Detection of toxigenic *Vibrio cholerae* with new multiplex PCR. J Infect Public Health 5(3):263–267

Huang Y-T, Chen S-U, Wu M-Z et al (2006) Molecular evidence for vertical transmission of listeriosis, Taiwan. J Med Microbiol 55:1601–1603

Huq A, Haley BJ, Taviani E et al (2013) Detection, Isolation, and Identification of *Vibrio cholerae* from the Environment. Cur Protoc Microbiol. https://doi.org/10.1002/9780471729259.mc06a05s26

Igbinosa IH, Igumbor EU, Aghdasi F et al (2012) Emerging Aeromonas species infections and their significance in public health. Sci World J:625023. https://doi.org/10.1100/2012/625023

Jakopanec I, Borgenk VL, Lund H et al (2008) A large waterborne outbreak of campylobacteriosis in Norway: the need to focus on distribution system safety. BMC Infect Dis 8:128. https://doi.org/10.1186/1471-2334-8-128

Janda JM, Abbott SL (2010) The genus *Aeromonas*: taxonomy, pathogenicity, and infection. Clin Microbiol Rev 23(1):35–73

Jensen AK, Bjorkman JT, Ethelberg S et al (2016) Molecular typing and epidemiology of human listeriosis cases, Denmark, 2002–2012. Emerg Infect Dis 22(4):625–633

Jiang Z, DuPont HL, Brown EL et al (2010) Microbial etiology of travelers' diarrhea in Mexico, Guatemala, and India: importance of enterotoxigenic *Bacteroides fragilis* and *Arcobacter* species. J Clin Microbiol 48(4):1417–1419

Jones KE, Patel NG, Levy MA et al (2008) Global trends in emerging infectious diseases. Nature 451(7181):990–993

JotScroll (2017) Mycobacterium tuberculosis morphology, characteristics, acid fast stain and culture media. Retrieved from JotScroll: http://www.jotscroll.com/forums/11/posts/159/mycobacterium-tuberculosis-morphology-acid-fast-stain-characteristics.html

Kaakoush NO, Castaño-Rodríguez N, Mitchell HM, Man SM (2015) Global epidemiology of *Campylobacter* infection. Clin Microbiol Rev 28(3):687–740

Kain KC, Kelly MT (1989) Clinical features, epidemiology, and treatment of *Plesiomonas shigelloides* diarrhea. J Clin Microbiol 27(5):998–1001

Kali A, Kalaivani R, Charles P, Seetha KS (2016) Aeromonas hydrophila meningitis and fulminant sepsis in preterm newborn: a case report and review of literature. Indian J Med Microbiol 34(4):544–547

Kassa T, Gebre-selassie S, Asrat D (2005) The prevalence of thermotolerant *Campylobacter* species in food animals in Jimma Zone, Southwest Ethiopia. Ethiop J Health Dev 19(3):225–229

Keddy KH. Sooka A, Parsons MB, et al (2013) Diagnosis of Vibrio cholerae O1 Infection in Africa J Infect Dis 208(Suppl_1):S23-S31.

Kemp R, Leatherbarrow AJH, Williams NJ et al (2005) Prevalence and genetic diversity of *Campylobacter* spp. in environmental water samples from a 100-square-kilometer predominantly dairy farming area. Appl Environ Microbiol 71(4):1876–1882

Khan IU, Hill S, Nowak E, Edge AT (2013a) Effect of incubation temperature on the detection of thermophilic *Campylobacter* species from freshwater beaches, nearby wastewater effluents, and bird fecal droppings. Appl Environ Microbiol 79(24):7639–7645

Khan IUH, Hill S, Nowak E et al (2013b) Investigation of the prevalence of thermophilic *Campylobacter* species at Lake Simcoe recreational beaches. Inland Waters 3:93–104

Khor WC, Puah SM, Tan JA et al (2015) Phenotypic and genetic diversity of *Aeromonas* species isolated from fresh water lakes in Malaysia. PLoS One 10(12):e0145933. https://doi.org/10.1371/journal.pone.0145933

Kim J, Oh E, Banting GS et al (2016) An improved culture method for selective isolation of *Campylobacter jejuni* from wastewater. Front Microbiol 7:1345

Kist M (1986) Who discovered *Campylobacter jejuni/coli*? A review of hitherto disregarded literature. Zentralbl Bakteriol Mikrobiol Hyg A 261(2):177–186

Knobler S, Mahmoud A, Lemon S, Pray L (2006) A world in motion: the global movement of people, products, pathogens, and power. In: Knobler S, Mahmoud A, Lemon S, Pray L (eds) The impact of globalization on infectious disease emergence and control: exploring the consequences and opportunities: Workshop Summary. The NAtional Academic Press, Washighton, DC, p 228. (21–48)

Korczak BM, Zurfluh M, Emler S et al (2009) Multiplex strategy for multilocus sequence typing, *fla* typing, and genetic determination of antimicrobial resistance of *Campylobacter jejuni* and *Campylobacter coli* isolates collected in Switzerland. J Clin Microbiol 47(7):1996–2007

Kwan PSL, Barrigas M, Bolton FJ et al (2008) Molecular epidemiology of *Campylobacter jejuni* populations in dairy cattle, wildlife, and the environment in a farmland area. Appl Environ Microbiol 74(16):5130–5138

La Rosa G, Fontana S, Di Grazia A et al (2007) Molecular identification and genetic analysis of norovirus genogroups I and II in water environments: Comparative analysis of different reverse transcription-PCR assays. Appl Environ Microbiol 73(13):4152–4161

La Rosa G, Fratini M, Libera SD, Iaconelli M, Muscillo M (2012) Emerging and potentially emerging viruses in water environments. Ann Ist Super Sanita 48(4):397–406

Lang P, Lefebure T, Wang W et al (2010) Expanded multilocus sequence typing and comparative genomic hybridization of *Campylobacter coli* isolates from multiple hosts. Appl Environ Microbiol 72(6):1913–1925

Lau S, Woo P, Teng J, Leung K, Yuen K (2002) Identification by 16s ribosomal DNA gene sequencing of *Arcobacter butzleri* bacteraemia in a patient with acute gangrenous appendicitis. Mol Pathol 55:182–185

Lee C, Agidi S, Marion JW, Lee J (2012) *Arcobacter* in Lake Erie beach waters: An emerging gastrointestinal pathogen linked with human-associated fecal contamination. Appl Environ Microbiol 78(16):5511–5519

Lengerh A, Moges F, Unakal C, Anagaw B (2013) Prevalence, associated risk factors and antimicrobial susceptibility pattern of *Campylobacter* species among under five diarrheic children at Gondar University Hospital, Northwest Ethiopia. BMC Paediatr 13:82–90

Levesque S, Fournier E, Carrier N et al (2013) Campylobacteriosis in urban versus rural areas: A case-case study integrated with molecular typing to validate risk factors and to attribute sources of infection. PLoS One 8(12):e83731. https://doi.org/10.1371/journal.pone.008373

Lu J, Ryu H, Vogel J, Domingo JS, Ashbolt NJ (2013) Molecular Detection of Campylobacter spp. And Fecal Indicator Bacteria during the Northern Migration of Sandhill Cranes (Grus canadensis) at the Central Platte River. Appl Environ Microbiol 79(12):3762–3769

Li F, Wang W, Zhu Z et al (2015) Distribution, virulence-associated genes and antimicrobial resistance of *Aeromonas* isolates from diarrheal patients and water, China. J Infect 70(6):600–608

Lippi D, Gotuzzo E (2014) The greatest steps towards the discovery of *Vibrio cholerae*. Clin Microbiol Infect 20(3):191–195

Lorber B (1997) Listeriosis. Clin Infect Dis 24:1–9

Louten J (2016) Chapter 16 – Emerging and re-emerging viral diseases. In: Louten J (ed) Essential human virology. Academic Press, Kennesaw, GA, pp 291–310. https://doi.org/10.1016/B978-0-12-800947-5.00016-8

Low JC, Donachie W (1997) A review of Listeria monocytogenes and listeriosis. Vet J 153(1):9–29

Lyon WJ (2001) aqMan PCR for detection of *Vibrio cholerae* O1, O139, Non-O1, and Non-O139 in pure cultures, raw oysters, and synthetic seawater. Appl Environ Microbiol 67(10):4685–4693

MacDonald E, White R, Mexia R et al (2015) Risk factors for sporadic domestically acquired *Campylobacter* infections in Norway 2010-2011: a national prospective case-control study. PLoS One. https://doi.org/10.1371/journal.pone.0139636

Man SM (2011) The clinical importance of emerging *Campylobacter* species. Nat Rev Gastroenterol Hepatol 8:669–685

Manchanda V, Gupta S, Bhalla P (2006) Meningococcal disease: History, epidemiology, pathogenesis, clinical manifestations, diagnosis, antimicrobial susceptibility and prevention. Indian J Med Microbiol 24:7–19

Marchiafava E, Celli A (1884) Spra i micrococchi della meningite cerebrospinale epidemica. Gazz degli Ospedali 5:59

Mathema B, Andrews JR, Cohen T et al (2017) Drivers of Tuberculosis Transmission. J Infect Dis 216(6):S644–S653

Matsuyama R, Kuninaga N, Morimoto T et al (2015) Isolation and antimicrobial susceptibility of *Plesiomonas shigelloides* from great cormorants (*Phalacrocorax carbo hanedae*) in Gifu and Shiga Prefectures, Japan. J Vet Med Sci 77(9):1179–1181

McCloskey B, Dar O, Zumla A et al (2014) Emerging infectious diseases and pandemic potential: status quo and reducing risk of global spread. Emerg Respir Tract Infect 14(10):P1001–P1010

Miyagi K, Hirai I, Sano K (2016) Distribution of *Aeromonas* species in environmental water used in daily life in Okinawa Prefecture, Japan. Environ Health Prevent Med 21(5):287–294

Moreno Y, Botella S, Alonso JL et al (2003) Specific detection of *Arcobacter* and *Campylobacter* strains in water and sewage by PCR and fluorescent in situ hybridization. Appl Environ Microbiol 69(2):118–1186

Morens DM, Fauci AS (2013) Emerging infectious diseases: threats to human health and global stability. PLoS Pathog 9(7):e1003467. https://doi.org/10.1371/journal.ppat.1003467

Morris GJ, Acheson D (2013) Cholera and other types of vibriosis: a story of human pandemics and oysters on the half shelf. Clin Infect Dis 37(2):272–280

Mottola A, Bonerba E, Bozzo G et al (2016) Occurrence of emerging food-borne pathogenic *Arcobacter* spp. isolated from pre-cut (ready-to-eat) vegetables. Int J Food Microbiol 236:33–37

Munita JM, Arias CA (2016) Mechanisms of antibiotic resistance. Microbiol Spectr 4(2). https://doi.org/10.1128/microbiolspec.VMBF-0016-2015.

Muriana P, Kushwaha K (2017) Food pathogens of concern: listeria monocytogenes. Oklahoma State University, Division of Agricultural Sciences and Natural Resources. Stillwater: Robert M. Kerr Food and Agricultural Products Center

Murray PR, Baron EJ, Jorgensen JH et al (eds) (2003) Manual of clinical microbiology, 8th edn. American Society for Microbiology, Herdon, VA

Musa MD, Ahmed WA (2017) Molecular detection of some *A.hydrophila* toxins and its antibiotics resistance pattern isolated from chicken feces in Thi-Qar Province (Iraq). Kufa. J Vet Med Sci 8(1):167–180

New York State Department of Health (2017) Retrieved October 10, 2018, from Listeriosis (*Listeria* infection): https://www.health.ny.gov/diseases/communicable/listeriosis/fact_sheet.htm

Nii-Trebi NI (2017) Emerging and neglected infectious diseases: insights, advances and challenges. Biomed Res Int 15. https://doi.org/10.1155/2017/5245021

Nwokocha ARC, Onyemelukwe NF (2014) *Plesiomonas shigelloides* diarrhea in Enugu area of south eastern Nigeria: incidence, clinical and epidemiological features. IOSR J Dent Med Sci 13(4):68–73

Obi CL, Potgieter N, Bessong PO, Matsaung G (2002) Assessment of the microbial quality of river water sources in rural Venda communities in South Africa. Water SA. Available on website http://www.wrc.org.za

Ohimian EI (2017) Emerging pathogens of global significance; priorities for attention and control. EC Microbiol 5(6):215–240

Okumura K, Shoji F, Yoshida M et al (2011) Severe sepsis caused by *Aeromonas hydrophila* in a patient using tocilizumab: a case report. J Med Case Rep 5(1):499

Oladokun MO, Okoh AI (2016) *Vibrio cholerae*: a historical perspective and current trend. Asian Pac J Trop Dis 6(11):895–908

Olcen P, Fredlund H (2001) Isolation, culture, and identification of Meningococci from clinical specimens. In: Pollard AJ, Maiden MJ (eds) Meningococcal disease: methods and protocols. Humana Press Inc, Totowa, NJ, pp 9–19

Olsvik O, Wachsmuth K, Kay B, Birkness KA, Yi A Sack A (1990) Laboratory observations on Plesiomonas shigelloides strains isolated from children with diarrhea in Peru. J Clin Microbiol 28(5):886–889

Pang M, Jiang J, Xie X et al (2015) Novel insights into the pathogenicity of epidemic *Aeromonas hydrophila* ST251 clones from comparative genomics. Sci Rep 5(1):9833

Pangallo D, Kaclikova T, Drahovaska H (2001) Detection of *Listeria monocytogenes* by polymerase chain reaction oriented to inlBgene. gene. New Microbiol 24:333–339

Patterson S, Drewe JA, Pfeiffer DU, Clutton-Brock TH (2017) Social and environmental factors affect tuberculosis related mortality in wild meerkats. J Anim Ecol 86(3):442–450

Paul R, Siitonen A, Karkkainen P (1990) *Plesiomonas shigelloides* bacteremia in a healthy girl with mild gastroenteritis. J Clin Microbiol 28(6):1445–1446

Persson S, Al-Shuweli S, Yapici S et al (2015) Identification of clinical *Aeromonas* species by rpoB and gyrB sequencing and development of a multiplex pcr method for detection of *Aeromonas hydrophila, A. caviae, A. veronii*, and *A. media*. J Clin Microbiol 53(2):653–656

Pham-Duc P, Nguyen H, Hattendorf J et al (2014) Diarrhoeal diseases among adult population in an agricultural community Hanam province, Vietnam, with high wastewater and excreta re-use. BMC Public Health 14:978–991

Prouzet-Mauléon V, Labadi L, Bouges N et al (2006) A*rcobacter butzleri*: Underestimated enteropathogen. Emerg Infect Dis 12(2):307–309

Public Health Agency of Canada (2011) Pathogen safety data sheets: infectious substances – Aeromonas hydrophila. Retrieved from Government of Canada: https://www.canada.ca/en/public-health/services/laboratory-biosafety-biosecurity/pathogen-safety-data-sheets-risk-assessment/aeromonas-hydrophila.html

Public Health England (2015) Identification of *Enterobacteriaceae*. UK standards for microbiology investigations. ID 16 4. https://www.gov.uk/uk-standards-for-microbiology-investigations-smi-quality-and-consistency-in-clinical-laboratories

Ranjbar R, Halaji M (2018) Epidemiology of *Listeria monocytogenes* prevalence in foods, animals and human origin from Iran: a systematic review and meta-analysis. BMC Public Health 18:1057–1068

Raquel MM, Megli CJ, Taylor RK (2010) Growth and Laboratory Maintenance of Vibrio cholerae. Curr Protoc Microbiol. https://doi.org/10.1002/9780471729259.mc06a01s17

Rashed SM, Mannan SB, Johura F-T et al (2012) Genetic characteristics of drug-resistant *Vibrio cholerae* O1 causing endemic cholera in Dhaka, 2006–2011. J Med Microbiol 61:1736–1745

Rasmussen-Ivey CR, Figueras MJ, McGarey D, Liles MR (2016) Virulence factors of *Aeromonas hydrophila*: in the wake of reclassification. Front Microbiol 7:1337

Razzolini MT, Günther WM, Martone-Rocha S, et al (2010). *Aeromonas* presence in drinking water from collective reservoirs and wells in peri-urban area in Brazil. Braz J Microbiol 41(3). doi:https://doi.org/10.1590/S1517-83822010000300020

Rolston KVI, Hopfer RL (1984) Diarrhoea due to *Plesiomonas shigelloides* in cancer patients. J Clin Microbiol 20(3):597–598

Rouphael NG, Stephens DS (2015) *N. meningitidis*: biology, microbiology, and epidemiology. Methods Mol Biol 799:1–20. https://doi.org/10.1007/978-1-61779-346-2_1

Salerno A, Deletoile A, Lefevre M et al (2007) Recombining population structure of *Plesiomonas shigelloides* (*Enterobacteriaceae*) revealed by multilocus sequence typing. J Bacteriol 189(21):7808–7818

Sandberg M, Nygård K, Meldal H et al (2006) Incidence trend and risk factors for *Campylobacter* infections in humans in Norway. BMC Public Health 2006(6):179–186

Santos N, Santos C, Valente T et al (2015) Widespread environmental contamination with *Mycobacterium tuberculosis* complex revealed by a molecular detection protocol. PLoS One 10(11):e0142079. https://doi.org/10.1371/journal.pone.0142079

Sebald M, Veron M (1963) Base DNA content and classification of vibrios. Ann Inst Pasteur (Paris) 105:897–910

Sellman J, Pederson P (2007) Emerging infectious diseases of immigrant patients. In: Walker PF, Barnett ED (eds) Immigrant medicine. Elsevier, pp 245–253. https://doi.org/10.1016/B978-0-323-03454-8.X5001-3

Shange N (2017) TimesLive. Retrieved October 11, 2018, from News: South Africa-Listeriosis, 10 things we know so far: https://www.timeslive.co.za/news/south-africa/2017-12-05-listeriosis-10-things-we-know-so-far/

Sharma S, Sachdeva P, Virdi JS (2003) Emerging water-borne pathogens. Appl Microbiol Biotechnol 61:424–428

Shepherd JG, Wang L, Reeves PR (2000) Comparison of O-antigen gene clusters of *Escherichia coli* (*Shigella*) sonnei and *Plesiomonas shigelloides* O17: sonnei gained its current plasmid-borne O-antigen genes from *P. shigelloides* in a recent event. Infect Immunol 68:6056–6061

Shigematsu M, Kaufmann ME, Charlett A et al (2000) An epidemiological study of *Plesiomonas shigelloides* diarrhoea among Japanese travellers. Epidemiol Infect 125:523–530

Silva J, Leite D, Fernandes M et al (2011) *Campylobacter* spp. as a foodborne pathogen: a review. Front Microbiol. https://doi.org/10.3389/fmicb.2011.00200

Soler L, Figueras MJ, Chacón MR et al (2002) Potential virulence and antimicrobial suscepti-bility of *Aeromonas popoffii* recovered from freshwater and seawater. FEMS Immunol Med Microbiol 32(3):243–247

Soler L, Yáñez MA, Chacon MR et al (2004) Phylogenetic analysis of the genus *Aeromonas* based on two housekeeping genes. Int J Syst Evol Microbiol 54(5):1511–1519

Sopwith W, Birtles A, Matthews M et al (2008) Identification of potential environmentally adapted *Campylobacter jejuni* strain, United Kingdom. Emerg Infect Dis 14(11):1769–1773

Sopwith W, Birtles A, Matthews M et al (2010) Investigation of food and environmental exposures relating to the epidemiology of *Campylobacter coli* in humans in Northwest England. Appl Environ Microbiol 76(1):129–135

Stainer R (1943) A note on the taxonomy of *Proteus hydrophilus*. J Bacteriol 46(2):213–214

Stephen DS (2009) Biology and pathogenesis of the evolutionarily successful, obligate human bacterium *N. meningitidis*. Vaccine 27(Suppl 2):B71–B77

Teka T, Faruque ASG, Hossain MI, Fuchs GJ (1999) *Aeromonas*-associated diarrhoea in Bangladeshi children: clinical and epidemiological characteristics. Anna Trop Paediatr 19(1):15–20

Theodoropoulos C, Wong TH, O'brien M, Stenzel D (2001) *Plesiomonas shigelloides* enters polar-ized human intestinal Caco-2 cells in an in vitro model system. Infect Immun 69(4):2260–2269

Tseng H, Liu CP, Li WC et al (2002) Characteristics of *Plesiomonas shigelloides* infection in Taiwan. J Microbiol Immunol Infect 35(1):47–52

Tulchinsky TH, Varavikova EA (2015) Communicable Diseases. In: Tulchinsky TH, Varavikova EA (eds) The New Public Health, 3rd edn. Academic Press, Cambridge, MA, pp 149–236

Ugarte-Ruiz M, Florez-Cuadrado D, Wassenaar TM et al (2015) Method comparison for enhanced recovery, isolation and qualitative detection of *C. jejuni* and *C. coli* from wastewater effluent samples. Int J Environ Res Public Health 12:2749–2764

United Nations Environment Programme Global Environment Monitoring System/Water Programme (UNEP/GEMS) (2008) Water quality for ecosystem and human health, 2nd edn. ISBN 92-95039-51-7 GEMS/Water website at http://www.gemswater.org/. Accessed 25 Nov 2016

Vallim DC, Hofer CB, Lisboa R et al (2015) Twenty Years of Listeria in Brazil: Occurrence of *Listeria* Species and *Listeria monocytogenes* Serovars in Food Samples in Brazil between 1990 and 2012. Biomed Res Int. https://doi.org/10.1155/2015/540204

Van den Abeele A, Vogelaers D, Van Hende J, Houf K (2014) Prevalence of *Arcobacter* Species among humans, Belgium, 2008–2013. Emerg Infect Dis 20(10):1731–1734

van der Werf MJ, Langendam MW, Huitric E, Manissero D (2012) Multidrug resistance after inap-propriate tuberculosis treatment: a meta-analysis. Eur Respir J 39(6):1511–1519

Vandamme P, De Ley J (1991) Proposal for a new family *Campylobacteraceae*. Int J Syst Bacteriol 41:451–455

Vandenberg O, Dediste A, Houf K, Ibekwem S, Souayah H, Cadranel S, Douat N, Zissis G, Butzler JP, Vandamme P (2004) *Arcobacter* species in humans. Emerg Infect Dis 10(10):1863–1867

Varoutas D, Katsianis D, Sphicopoulos T et al (2002) Economic viability of 3G mobile virtual net-work operators. In: Proceedings – 2002 International Conference on Third Generation Wireless and Beyond (Key Function of World Wireless Congress), vol 7, pp 60–63). Frontiers. https://doi.org/10.3389/fmicb.2016.01337

Vázquez-Boland JA, Kuhn M, Berche P et al (2001) *Listeria* pathogenesis and molecular virulence determinants. Clin Microbiol Rev 14(3):584–640

Vivekanandhan G, Savithamani K, Hatha AAM, Lakshmanaperumalsamy P (2002) Antibiotic resistance of *Aeromonas hydrophila* isolated from marketed fish and prawn of South India. Int J Food Microbiol 76(1–2):165–168

Von Graevenitz A, Bucher C (1983) Evaluation of differential and selective media for isolation of *Aeromonas* and *Plesiomonas* spp. from human faeces. J Clin Microbiol 17(1):16–21

Vouga M, Greub G (2016) Emerging bacterial pathogens: the past ad beyond. Clin Microbiol Infect 22(1):12–21

Wang Y, Zhao A, Zhu R et al (2012) Genetic diversity and molecular typing of *Listeria monocytogenes* in China. BMC Microbiol 12:119

Webb AL, Taboada EN, Selinger LB et al (2017) Prevalence and diversity of waterborne *Arcobacter butzleri* in southwestern Alberta, Canada. Can J Microbiol 63(4):330–340

Weichselbaum A (1887) Ueber die Aetiologie der akuten Meningitis cerebrospinalis. Fortschr Med 5:573–583

Wesley IV (1997) *Helicobacter* and *Arcobacter*: Potential human foodborne pathogens. Trend Food Sci Technol 8:293–299

WHO (2013) Emergencies, preparedness, response: Meningococcal disease: 2013 epidemic season in the African Meningitis Belt. Retrieved October 10, 2018, from World Health Organisation: http://www.who.int/csr/don/2013_06_06_menin/en/

WHO (2017) Prioritization of pathogens to guide discovery, research and development of new antibiotics for drug-resistant bacterial infections, including tuberculosis. World Health Organisation, Geneva. Retrieved from http://www.who.int/medicines/areas/rational_use/PPLreport_2017_09_19.pdf

WHO (2018a) Fact sheets; cholera. Retrieved October 10, 2018, from World Health Organisation: http://www.who.int/en/news-room/fact-sheets/detail/cholera

WHO (2018b) Emergencies, preparedness, response: Listeriosis-South Africa (Disease outbreak news). Retrieved October 10, 2018, from World Health Organisation: http://www.who.int/csr/don/02-may-2018-listeriosis-south-africa/en/

WHO (2018c) Media Centre: fact sheet- Listeriosis. Retrieved October 19, 2018, from World Health Organisation: https://www.who.int/mediacentre/factsheets/listeriosis/en/

WHO (2018d) *Meningococcal meningitis*. Retrieved October 23, 2018, from World Health Organisation: http://www.who.int/news-room/fact-sheets/detail/meningococcal-meningitis

WHO (2018e) *News: Campylobacter*. Retrieved October 28, 2018, from World Health Organisation: http://www.who.int/news-room/fact-sheets/detail/campylobacter

WHO (2018f) Fact sheets: antibiotic resistance. Retrieved October 30, 2018, from World Health Organisation: http://www.who.int/en/news-room/fact-sheets/detail/antibiotic-resistance

WHO (2018g) Managing epidemics: key facts about major deadly diseases. World Health Organisation, Geneva, Switzerland

Wirth T, Hildebrand F, Allix-Beguec C et al (2008) Origin, spread and demography of the *Mycobacterium tuberculosis* complex. PLoS Pathog 4(9):e1000160

Wong TY, Tsui HY, So MK et al (2000) *Plesiomonas shigelloides* infection in Hong Kong: retrospective study of 167 laboratory-confirmed cases. Hong Kong Med J 6(4):375–380

Woolhouse ME, Gowtage-Sequeria S (2005) Host range and emerging and reemerging pathogens. Emerg Infect Dis 11(12):1842–1847

World Health Organisation (2011) Immunization, vaccines and biologicals: meningococcal meningitis. Retrieved October 21, 2018, from World Health Organisation: https://www.who.int/immunization/topics/meningitis/en/

World Health Organisation (2013) The Global View of Campylobacteriosis. WHO Document Production Services, Geneva

World Health Organisation (2015) Global Action Plan on Antimicrobial resistance. WHO Document Production Services, Geneva, Switzerland. Retrieved from http://apps.who.int/iris/bitstream/handle/10665/193736/9789241509763_eng.pdf?sequence=1

Wybo I, Lindenburg F, Houf K (2004) Isolation of *Arcobacter skirrowii* from a patient with chronic diarrhea. J Clin Microbiol 42(4):1851–1852

Xia F, Liu P, Zhou Y (2015) Meningoencephalitis caused by *Plesiomonas shigelloides* in a Chinese neonate: case report and literature review. Italian J Paediatr 41:3–7

Yadava JP, Jain M, Goel AK (2013) Detection and confirmation of toxigenic *Vibrio cholerae* 01 in environmental and clinical samples by a direct cell multiplex PCR. Water SA 39(5):611–614

Yáñez MA, Catalán V, Apráiz D et al (2003) Phylogenetic analysis of members of the genus *Aeromonas* based on gyrB gene sequences. Int J Syst Evol Microbiol 53(3):875–883

Yang Y, Miao P, Li H et al (2018) Antibiotic susceptibility and molecular characterization of *Aeromonas hydrophila* from grass carp. J Food Saf 38(1):e12393

Young LS, LaForce FM, Head JJ et al (1972) A simultaneous outbreak of meningococcal and influenza infections. New Engl J Med 287:5–9

Zacharow I, Bystron J, WaBecka-Zacharska E et al (2015) Genetic diversity and incidence of virulence-associated genes of *Arcobacter butzleri* and *Arcobacter cryaerophilus* isolates from pork, beef, and chicken meat in Poland. Biomed Res Int. https://doi.org/10.1155/2015/956507

Zhang J, Yang Y, Zhao L, Li Y, Xie S, Liu Y, (2015) Distribution of sediment bacterial and archaeal communities in plateau freshwater lakes. Appl Microbiol Biotechnol 99 (7):3291–3302

Zhou X, Jiao X (2005) Polymerase chain reaction detection of Listeria monocytogenes using oligonucleotide primers targeting actA gene. Food Control, 16(2), 125–130

Chapter 3
Vibrio cholerae and Cholera: A Recent African Perspective

Wouter J. le Roux, Lisa M. Schaefer, and Stephanus N. Venter

3.1 Introduction

With an estimated burden of 1.4–4.3 million cases per year, cholera remains a major health concern worldwide (W.H.O. 2014). Much of the research addressing cholera has focused on regions of Asia and the Indian sub-continent, a region where cholera has been ever-present over the past century or more. However, since the introduction of the seventh pandemic strains into Africa in 1970, this continent has suffered severely, and African countries accounted for the majority of global cases reported to the World Health Organization (WHO) in many recent years (Bockemuhl and Schröter 1975; W.H.O. 2014). For this reason, there has been an increased research interest from both international and local researchers to conduct cholera-related research on African soil.

Hundreds of peer-reviewed journal articles have been published on this topic within the last decade, and it is challenging for even the most dedicated professionals to stay abreast of the vast collection of knowledge. To facilitate knowledge uptake, this chapter aims to investigate how local research contributed to the understanding of the disease (its impact, spread, endemicity and management) and the cholera causative agent, *Vibrio cholerae*. It summarizes the findings of a decade of research in a structured format to highlight not only the successes but also the potential research gaps that may have acted as hurdles in the fight against cholera.

Information from hundreds of publications was used to compile this chapter. The data represents a decade of research outputs, starting in 2009 and ending in 2018,

W. J. le Roux (✉) · L. M. Schaefer
Water Centre, Council for Scientific and Industrial Research (CSIR), Pretoria, South Africa
e-mail: wleroux@csir.co.za

S. N. Venter
Department of Microbiology and Plant Pathology, University of Pretoria, Pretoria, South Africa

© Springer Nature Switzerland AG 2020
A. L. K. Abia, G. R. Lanza (eds.), *Current Microbiological Research in Africa*,
https://doi.org/10.1007/978-3-030-35296-7_3

and this chapter focusses on peer-reviewed journal articles that report on cholera or *Vibrio cholerae* in the African context. Every effort was made to include all the relevant publications within the selected timeframe (though this cannot be guaranteed due to the large number of publications screened). Keyword searches using 'cholera', '*V. cholerae*' and 'Africa' were performed on three different platforms (PubMed©, ScienceDirect© and Scopus©).

For clarity, the data were organized into two broad groupings, those outputs that relate mostly to the disease and those that relate mainly to the causative agent. In each of these groups, the research outputs are further discussed under different subtopics. The general structure of this chapter (and the number of publications used as the input for each topic) is given in Fig. 3.1.

3.2 Cholera: The Disease

This section covers recent publications that focus on cholera cases, outbreaks or epidemics in Africa. Researchers investigated aspects such as epidemiology, vaccination campaigns, host-related drivers of spread, clinical diagnosis and risk factors (amongst others).

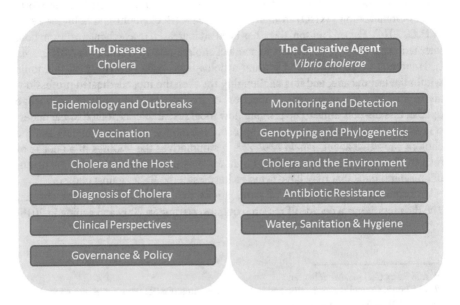

Fig. 3.1 The structure of the groupings and topics as used in this chapter

3.2.1 Epidemiology and Outbreaks

In the African cholera context, the bulk of peer-reviewed outputs dealt with outbreaks, and these publications often contained an epidemiological perspective. The majority reported on recent disease outbreaks, with a small number also looking at historical data in order to gain new insights. Gauzere and Aubry (2012) discuss pre-seventh pandemic cholera epidemics on Reunion Island during the nineteenth century, thereby showing that Africa suffered under the burden of this disease long before the introduction of the seventh pandemic strains in the 1970s. However, the seventh pandemic affected Africa more severely than any other continent and Mengel et al. (2014) calculated that 3.2 million suspected cholera cases were reported to the World Health Organization between 1970 and 2011. This accounted for nearly half of the cases reported worldwide in the same time frame. In their publication titled 'Cholera outbreaks in Africa', the authors provide a valuable review on local epidemics over the last half-century and point out how Africa was disproportionally affected (Mengel et al. 2014).

According to data from the World Health Organization, 90% of all reported cases between 2001 and 2009 were from African countries. However, this changed with the advent of the Haiti outbreak and a coinciding reduction of cases from the continent, allowing the proportion of African cholera cases to decline to less than 50% of all globally reported cases (2010–2013). Even though African cholera cases have declined in recent years, the cholera fatality rate (CFR) in Africa remains higher than that of outbreaks reported elsewhere in the world. The average African CFR (2.28%) was nearly double that of the combined CFR of other regions between 2009 and 2013 (W.H.O. 2011, 2012, 2013, 2014).

The severity of African cholera outbreaks warranted considerable scientific interest, and it is therefore understandable that so many publications focussed on this research area in the last decade. Most of the peer-reviewed outputs under this subheading reported on cholera case data covering a specific outbreak and for a specific country, region or cohort of people. A literature review discovered peer-reviewed publications from 16 African countries, many with multiple publications during the review period (2009–2018) (Fig. 3.2).

Epidemiological reports from Zimbabwe, Ghana, Kenya and Uganda dominated the literature, each boasting more than five publications over the last decade (Shikanga et al. 2009; Ahmed et al. 2011; Adagbada et al. 2012; Mbopi-Keou et al. 2012; Mahamud et al. 2012; Onyango et al. 2013; Morof et al. 2013; Cartwright et al. 2013; Djomassi et al. 2013; Loharikar et al. 2013). Reports from Benin, Burkina Faso, Cameroon, Democratic Republic of Congo, Ethiopia, Guinea-Bissau, Ivory Coast, Mozambique, Nigeria, South Africa, Tanzania and Togo made up the remainder of the literature (Bartels et al. 2010; Gbary et al. 2011a, b; Kyelem et al. 2011; Kelvin 2011; Cummings et al. 2012; Opare et al. 2012; Mayega et al. 2013). Many of the studies had similar findings, for instance, that men were disproportionately affected by cholera outbreaks and faced a higher mortality risk (Bartels et al. 2010; Morof et al. 2013). Concern was often expressed over the high fatality rate

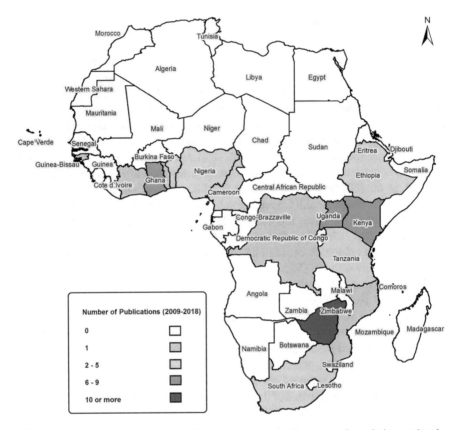

Fig. 3.2 Number of peer-reviewed publications per country that reported on cholera outbreaks (2009–2018)

experienced in Africa, and reference was made to the possible reasons for this (Shikanga et al. 2009; Ahmed et al. 2011; Onyango et al. 2013; Cartwright et al. 2013; Djomassi et al. 2013). Possible reasons that were identified included inadequacies in the healthcare services sector. These services were often too far away, too few, or understaffed and poorly equipped (Mason 2009; Ahmed et al. 2011; Cartwright et al. 2013).

When uncharacteristically low case fatality rates were found during cholera outbreaks, or when the CFR data showed a rapid temporal decline, researchers often cited the availability of quality medical care, free treatment kits, proper hygiene measures, patient compliance with treatment plans, and proactive planning by governments as the main reasons (Gbary et al. 2011b; Mbopi-Keou et al. 2012). Whereas the role of healthcare facilities was found to play a significant part in case fatality rates, the reasons cited for the rapid and uncontained spread of epidemics once an outbreak ignited were often linked to the lack of safe drinking water and poor standards of sanitation and hygiene (Mason 2009; Bartels et al. 2010; Ahmed et al. 2011; Adagbada et al. 2012; Cummings et al. 2012).

Some studies attempted to find the index case(s) or endeavoured to understand the spread pattern and drivers that fuelled the outbreak (George et al. 2016; Sule et al. 2017; Ohene-Adjei et al. 2017). One study in Ghana evaluated a region's cholera surveillance system (Osu Klottey District) and found it to be sensitive, simple, stable, flexible and acceptable (Adjei et al. 2017). They believed that the effectiveness of this system helped Ghana achieve a country-wide CFR of less than 1%. Two different outbreak reports from other regions in Ghana recorded CFR values higher than the 1% target suggested by the WHO (Acquah et al. 2016; Noora et al. 2017). It may be that these two regions lack the advantageous surveillance system present in the Osu Klottey District. The plight of vulnerable populations was highlighted by two different reports, both from Uganda. In the first study, it was reported that a cholera outbreak in a psychiatric hospital led to a CFR of 28%, and in the second it was shown that fishermen's villages were disproportionally affected as they contributed to more than 50% of Uganda's cholera morbidity and mortality numbers (even though they added up to less than 10% of the country's total population) (Bwire et al. 2015, 2017a).

The literature covered above has one thing in common; they all looked at specific outbreaks or epidemics (typically one outbreak present in one region). Some researchers expanded their coverage beyond this and investigated cholera case data over more extended time frames in order to identify temporal trends, or analysed data from different regions (and therefore possibly different outbreaks) (Ekra et al. 2009; Bwire et al. 2013a; Essoya et al. 2013; Gujral et al. 2013; Mutonga et al. 2013). Using such an approach allowed for the identification of high- and low-risk regions in Uganda (Bwire et al. 2013b). The authors feel that surveillance activities with a broader scope (longer time frames and covering larger regions) would help to inform preventative efforts such as vaccination, as it would help to prioritize high-risk regions/populations (Bwire et al. 2013b). Results from an assessment in two districts in Ghana suggested that the country's epidemic control activities were effective, but there was a need to strengthen preparedness planning by improving community surveillance systems and creating awareness about the risk of cholera so that affected people may seek prompt healthcare assistance (Ohene et al. 2016). Sadly, it was precisely a delay in seeking healthcare that led to higher CFRs in the 2010 Zimbabwean outbreak in Kadoma City compared to an outbreak in the same area in 2008 (Maponga et al. 2015). Sauvageot et al. (2016) investigated cholera incidence and mortality in sub-Saharan Africa by analysing data from the African Cholera Surveillance Network (2011 to 2013) (Sauvageot et al. 2016). They found that substantial variation occurred in cholera incidence, age distribution, clinical presentation, culture confirmation, CFR and testing frequency between different African countries.

3.2.2 Vaccination

The adage that 'prevention is better than cure' applies well to the cholera conundrum, and with this in mind, the relatively recent development of efficient oral cholera vaccines (OCVs) show great promise. Until recently, safe and effective cholera vaccines were not readily available. However, that has changed with the recent introduction of two different World Health Organization pre-qualified oral cholera vaccines (OCVs) named Dukoral and Shanchol (W.H.O. 2014). Both these vaccines have been proved to be safe, immunogenic and effective and have been used in mass vaccination campaigns. Meanwhile, the production of a previously licensed oral live attenuated single-dose vaccine has been discontinued, while a former parenteral cholera vaccine never received WHO approval (W.H.O. 2014).

Dukoral consists of killed whole-cell *V. cholerae* O1 and purified recombinant B-subunit cholera toxin (Valneva 2019). Depending on the patient's age, the OCV is administered in two or three doses, with booster doses given 1–6 weeks after the initial dose. In Bangladesh and Peru, the vaccine provided 85% protection during the first few months, but the immunity declined rapidly in younger children after 6 months. In older children and adults, the level of protection remained around 60% for up to 2 years. Shanchol is a bivalent OCV that contains killed whole-cell *Vibrio cholerae* O1 and O139 but lacks a recombinant B-subunit cholera toxin component (I.V.I., International Vaccine Institute 2015). This OCV has the practical advantage that it does not require reconstitution in a buffer solution before administration. It is administered in two doses, with 2 weeks in between the first and last administration. In a field trial, Shanchol provided 65% protection, and more importantly the protection was sustained for at least 5 years without requiring additional booster doses. Both OCVs (Dukoral and Shanchol) are not approved for use in infants (<2 years and < 1 year of age, respectively) (W.H.O. 2013, 2014).

Some studies recently reported on the effectiveness, predicted effectiveness or tools to calculate the effectiveness of the WHO approved OCVs in Africa. In two different studies that used existing data from cholera outbreaks in Guinea-Bissau (Azman et al. 2012) and Zimbabwe, India and Tanzania (Reyburn et al. 2011a) researchers estimated the effect that reactive cholera vaccination campaigns would have had. Both groups found that reactive vaccination would likely have averted a large portion of cases and saved many lives. The same conclusion was reached when two groups of researchers studied recent data from an outbreak in South Sudan; the results suggest that vaccination halted the spread of cholera and that vaccinated individuals were 4.5-fold less likely (than non-vaccinated individuals) to contract the disease (Azman et al. 2016b; Bekolo et al. 2016).

Gabutti et al. (2012) estimated the efficacy of the oral anticholeric vaccine Dukoral in subjects travelling to high-risk areas (often in African countries). Their results showed that Dukoral was effective in reducing the risk of contracting diarrhoea in travellers exposed to high-risk conditions. No serious adverse effects were reported by the vaccinated cohort (Gabutti et al. 2012). Initially OCVs were not approved for use in pregnant women. However, recent studies have not shown any significant adverse effects and inclusion of prenatal patients in cholera vaccination efforts is now encouraged (Grout et al. 2015; Moro and Sukumaran 2017). In Zanzibar, Khatib et al. (2012) found that Dukoral conferred a 79% protection against cholera in participants who received two vaccine doses (Khatib et al. 2012). Indirect protection (often referred to as 'herd protection') was also demonstrated by a decline in the cholera risk in non-vaccinated residents within a neighbourhood as the vaccine coverage increased. Based on the findings, the authors suggested that the oral cholera vaccine offered both direct and indirect (herd) protection in the sub-Saharan African setting (Khatib et al. 2012). Reports on the use of Shanchol from Ethiopia, Sudan and Guinea were also encouraging (Ciglenecki et al. 2013; Luquero et al. 2013, 2014; Desai et al. 2014). In two different reports, Ciglenecki et al. (2013) and Luquero et al. (2013) reported on the first large-scale use (>300,000 doses) of an oral cholera vaccine as an outbreak control measure in Africa (Ciglenecki et al. 2013; Luquero et al. 2013). The study was carried out in Guinea, and this was also the first time Shanchol was used in Africa. Their study addressed concerns about feasibility, timeliness and acceptability by the population, all factors which may discourage the use of OCVs during epidemics. Despite logistical challenges (short time frames, the requirement of a two-dose schedule, the remote setting and dealing with a highly mobile population) the campaign was well accepted, and high vaccination coverage was achieved (Ciglenecki et al. 2013; Luquero et al. 2013). A matched case study determined that vaccination with Shanchol in Guinea resulted in effective protection (calculated at 86.6%) of at-risk members in households (these were selected based on having at least one resident cholera patient) (Luquero et al. 2014). The results showed that the OCV was effective when used in response to a cholera outbreak. In Sudan, Shanchol was administered to refugees within the challenging context of a humanitarian crisis after conflict erupted in South Sudan's capital city, forcing large numbers of displaced people into temporary camps (W.H.O. (World Health Organization) 2014). Despite the challenges, the pre-emptive campaigns were successful in the internally displaced people (IDP) camps, with valuable lessons learnt along the way. In Rwanda, a cholera outbreak occurred in a refugee camp and provision was made to administer Shanchol to approximately 10,000 refugees (Binagwaho et al. 2012). However, authorities managed to contain the spread of the disease by proactively applying sanitation measures (Binagwaho et al. 2012). In a randomized, controlled, double-blinded, placebo-controlled trial carried out in Ethiopia, a seroconversion rate of 81% and 77% was seen after a two-dose application of Shanchol in adults and children, respectively (Desai et al. 2014). No serious adverse effects were reported by participants during this study. Similar seroconversion rates were seen during a more recent outbreak in South Sudan, and the analyses suggested that a single dose of Shanchol may provide adequate protection (instead of the generally accepted two-dose applications) (Iyer et al. 2016). Single-dose applications of OCVs were further investigated by several authors, mainly using Sudan and Zambia data, and all showed that this approach could be more cost-effective while still providing sufficient protection (Azman et al. 2016a; Parker et al. 2017; Poncin et al. 2018).

In a fascinating study by Kim et al. (2016), researchers estimated the impact that vaccination campaigns can have on future cholera case numbers. Using mathematical modelling, they calculated that periodic OCV vaccination (every 3 or 5 years) can significantly reduce the global burden of cholera (up to 80% fewer cases in some African countries), but that it is unlikely to eradicate the disease. The same study also mentions a possible shift in cholera transmission mechanisms (away from water>human towards human>human transmission) as the impact of vaccination increases (Kim et al. 2016).

Several researchers investigated the cost versus benefit of reactive cholera vaccination (specifically OCV) campaigns. Jeuland et al. (2009) found that the exclusion of indirect effects (i.e. herd protection) may lead to the underestimation of the cost-effectiveness of vaccination programs with oral cholera vaccines (Jeuland et al. 2009). They showed the cost-effectiveness of OCV campaigns using data from some countries (including one African country, Mozambique). Using data from the recent large-scale cholera outbreak in Zimbabwe, Kim et al. (2011) concluded that reactive vaccination campaigns have the potential to be cost-effective in the control of cholera outbreaks. In their model, the cost-effectiveness was closely linked to the number of case fatalities, per dose vaccination price and the size of the outbreak. In light of the relatively high OCV prices during a Zanzibar mass vaccination campaign, and the relatively low numbers of people at risk (low incidence rates), Schaetti et al. (2012b) concluded that this specific campaign was not cost-effective. The studies mentioned above all looked at the cost of the vaccine campaign versus the benefit in the reduction of cases, fatality rates or DALYs, but did not include costs incurred by residents who need to travel to vaccination centres. Using a travel-cost approach, and data from Beira (Mozambique), Jeuland et al. (2010) reported on a households' demand for 'no cost' cholera vaccination. Their results indicated that, in contrast to wealthier households, poor households were often unable to spend the necessary time and money in order to receive vaccinations, mainly because they were situated farther away from OCV administration centres than wealthier households. Poorer households were also (in general) less well informed about the OCV campaign. In Malawi, the financial cost of vaccination (Shanchol, two doses) was calculated using a standardized cholera cost tool, Choltool, at USD 7.14 per person vaccinated. The total economic cost (including indirect patient-related costs) was calculated as USD 8.75 (Ilboudo and Le Gargasson 2017). The researchers found that the most significant proportion of the cost went towards the procurement and shipment of the vaccines, but that human resource expenses related to the roll-out of the campaign contributed significantly to the overall cost. In certain settings, cholera vaccination campaigns can be rolled out in a more cost-effective way, as Poncin et al. (2018) showed in Zambia where a campaign cost was calculated at USD 2.31 per dose. However, indirect costs were not included in this calculation. The successful implementation of the Global Oral Cholera Vaccine Emergency Stockpile in 2013 and first African use thereof in Sudan in 2014 may have helped to drive down costs and increase availability. Even though this has helped to streamline the process, vaccine availability is still inadequate in most outbreak situations.

Due to the global shortage of OCVs vaccination campaigns and the number and location of vaccine delivery sites must be carefully planned in order to effectively use the available doses and to keep travel and queuing costs low (Jeuland et al. 2010). The role of effective logistics in ensuring good vaccine coverage was also mentioned by Schaetti et al. in a study that looked at the lessons learnt from the OCV campaigns in Zanzibar (Schaetti et al. 2012a). Using data from Niger, Guerra et al. (2012) made a case for using cholera surveillance data to prioritize areas under consideration for vaccination. Prioritization of high-risk areas, as shown in the simulated study done in Guinea-Bissau, could be vital in limiting the size and spread of outbreaks (Azman et al. 2012). Insufficient supplies of WHO pre-qualified vaccines prohibit their widespread and indiscriminate use in Africa, and Deen et al. (2016) called for the development of prioritization guidelines. They also showed how a scenario approach could help in the ranking process of cholera epidemics, thereby allowing for prioritized selection of regions during vaccination campaigns.

In a cross-sectional study done in the Democratic Republic of Congo (DRC) anticipated acceptance of cholera vaccines (specifically OCV) was investigated (Merten et al. 2013a). The results suggested a high motivation to use an OCV, but only if it seemed affordable. This study highlights the role that cost (direct and indirect) plays in the potential success of a cholera vaccination campaign (Merten et al. 2013a). Though there are many technical and logistical reasons why cholera vaccination campaigns have not always been successful in Africa, it is the opinion of von Seidlein et al. (2013) that the most significant barrier has been that populations affected by cholera outbreaks are underprivileged and lack a strong political voice.

Many aspects of cholera outbreaks are dependent on human behaviour, and community acceptance of OCVs (based on unique social and cultural landscapes) is not always a given. In the Democratic Republic of Congo (D.R.C.), Western Kenya and Zanzibar, the anticipated acceptance of OCVs were found to be high (>90%), on condition that the vaccine was provided at no or a low cost (Schaetti et al. 2011; Merten et al. 2013a, b; Sundaram et al. 2013). Cholera-specific symptoms, income, and education were positively associated with OCV acceptance in Zanzibar and Kenya (Schaetti et al. 2011; Sundaram et al. 2013). Interestingly the study carried out by Schaetti et al. (2011) to determine anticipated OCV acceptance was followed up by a study on the actual uptake of OCVs after a mass vaccination campaign was completed (later) in the same study area (Schaetti et al. 2012a). They reported that the OCV uptake was lower than the anticipated acceptance, with only 49.7% of study respondents having taken the two doses required for effective vaccination (Schaetti et al. 2012a). They found that female gender, rural residency, and advanced age were positive determinants of OCV uptake, while men were less inclined to make use of the protection offered by the oral vaccines. Serious cholera-specific symptoms and fear within the community that the local healthcare system would be overwhelmed by the outbreak were also positively correlated with OCV uptake (Schaetti et al. 2012a). In some African countries, the lack of trust that was associated with perceived institutional negligence (especially by Governments) caused hesitancy in the acceptance of cholera vaccination campaigns (Démolis et al. 2018). The authors highlighted the pivotal role that community leaders can play in alleviating politically motivated resistance to vaccination efforts. Cholera vaccination campaigns are often well accepted in African countries, although some populations remain hesitant. For example, 97.6% acceptance was recorded in Malawi during one specific campaign (Msyamboza et al. 2016). To increase acceptance, it may be important to shape specific delivery strategies for unique and at-risk populations, in the same manner that fishers in Malawi were targeted in a recent cholera vaccination campaign

(Sauvageot et al. 2017; Heyerdahl et al. 2018). This campaign used simple but effective tailor-made approaches to reach high levels of vaccine coverage in hard-to-reach populations. A summary of the most often cited reasons for cholera vaccine acceptance or hesitancy is given in Table 3.1 (Choi et al. 2010b; Nyambedha et al. 2013; Desai et al. 2014; Msyamboza et al. 2016; Deen et al. 2016; Sundaram et al. 2016; Peprah et al. 2016; Démolis et al. 2018; Pugliese-Garcia et al. 2018).

Table 3.1 Cholera vaccine acceptance considerations in African countries (2009–2018)

Associated with an increase in acceptance
• Vaccination is free (no cost)
• Higher income (only associated with increased uptake in some countries)
• Cholera perceived to be a severe disease
• Populations who perceive themselves to be very vulnerable to cholera
• Involvement of community leaders during vaccination campaign roll-out
• Visible success by public health officials (earlier outbreaks or other health emergencies)
• Populations with less access to public health facilities (and therefore deem themselves to be more vulnerable)
• Awareness of other cholera prevention methods (like good sanitation and hygiene)
• The trust that displaced populations have in aid organizations (like the United Nations)
• A higher level of education
• Easy access to vaccination centres (transport to centres, or mobile centres coming to the affected areas)
Associated with a decrease in acceptance
• Payment required for vaccination (cost of vaccination)
• Low perceived need for immunization
• Rumours related to vaccine safety (and fear of infection due to vaccination)
• Fear of side effects
• Negative experiences following routine childhood immunization
• Fear of injections and pain (wrongly with cholera vaccines)
• Fear of taking the vaccine alongside other medication or with alcohol
• Distrust in national institutions (and perceived institutional negligence)
• Politically motivated resistance
• Fears of ethnic persecution (for marginalized communities (refugees))
• Higher age
• Larger households
• Lack of education
• High level of social vulnerability
• Religious beliefs that are in contrast to healthcare practices
• Distrust towards western medicine
• Reliance on traditional remedies
• Lack of time
• Men are less likely to accept vaccination than women

3.2.3 Cholera and the Host

Vibrio cholerae (the species of bacterium that causes cholera) is well known for its ability to survive in two completely different settings: the natural aquatic environment and the human host environment. This section looks specifically at African research outputs that focussed on host-related aspects of cholera over the last decade, and how this contributed to our understanding of outbreaks in terms of spread and recurrence. Studies are examined that describe behavioural factors that increase (or decrease) the risk of outbreaks of diseases in specific populations or areas, and briefly cover African studies that look at the impact of cholera on people's lives.

The role of human movement in the transmission and spread of cholera epidemics was emphasized in several studies, with the role of specific factors highlighted by some, while others reported on heterogeneity and complexity of transmission dynamics (Mari et al. 2012; Njagarah and Nyabadza 2014; Njagarah 2015; Mukandavire and Morris 2015; Weill et al. 2017). In a study that looked at the control of cholera in communities that are linked by migration, researchers found that the implementation of controls (such as proper hygiene, sanitation and vaccination in both affected communities) may mitigate cholera outbreaks within half the time required for self-limitation (Njagarah 2015). Using a novel approach to track human movement (by looking at mobile phone usage) in Senegal during the 2005 epidemic, researchers showed how a mass gathering that occurred early in the outbreak had a significant influence on the course of the epidemic (Finger et al. 2016). Mukandavire et al. (2011) noted the heterogeneity of cholera transmission dynamics in different provinces of Zimbabwe during the country's recent epidemic. The heterogeneity was ascribed to differences in environment, socio-economic conditions and cultural practices (Mukandavire et al. 2011). Osei et al. (2012) also observed spatial variation in the risk of cholera infection upon examining epidemiological data from Ghana; a higher risk of cholera was associated when a community contained slum settlements, had high population density and was near refuse dumps.

Cameroon is an African country that has experienced many cholera outbreaks in the last two decades (many with a high case fatality rate), and it is here that Djouma et al. (2016) investigated risk factors that may have contributed to the disease burden. The researchers found that non-use of cholera treatment centres, delays in deciding to go to these treatment centres or choosing to go to temporary treatment centres instead of hospitals were all risk factors for cholera-related mortality. Another study from Cameroon found that age below 21 years, eating outside of the home and poor food preservation practices were independent risk factors of cholera (Nsagha et al. 2015). In South Sudan, Ujjiga et al. (2015) identified travelling and eating away from home as cholera risk factors, while treating water and receiving oral cholera vaccination were protective. A geospatial assessment of cholera in a rapidly expanding urban area in Nigeria revealed that living close to waste dump sites or a market place was associated with an increased cholera risk (Olanrewaju and Adepoju 2017). In Kenya, cholera incidence was linked to open defecation, poverty, the use of unimproved water sources and to a low concentration of healthcare facilities (Cowman et al. 2017).

Using data from a 2015 outbreak in Ghana, a study investigated the complex interplay that exists between everyday risks (like poor sanitation) and disaster risks (like flooding events) (Songsore 2017). The insights of the study can help with the planning and thought processes necessary to address groups of overlapping risks.

Hove-Musekwa et al. (2011) found that numerical simulations suggest an increase in susceptibility to cholera due to malnutrition, and suggested that nutritional issues should be addressed in impoverished communities affected by cholera. Poverty was also reported by Penrose et al. (2010) to be a significant driver of cholera in Dar es Salaam, Tanzania. They found that cholera incidence was closely associated with informal settlements (Penrose et al. 2010).

According to Roy et al. (2014), epidemic cholera propagate in similar ways to aggressive wildfires (in non-endemic areas) which have implications for the effectiveness of control measures and the mechanisms that may ultimately limit the size of outbreaks. A detailed understanding of the interconnection of potential host-related drivers of transmission within Africa is essential in the fight against the disease.

In order to understand how communities react and respond to outbreaks, the social and cultural features of cholera need to be taken into account. Using a vignette-based semi-structured interview approach, researchers studied sociocultural aspects of the disease in regions located in the Democratic Republic of Congo (DRC), Kenya, and Zanzibar (Schaetti et al. 2010, 2013; Nyambedha et al. 2013). The results from the three regions were summarized in a systematic comparison (Schaetti et al. 2013). Cholera was reported to have a significant impact on the respondent's lives in all three areas, with the social impact being mainly characterized by financial concerns (for instance, loss of income). The economic cost of cholera (carried by households) was determined by a team of researchers that analysed data from a severe outbreak in Ghana (Awalime et al. 2017). They found that the direct cost to the household, that had at least one member falling ill with cholera, was between US$ 62 and US$ 107, and that costs were higher for poor households that were situated in high incidence areas. Similar findings were reported in Malawi, where the cost of cholera per affected household was found to be US$ 65.6 on average (Ilboudo et al. 2017). The researchers included direct and indirect costs in the equation. They also looked at the cost that treatment facilities had to bear, and found that the direct costs were on average US$ 59.7 per cholera patient treated (Ilboudo et al. 2017). Both studies showed that a reduction in cholera cases could lead to a substantial potential cost saving at the household and national level.

Several publications have reported on the level of awareness and preparedness for cholera outbreaks in different communities. Knowledge can increase a population's resilience against cholera as it may equip household members to prevent infection and respond appropriately to illness. In one study that looked at such perceptions and practices in multiple countries, the researchers found that clean water (and food) along with vaccination was perceived to be sound preventative measures (in DRC), while elsewhere safe water, sanitation and health education was highlighted as being important considerations (in Kenya and Zanzibar). Rehydration (using oral rehydration solutions) was highlighted by all the countries included in the study, and healthcare facilities were universally seen as a source of help. Though many similarities were noted regarding sociocultural features between the study areas, there were also differences aplenty (in particular vulnerability perceptions). The results emphasized the need for similar studies elsewhere (Schaetti et al. 2013). Cultural beliefs and practices that may influence cholera spread and transmission were studied in Cameroon using a qualitative approach (Ngwa et al. 2017). During group discussions, participants said they believed cholera to be reprimand from God and that transmission occurred through the air, but some participants also felt that mass gatherings (like funerals and weddings), poor sanitary practices and mountaintop burials might contribute to cholera outbreaks. Many cultural beliefs clashed with the good practices of western medicine (for instance, going to the hospital); this conflict was termed by the researchers as a 'rural-urban mentality confrontation'. The authors found that many other sociocultural factors also play a role in cholera transmission and call for a better understanding of the sociocultural context before launching an outbreak response. A quantitative case-control study that was also carried out in Cameroon found that most respondents thought that poor hygienic practices and contaminated water were the main transmission routes of cholera (Nsagha et al. 2015). In South Africa, an assessment of the knowledge, attitudes and practices regarding cholera preparedness was conducted within a village that was affected by a cross-border cholera outbreak in 2008 and 2009. A large portion of community members indicated that they knew how cholera was contracted (via contaminated water) and that they knew how to prevent getting ill (Ncube et al. 2016). Respondents also knew about treatment for cholera at healthcare facilities, though fewer knew about oral rehydration solutions that can be used in home-care. However, many undesirable practices were witnessed in the village (like poor hygiene and sanitation), indicating that the sound knowledge regarding cholera prevention and response may still need to be converted from theory into standard practice. The insights gained from community studies may prove to be invaluable in the planning of programs and initiatives directed at cholera control and elimination.

In Nigeria, a study focusing on the role that healthcare workers (HCW) play during cholera emergency response situations showed that HCWs were often too few (Oladele et al. 2012). They frequently lacked adequate training and often experienced a shortage of emergency response kits. Questionnaire data also suggested that HCWs, even though vigilant of personal safety precautions, often contracted cholera during epidemics suggesting that they may be at higher risk than the general population (Oladele et al. 2012). In Kenya, Date et al. (2013) evaluated the effectiveness of a cholera education programme that was rolled out in response to recurring cholera outbreaks. The authors found that attendance of a cholera education event may have (in some instances) positively influenced household water treatment practices, but that there was a gap between the knowledge and practice of water treatment (Date et al. 2013). Interpretation of the study's results was complicated by the fact that respondents in the comparison group, albeit to a lesser extent than the intervention group, recalled attending other cholera education events (unrelated to the current education intervention) (Date et al. 2013).

3.2.4 Diagnosis of Cholera

Keddy et al. (2013) discussed the challenges and options relating to the diagnosis of cholera in an African setting. They provided insights into the state diagnostic testing, called for the standardization of diagnostic methods, reviewed newer technologies that may in future become more widely used and discussed some of the unique African challenges (Keddy et al. 2013). A reliable diagnosis is a critical first step and is required to initiate effective healthcare solutions at the onset of cholera outbreaks. Three papers evaluated the performance of a rapid diagnostic method (Crystal VC Rapid Dipstick) in stool samples (Harris et al. 2009; Ley et al. 2012; Martinez-Pino et al. 2013). The Crystal VC test was reported to have a high sensitivity at 93% or higher (comparable to that of standard culture methods) but exhibited poor specificity (49.2% in one study, and 71–76% in another) (Harris et al. 2009; Ley et al. 2012). Ley et al. (2012) conclude that the current version of the Crystal VC dipstick could potentially be used as a screening tool in the field, but that the high proportion of false positive results would necessitate the confirmation of results by stool culture. Another potential problem of using the Crystal VC test is that individuals vaccinated with Shanchol often test positive for *Vibrio cholerae* O139 for up to 5 days after ingesting the oral cholera vaccine (however, the cross-reactivity was not seen with *Vibrio cholerae* O1) (Martinez-Pino et al. 2013). As such the Crystal VC test could be used without delay during and after reactive oral cholera vaccination campaigns in outbreaks where *Vibrio cholerae* O1 is the causative agent, but the specificity limitations of the test may still pose a challenge. To address the specificity challenges, two studies investigated the use of an enrichment step before identification/detection with the Crystal VC dipstick (Ontweka et al. 2016; Bwire et al. 2017b). Both studies reported an increase in specificity, and that the optimized performance of the dipstick (when used in conjunction with pre-enrichment) makes it a viable option for rapid case detection in rural African areas.

3.2.5 Clinical Perspectives

Oral rehydration is still widely promoted as an effective treatment option for cholera patients. Musekiwa and Volmink (2011) investigated the use of oral rehydration salt solutions (ORS) with an osmolarity of ≤270 mOsm/L versus an osmolarity of ≥310 mOsm/L. In people treated for cholera using ORS with an osmolarity of ≤270 mOsm/L, biochemical hyponatraemia was sometimes seen, compared to patients treated with ORS with an osmolarity of ≥310 mOsm/L although the hyponatraemia did not appear to be associated with any serious consequences.

Although rehydration plays a pivotal role in reducing cholera-related mortalities, antibiotics have traditionally been used to reduce the duration of disease, lessen organism shedding and treat severe cases. However, the widespread use of antibiotics in prophylaxis and treatment has resulted in the emergence of antibiotic-resistant strains, with many African countries reporting multidrug resistant strains (Kacou-N'douba et al. 2012; Marin et al. 2013). The surveillance of antibiotic susceptibility patterns in clinical and environmental strains (the latter could serve as a genetic reservoir) continues to be necessary to ensure that antibiotics remain an effective treatment option. It is also imperative that new antimicrobial drugs be discovered and developed. Tebou et al. (2017) found that flavonoids extracted from a small evergreen tree (*Maytenus buchananii*) exhibited antimicrobial qualities that were effective against *Vibrio cholerae*. The leaves of this tree have also been traditionally used to treat other ailments (Tebou et al. 2017). Plants from the *Piper* genus have also traditionally been used for medicinal purposes in Africa, and bioactive compounds extracted from them have recently been shown to have antimicrobial activity against *V. cholerae* (Mgbeahuruike et al. 2017). In both cases, further investigation, testing and development will be required to elucidate the full potential for clinical applications (Mgbeahuruike et al. 2017; Tebou et al. 2017). It may still be a long path before any of these promising compounds are cleared for therapeutic use against current antibiotic-resistant cholera clones.

Okosun and Makinde (2014) investigated the synergistic relationship between malaria and cholera using a modelling approach. They found that malaria infection may be associated with an increased risk of cholera. However, cholera infection was not found to be associated with an increased risk for malaria. In an article that focussed on the HIV immune suppression and how that could potentially predispose individuals to cholera, Mushayabasa and Bhunu (2012) developed a simple mathematical model to assess whether HIV infection was associated with an increased risk for cholera. The findings suggested that HIV infection was associated with an increased risk for cholera in cholera-endemic areas.

As seen with HIV, the human immune system plays a vital role in combatting and preventing infections like cholera. Ssemakalu et al. (2014) reviewed literature about the influence of solar water disinfection on immunity against cholera. In the solar water disinfection method (SODIS), heat and ultraviolet radiation from sunlight are used to inactivate pathogens within water intended for consumption (McGuigan et al. 2012). The inactivated pathogens, unlike in the case of filtration, are still consumed during drinking, and these antigens may interact with determinants of human mucosal immunity. The review highlighted gaps in the current body of knowledge as it pertains to the ability of SODIS to promote cholera immunity, and Ssemakalu et al. (2014) suggested further study to clarify this.

3.2.6 Governance and Policy

In response to cholera outbreaks, governments often launch a healthcare intervention (or a suite of interventions). Makoutode et al. (2010) assessed the quality of an outbreak response following the 2008 cholera outbreak in Benin. The authors concluded that the national response plan for the mitigation of cholera outbreaks worked well in Benin, as the case fatality rate was kept low during this recent outbreak. In recent times there has been a shift from a reactive response to a more proactive approach, with the ongoing burden of cholera in Africa prompting several initiatives aimed at the elimination and reduction of the disease (Muyembe et al. 2013; Bompangue 2014; Haaser 2014; Hessel 2014). In the Democratic Republic of Congo (DRC), national policies and plans were drafted to combat cholera (Muyembe et al. 2013; Bompangue 2014). The government of the DRC committed to cholera elimination through the adoption of a new national policy built on improved alert, response, case management and prevention strategies (Muyembe et al. 2013). The authors used historical epidemiological data (2000–2011), along with extensive research (2005 onwards), to understand the spatial and temporal dynamics of cholera in the DRC. The study showed that cholera epidemics were focused in and around the areas of a small lake, being more prevalent in the dry season (Bompangue 2014). It was recommended that these areas (especially in the dry season) be prioritized. Based on the findings, the DRC adopted a multisectoral plan to eliminate cholera as a public health threat by 2017. In conjunction with the national plan, a grouping of governmental, non-governmental and international organizations used a geographical information systems approach to study epidemiology, sanitation and hygiene to determine the location of cholera 'hotspots' (Haaser 2014). The study identified eight cholera 'hotspots' associated with cholera emergence in the Great Lakes Region of the DRC.

The importance of diverse preventative interventions (like clean water provision and improved sanitation) and a move from cholera response and treatment to one of proactive cholera prevention have been emphasized (Haaser 2014; Hessel 2014). Subsequently, the Global Alliance against Cholera (GAAC) was established in 2010. The primary objective of the alliance is to advocate how a sustainable multisectoral approach (as shown in the DRC) can be applied to reduce and eliminate cholera in other parts of the world (Haaser 2014). The role of another international initiative 'The Initiative against Diarrheal and Enteric diseases in Africa and Asia' (IDEA) in fighting cholera was reported by Hessel (2014). This initiative is a network of independent, multidisciplinary and multinational health professionals involved in enteric disease and cholera control and prevention. IDEA shares information and experiences for the improved control and prevention of cholera and other enteric diseases. It is believed that the interdisciplinary, inter-sectoral and cross-border approaches create an environment conducive to cross-fertilization of knowledge and practices, thereby aiding the design and formulation of novel cholera prevention strategies (Hessel 2014). Even with the existence of platforms like IDEA and GAAC cross-border cholera outbreaks remain a challenge. This was again highlighted in a recent study that investigated cholera cases on the Uganda-DRC and the Malawi-Mozambique borders (Bwire et al. 2016). The researchers found that outbreak control efforts primarily involved unilateral measures implemented by only one of the two affected countries. The authors call for guidelines and strategies that assign clear roles and responsibilities for cholera actors on both sides of the fence. This will benefit bordering countries in a mutualistic way and decrease the cholera burden of marginalized peri-border communities.

3.3 Vibrio Cholerae: The Causative Agent

3.3.1 Monitoring and Detection

Efforts to manage and mitigate cholera outbreaks often start with bacteriological monitoring of aquatic ecosystems. This is done because proactive monitoring of *V. cholerae* can help to prevent dissemination of the etiological agent *V. cholerae* before the onset of epidemics (Faruque et al. 2006). In Africa, many studies focused on the clinical detection of cholera cases (this is evident from literature covered earlier in this chapter), while less effort has been given to the monitoring and surveillance of the etiological agent, *V. cholerae*. Kaboré et al. (2018) mentioned that very little is known about the occurrence of these cholera-causing bacteria in African freshwater resources (Saidou et al. 2017).

In Africa, *V. cholerae* was occasionally isolated from environmental samples using only conventional methods like selective enrichment and cultivation of bacteria (Yirenya-Tawiah et al. 2018). Many other studies used molecular methods like polymerase chain reaction (PCR) (sometimes in conjunction with culture-based methods) to detect *V. cholerae* (du Preez et al. 2010; Madoroba and Momba 2010; Keshav et al. 2010; Ntema et al. 2010; le Roux and van Blerk 2011; Teklehaimanot et al. 2015; Gdoura et al. 2016). Polymerase chain reaction was often utilized to detect and characterize *V. cholerae*, distinguish between O1, non-O1 and O139 serogroup isolates, and determine the presence/absence of virulence factors (like cholera toxin) (Dalusi et al. 2015; Bwire et al. 2018a). In some studies, quantification was carried out using a qPCR or Most Probable Number PCR approach (Machado and Bordalo 2016; Saidou et al. 2017). A direct fluorescent antibody (DFA) method was also used in one published study, though the researchers reported poor specificity of the antibody (du Preez et al. 2010). Biochemical identification, which was commonly utilized a few years ago for microbiological confirmation of presumptive isolates, requires the use of pure cultures. This disadvantage, along with the widely known difficulties in isolating *V. cholerae* from environmental samples, may have led to this method being largely abandoned in recent times. This is evident by the lack of African *V. cholerae* monitoring studies that relied on biochemical confirmation in recent years. PCR-based detection was favoured instead, even though PCR has the disadvantage of not being able to distinguish between live or dead cells. Making use of a selective enrichment or a growth step before employing PCR can provide some information on viability. Diagnostic dipsticks are sometimes used for the detection of *Vibrio cholerae* O1 or O139 in clinical stool samples, but not for environmental monitoring due to their low sensitivity. However, two recent studies evaluated the use of the Crystal VC dipstick (made in India) using a pre-enrichment step in alkaline peptone water (Ontweka et al. 2016; Bwire et al. 2017b). The low-cost method employs sterile gauze filtration to concentrate cholera bacteria, and the dipsticks performed very well when used in conjunction with an enrichment step (a specificity of >90% compared to PCR results was reported). The authors also provide a solution for DNA preservation using filter paper, a cost-effective and straightforward method to allow long-term storage and transport of samples when delayed PCR analysis is required (Ontweka et al. 2016).

In Africa, *Vibrio cholerae* was often detected in water samples. However, detection rates vary considerably between studies. There are many possible reasons for the difference in detection rates. Foremost is the diverse range of water sources that were sampled. Some studies looked at stored drinking water, others focused on freshwater resources (like lakes and rivers), and some investigated treated wastewater effluents. Treated wastewater effluents in South Africa often harboured *V. cholerae*, with over 90% of samples containing presumptive isolates (although PCR confirmation excluded about half of these) (Teklehaimanot et al. 2015). No cholera bacteria from the O1 and O139 serogroups were detected in this study. In Uganda's Greater Lakes region, *V. cholerae* was detected in 10.8% of surface water samples; however, no enterotoxigenic strains were detected (genes coding for cholera toxin were absent) (Bwire et al. 2018a). Interestingly, the researchers found a limited number of atypical serogroup O139 isolates during their study. The first reports of *V. cholerae* O139 in African waters came from Mozambique in 2010 and also from Tanzania in 2015; this serogroup has very rarely been found on this continent (du Preez et al. 2010; Dalusi et al. 2015). The *V. cholerae* detection rate in Tanzanian coastal water (at 10%) was found to be very similar to that reported in the Uganda freshwater study, with the difference being that many of the Tanzanian strains harboured cholera toxin and could, therefore, be regarded as enterotoxigenic (Dalusi et al. 2015). In Burkina Faso, 35.8% of surface water reservoirs (lakes and dams) were found to contain *V. cholerae*, but O1 or O139 serogroup organisms were not detected, nor did any of the strains harbour the cholera toxin (Saidou et al. 2017). Drinking water samples that were obtained from household storage systems in Ghana's Accra Metropolitan Area frequently contained presumptive *V. cholerae* (83.8% detection rate) (Yirenya-Tawiah et al. 2018). A small percentage of the isolates (1.5%) were found to be of serogroup O1, which is essential because it is frequently implicated in cholera epidemics in Africa. *Vibrio cholerae* of the serotype O1 was also infrequently detected in environmental samples from two other studies by Keshav et al. (2010) in livestock faecal samples and du Preez et al. (2010) in marine/estuarine water samples from Beira, Mozambique. In South Africa and Guinea Bissau, the presence of enterotoxigenic strains was also noted to be much lower than the overall detection rate of *V. cholerae* in environmental water samples (le Roux and van Blerk 2011; Machado and Bordalo 2016). In Cameroon, a large number of environmental water samples (n = 1011) were tested for the presence of cholera bacteria (Debes et al. 2016a). The researchers found that 24.1% of the samples contained *V. cholerae*. However, no O1 or O139 serogroup bacteria were amongst those detected. The researchers stated that hospitals in the study area recorded significant disease caused by non-O1 and non-O139 *V. cholerae* strains during the study period. This emphasizes the importance of environmental monitoring, even if enterotoxigenic O1 and O139 strains are not prevalent.

Though most of the environmental surveillance was carried out on water samples, one study also looked at faecal matter obtained from livestock in rural areas of South Africa. They found that 32.2% of all samples tested positive for the presence of *V. cholerae* and that a small percentage of the *V. cholerae* isolates (approximately 5%) belonged to the O1 serogroup (Keshav et al. 2010). Interestingly excreta from cows contained these bacteria far more often than that of donkeys, goats, pigeons, pigs or chicken and this may indicate that certain animals are disproportionally affected by *V. cholerae*. Elsewhere in Africa, cholera was also found to be a significant disease in Zoo-kept birds (Emikpe et al. 2016). In Egypt, 225 crustacean samples were analysed for the presence of pathogenic *Vibrio* species, and it was found that 0.9% of these samples contained *V. cholerae* (Ahmed et al. 2018). In Tanzania, *Vibrio cholerae* was often detected in wastewater, vegetables (irrigated with wastewater) and fish samples (from stabilization ponds), with detection rates of 36.7, 21.7 and 23.3%, respectively (Hounmanou et al. 2016).

In South Africa, riverbed sediments were analysed for the presence of *V. cholerae* virulence-associated genes (VAGs), and in this study, the haemolysin gene (*hlyA*) was regularly detected. The presence of *V. cholerae* VAGs in sediments suggests that riverbeds may act as reservoirs for these potentially pathogenic bacteria (Abia et al. 2017). In an earlier study, Abia et al. also showed that exposure to river water under conditions of riverbed sediment resuspension could be associated with an increased public health risk. *Vibrio cholerae* was the most detected pathogen in the sediments of this peri-urban river (Abia et al. 2016a).

Smith et al. (2012) highlighted how important reliable and trustworthy results are when it comes to the detection of environmental sources of epidemic diseases like cholera. They reported on a public health response that followed a pseudo-alert of *Vibrio cholerae* O1 in a river water sample (Smith et al. 2012). A laboratory incorrectly reported the presence of *V. cholerae* O1 after they likely contaminated the sample with a *V. cholerae* O1 reference strain (Smith et al. 2012). False alarms like the example above can lead to the costly deployment of limited public health resources.

3.3.2 Genotyping and Phylogenetics

A current focus area in cholera research is the evolution and global spread of *V. cholerae* strains, especially atypical strains, using genomic approaches. It has been shown that related *Vibrio cholerae* O1 strains often circulate between the African and Asian continents and these insights have broadened (and continue to expand) our understanding of waterborne disease epidemiology (Mutreja et al. 2011).

The current Seventh pandemic is caused by *Vibrio cholerae* O1 strains of biotype El Tor, with the genes encoding the cholera toxin (*ctxAB*) contained on the genome of a CTX prophage. However, El Tor biotype strains with classical biotype characteristics (hybrid strains) and El Tor strains that harbour the classical type cholera toxin (altered strains) have been reported since the early 1990s (Nair et al. 2006). In this chapter, the two variant types will collectively be referred to as atypical strains. Reports on the occurrence of atypical strains in Africa suggest that these variants have become widespread on the continent (Choi et al. 2010a, b; Ceccarelli et al. 2011;

Ismail et al. 2011; Bhuiyan et al. 2012; Naha et al. 2013; Oyedeji et al. 2013; Saidi et al. 2014; Sambe-Ba et al. 2017). One of the earliest examples of the occurrence of atypical strains in Africa was the discovery of strain B-33 in Mozambique in 2004 by Faruque et al. (2007). They described an El Tor strain that carries the *rstR* gene of Classical biotype (*rstR*[Class]). The CTX prophage was integrated into the small chromosome of this strain's genome, rather than being present on the large chromosome as is typical for the seventh pandemic *Vibrio cholerae* O1 El Tor isolates. Also, RS1 (a CTX phage-related element) seemed to be absent from the genome of the Mozambique strain (B-33). It was later shown by Choi et al. (2010b) that the strain carries a new type of the RS1 element on the large chromosome, along with a *ctxB* gene of classical origin. Based on the genome-wide similarity to that of other seventh-pandemic El Tor strains, it has been proposed that the atypical Mozambique strain evolved from a seventh pandemic progenitor (Faruque et al. 2007; Halder et al. 2010). An atypical El Tor strain was isolated during an outbreak in Angola in 2006, and interestingly this isolate carried ctxB[Class] on the large chromosome (Ceccarelli et al. 2011). Choi et al. (2010a) analysed several atypical *Vibrio cholerae* strains (isolated from various regions in Asia and Africa) using multilocus variable-number tandem repeat analyses (MLVA). They identified three MLVA profile groups amongst the atypical strains and conclude that the origin and diversification of these altered and hybrid strains require further study. Various authors reported that atypical *Vibrio cholerae* strains express cholera toxin at higher levels under in vitro conditions (Choi et al. 2010a; Ssemakalu et al. 2013; Naha et al. 2013); however, the link between up-regulated expression of cholera toxin and virulence needs to be investigated further.

   The emergence of atypical *Vibrio cholerae* El Tor strains received much attention in recent African studies, with authors often using molecular techniques to compare novel variants to other endemic and historical strains. Pulsed field gel electrophoreses (PFGE) and Multiple-locus variable number tandem repeat analysis (MLVA) were the phylogenetic techniques most often used (Choi et al. 2010a; Reimer et al. 2011; Bhuiyan et al. 2012; Mohamed et al. 2012; Ismail et al. 2012, 2013; Naha et al. 2013; Rebaudet et al. 2014; Saidi et al. 2014; Adewale et al. 2016). Using MLVA Choi et al. (2010a) showed that atypical *Vibrio cholerae* strains (isolated from various regions in Africa and Asia) could be grouped into three different groups and that they were distant from the Seventh Pandemic El Tor strain. Using the same technique, Mohamed et al. (2012) showed that multiple genetic lineages of *Vibrio cholerae* were simultaneously infecting people in Kenya. Rebaudet et al. (2014) demonstrated the inter-epidemic progressive genetic diversification of strains from a single genotype (from a single index case) in Guinea. Ribotyping and random amplified polymorphic DNA (RAPD) was used in Somalia to compare outbreak strains isolated in the late 1990s, which confirmed that this group of strains had a clonal origin (Scrascia et al. 2009). Researchers in Zambia and Zanzibar reported that atypical El Tor strains formed separate PFGE clusters when compared to the seventh pandemic El Tor isolates. It was also shown that the atypical Zambian strains were closely related to Bangladesh altered strains, and less so to Mozambique variants (Bhuiyan et al. 2012; Naha et al. 2013). In many countries, the atypical strains of that country or region grouped closely together, indicating a single clonal lineage (Bhuiyan et al. 2012; Ismail et al. 2013; Naha et al. 2013; Saidi et al. 2014). In Nigeria *V. cholerae* isolates from outbreaks from 2007 to 2013 were analysed with PFGE and the results showed that O1 serogroup bacteria from clinical and water samples clustered together, but that some isolates grouped per region (Adewale et al. 2016). The simultaneous presence of two distinct *ctxB* alleles was also found, with Ogawa serotype strains displaying the Haitian type and Inaba serotype strains carrying the classical *ctxB* allele.

For a few years, publications that focused on atypical strains dominated literature space within the field of African *V. cholerae* phylogenetic research (though there were a limited number of exceptions during this review period). Recently the interest has shifted back to phylogenetics with the aim of understanding transmission and dissemination of cholera bacteria, be it locally and/or globally. In Kenya, it was found that O1 *Vibrio cholerae* El Tor strains carrying an SXT/R391-like element did not significantly change over 13 years (Kiiru et al. 2009). Ghanian *Vibrio cholerae* strains isolated in two periods (1970-1980, 2006) were compared using multilocus sequence analysis (MLSA), gene comparisons, and PFGE by Thompson et al. (2011). The MLSA results revealed two major clusters: El Tor and Amazonia/ Ghana, with the latter occurring only in the late 1970s and found to be closely related to South American strains (Thompson et al. 2011). PFGE was used to characterize a subset of the 1970s samples, where the higher resolution of this method revealed genetic diversification within the strains from Ghana/Amazonia and El Tor clusters. Ismail et al. (2012) used PFGE data to described how an O1 isolate was imported to South Africa from India. In an attempt to identify the source of the Haiti outbreak strains, Reimer et al. (2011) used an assortment of techniques (including PFGE and Single Nucleotide Polymorphisms (SNPs)); however, the origins of the outbreak strain could not be traced. Nevertheless, data from another study using SNPs as high-resolution markers suggested that the Haitian strain may have had its origins in southern Asia (Mutreja et al. 2011). A comparative characterization, using PFGE amongst other methods, revealed a complex relationship between human isolates from five sub-Saharan African countries (Smith et al. 2015). The isolates were collected between 2010 and 2013 and came from Guinea, DRC, Togo, Ivory Coast and Mozambique. In some instances, isolates clustered together by country, but other clusters were made up of isolates that were shared between countries—indicating that the same genetic lineage was present in multiple nations. In an in-depth phylogenetic study that investigated Seventh Pandemic *V. cholerae* strains from 45 African countries (over 49 years), it was found that most major epidemics were caused by a single expanded lineage (Weill et al. 2017). This lineage was introduced into Africa at least 11 times since 1970 and caused epidemics that lasted up to 28 years. According to the authors, their results indicate that human-related factors are more important than climatic and environmental factors in the dynamics of cholera in Africa (Weill et al. 2017).

Many African countries lack the laboratory capabilities to preserve, store and transport samples from cholera-affected areas to off-site (often far off) analytical laboratories as would be required for phylogenetic studies. For this reason, Debes et al. (2016b) evaluated the use of filter paper for simplified sample preservation for delayed molecular characterization. They tested the method on samples obtained from Cameroon and found that it provided a viable low-cost solution. Their MLVA results also show that outbreak strains from Cameroon in 2014 were closely related but could still be classified into two distinct clonal complexes. Another study reported on a potential solution for difficulties that are often experienced when trying to type *V. cholerae*; the difficulties include low cell numbers in environmental water and poor isolation efficiency due to the viable-but-non-culturable (VBNC) state (Vezzulli et al. 2017). The authors evaluated a whole-genome enrichment (WGE) approach for use with next-generation sequences. In short, the approach uses biotinylated RNA baits for the target enrichment of *V. cholerae* DNA by utilizing hybridization. The approach produced next-generation sequencing results from Tanzanian environmental samples that exhibited a 2500 times increase in sequencing coverage compared to theoretical calculations of coverage without the use of WGE (Vezzulli et al. 2017). The study provided a 'proof of concept' for this new approach that may aid future phylogenetic efforts.

Phylogenetic analysis based on whole-genome sequences (WGS) was used to study the lineages of *Vibrio cholerae* strains in two comprehensive studies: one with a focus on the global spread of seventh pandemic strains and the other investigating the geophylogeny of clinical and environmental strains in Kenya (Mutreja et al. 2011; Kiiru et al. 2013). To understand the evolution of the seventh pandemic lineage, Mutreja et al. (2011) identified SNPs in 154 whole-genome sequences of globally and temporally representative *Vibrio cholerae* isolates. Using this approach, they showed that the seventh pandemic spread from the Bay of Bengal (with a common ancestor in the 1950s) in a series of overlapping waves, and they also identified several trans-continental events. The concept of periodic radiation from a single source followed by local evolution, and in some cases, local extinction in non-endemic areas, was well supported by their results. The clonal expansion of this lineage exhibited a strong temporal signature and the authors calculated a consistent SNP accumulation rate of 3.3 SNPs per year. The estimated rate of mutation was found to be 2.5 to 5 times slower than that estimated for the recent clonal expansion of some other bacterial pathogens (Harris et al. 2010; Croucher et al. 2011). Kiiru et al. (2013) conducted a phylogentic study on Kenyan *V. cholerae* isolates using an approach similar to that of Mutreja et al. (2011). They found that many Kenyan *V. cholerae* environmental isolates were distinct from the monophyletic seventh pandemic lineage of *V. cholerae* O1 El Tor. However, several *V. cholerae* O1 isolates from environmental sources were closely related to the seventh pandemic strains, indicating that clinically significant strains co-exist with environmental strains in the environment. This presents the possibility of genetic recombination events between these distinct populations. It also shows that the level of environmental contamination by clinically relevant strains is significant and suggests that environmental reservoirs play an important role in outbreaks. They found that the genomes of the Kenyan *Vibrio cholerae* O1 El Tor isolates are clonally related to other seventh pandemic El Tor strains elsewhere in the world, even though these isolates could be assigned into two sub-clades. Interestingly, the authors of these two articles did not make mention of atypical/alternative/hybrid

Vibrio cholerae strains, a phenomenon that some believe may signal the beginning of the eighth pandemic (Mutreja et al. 2011; Kiiru et al. 2013). In Tanzania, cholera strains that have been causing outbreaks since 1974 were not adequately studied, but that changed when Kachwamba et al. (2017) investigated 96 isolates from recent outbreaks in the country using a combination of MLVA and whole-genome sequence analysis. They found that Tanzanian *V. cholerae* strains often cluster together when from the same outbreak, but that distinct strains also circulated simultaneously. A study done in Cameroon, also using WGS, investigated strains from the 2010/2011 outbreak and found these strains to be clonal, and they also clustered distantly from other African strains (Kaas et al. 2016). The authors suggest that *V. cholerae* is endemic to the Lake Chad basin in Cameroon and that the local strains are different from other African strains. The findings from the Cameroon study agree well with a study that analysed isolates from the Democratic Republic of Congo (DRC), Zambia, Guinea and Togo (Moore et al. 2015). In this multi-country study that looked at the African Great Lakes and West Africa regions (the foci of cholera in Africa), MLVA was used to infer phylogenetic relationships of 337 isolates. They found that isolates from the Great Lakes Region clustered closely together and that West African isolates formed another distinct and separate cluster. At a country-level scale, their analysis showed that some clusters persisted in their specific countries for several years, causing expansive epidemics from time to time (Moore et al. 2015). Similar findings emerged from Uganda where MLVA and WGS analysis of 2014 to 2016 outbreak isolates showed local and regional transmission of outbreak strains (Bwire et al. 2018b). The result suggests that region-specific mechanisms of cholera emergence play a significant role in the African cholera paradigm. Another factor in cholera emergence is the role that global interconnectivity (travel) and the spread of seventh pandemic strains play in driving cholera outbreaks and epidemics. Two different phylogenetic studies in Mozambique studied outbreak isolates spanning a decade or more (1997–2014 and 2002–2012), and both studies found that pandemic strains heavily influenced the local cholera scene (Langa et al. 2015; Garrine et al. 2017). It is interesting that Garrine et al. found that a dominant clonal lineage persisted in Mozambique for at least 8 years, suggesting that this strain had an environmental or human population reservoir for the duration of its presence (Garrine et al. 2017).

According to Mutreja et al. (2011), the Classical and El Tor clades did not originate from a recent common ancestor but instead were likely independent derivatives with distinct phylogenetic histories. Though reference was made to different *ctxB* sequences in both articles, it was not immediately evident how their results relate to the current discussion on variant *V. cholerae* strains that carry classical CTX prophages. Elsewhere (in Angola) Valia et al. (2013) studied *Vibrio cholerae* O1 epidemic variants, isolated between 1992 and 2006. They investigated mobile genetic elements (MGE), Integrative Conjugative Elements (ICEs), VSP-II regions and genomic islands (GIs) and compared the profiles between Angola strains, as well as to reference strains from elsewhere. Their results identified variability within the 1990s epidemic strains, showing different rearrangements in a dynamic part of the genome not present in the prototypical *V. cholerae* O1 N16961. The 2006 strains also differed from the current pandemic *V. cholerae* O1 strain. The research emphasized the role of horizontal gene transfer (HGT) in the diversification of epidemic *V. cholerae* strains.

3.3.3   Cholera and the Environment

Rising temperatures along with changes in rainfall patterns (as a result of climate change) may have potentially severe consequences for human health, with an increased risk of diarrhoeal diseases being widely anticipated (Paz 2009; Trærup et al. 2011). Recently, several researchers studied the role that environmental factors (temperature and rainfall) play in cholera outbreaks. In some cases, the aim was to prepare for climate change scenarios, whereas others reported on attempts to predict regional cholera outbreaks in the short term, or better understand environmental drivers of African cholera outbreaks.

Using historical data, Paz (2009) studied the association between cholera rates and the annual variability of air temperature and sea surface temperature (SST) in South-Eastern Africa between 1971 and 2006. It was shown that the annual mean air temperature and SST (at the local scale as well as at hemispheric scales) had a significant impact on cholera incidence during the study period. Similarly, results from a study by Trærup et al. (2011) showed a significant relationship between temperature and the incidence of cholera in Tanzania. They calculated that with a 1 °C temperature increase, the initial relative risk of cholera increased by 15 to 29% and that by 2030, the total cost of cholera (based on a one- to two-degree increase in temperature) could be in the range of 0.32–1.4% of GDP (Trærup et al. 2011).

In Zanzibar, data from cholera surveillance reports (1997–2006) was correlated with temperature and rainfall statistics (Reyburn et al. 2011b). They found that a one-degree Celsius rise in temperature at 4 months lag resulted in a twofold increase of cholera cases. An increase of 200 mm in rainfall at 2 months lag resulted in a 1.6-fold increase of cholera cases. The interaction between temperature and rainfall also bore a significant positive association with cholera ($p < 0.04$), with a 1-month lag (Reyburn et al. 2011b). Using data from three cholera outbreaks in Zambia (2003–2006) Luque Fernandez et al. (2012) found that temperature and rainfall increases before the start of the rainy season could predict elevated cholera cases (within 3 weeks). In Nigeria, temperature and rainfall (along with poverty, and population density) were positively correlated with increased cholera cases and deaths (Leckebusch and Abdussalam 2015). Hydroclimatic conditions (related to anomalies of temperature and precipitation) were found to have helped trigger and drive a 2008 cholera outbreak in Zimbabwe (Jutla et al. 2015). These and other findings strengthen the notion that cholera incidence is influenced by factors that are directly or indirectly linked to climatic events (like deteriorating sanitation, elevated temperatures and heavy rainfall) (Dangbé et al. 2018). Abia et al. (2016b) showed that *V. cholerae* is well adapted to survival in riverbed sediments at higher temperatures (more so than *Escherichia coli*). Increases in African water and sediment temperatures that are due to climate change may, therefore, drive the geographical expansion of environmental *V. cholerae* reservoirs.

In Senegal, de Magny et al. (2012) compared rainfall patterns between 2002 and 2005, the relationship between the Sea Surface Temperature (SST) gradient in the tropical Atlantic Ocean, and precipitation over Senegal for 2005. A specific pattern of rainfall was seen throughout the Dakar region during August of 2005, and the associated rainfall anomaly coincided with an intensifying of the cholera epidemic. It was suggested that high-resolution rainfall forecasts at sub-seasonal time scales could support efforts towards cholera early warning systems. Studying the correlation of large-scale weather patterns with cholera data in several central African countries, Nkoko et al. (2011) found that cholera cases significantly increased during abnormally warm El Nino events, but declined or remained stable between these events. Some other factors, such as the role that location, plankton levels and fishing activities play, were investigated and the researchers stated that links between cholera outbreaks, climate and lake environments require additional multidisciplinary study.

Interestingly a positive association between water-hyacinth coverage in Lake Victoria and cholera cases reported in Nyanza Province (Kenya) was observed by Feikin et al. (2010). The results suggested that water-hyacinths associated with freshwater lakes might contribute to initiating cholera outbreaks and causing sporadic disease in East Africa. Another life form that is present in freshwater lakes, midges (also called lake flies), have larval life stages that support the proliferation of *V. cholerae*. All *Chironomus transvaalensis* larvae and exuviae samples that were collected and examined from Lake Manzala (Egypt) during a recent study contained *V. cholerae* bacteria, though none of the isolates belonged to the epidemic serogroups (O1 and O139) (Lotfi et al. 2016). In a study that considered the spatial distribution of cholera cases in Harare (Zimbabwe), it was found that for every 100 meters of increase in elevation, the cholera risk decreased by 30% (Luque Fernandez et al. 2012). It may be that elevation is a good indicator of the distance between contaminated surface water flows and susceptible individuals.

Another environmental factor that could potentially be impacting on *Vibrio cholerae* pathogenicity is sunlight. Early reports showed that sunlight, and the accompanying ultraviolet radiation, may increase cholera toxin secretion and induce the propagation of the *ctx* phage (Faruque et al. 2000; Quinones et al. 2006a, b). However, the research was conducted with *V. cholerae* in nutrient-rich environments, and at short exposure periods. Ssemakalu et al. (2013) investigated the impact of solar irradiation on cholera toxin secretion in water (a nutrient-poor environment) over an extended period. The research was carried out to determine if the SODIS method (solar disinfection) would increase cholera toxin secretion in SODIS-treated waters. They found that exposure to sunlight over 7 h, and also over 31 h, damaged the genes coding for cholera toxin, and more importantly that sunlight did not increase the expression of cholera toxin under SODIS conditions (Ssemakalu et al. 2013). However, it remains unclear to what extent sunlight may influence *V. cholerae* virulence at sublethal doses as encountered in aquatic environments.

It is undeniable that environmental factors play a role in cholera outbreaks and influence the start and progression of epidemics. However, the Indian/Asian cholera paradigm (wherein the epidemiology of this prototypical waterborne disease is considered to be driven mostly by climate-induced variations in coastal cholera reservoirs) cannot fully explain cholera in Africa (Rebaudet et al. 2013a). In a systematic review of factors that drive cholera in coastal Africa, Rebaudet et al. (2013a) showed that coastal epidemics constitute a minor part of the African cholera burden and that there is little evidence to demonstrate a perennial coastal reservoir of toxigenic *V. cholerae* around the African continent. From 2009 to 2011, three-quarters of all cholera cases in Africa occurred in inland regions (Rebaudet et al. 2013b). Instead of climate events influencing a coastal reservoir, the authors argue that climate events (like rainfall) may exacerbate poor sanitation practices and lacking infrastructure problems, thereby creating urban environments that favour the spread of cholera. Additionally, human activities like fishing or travelling (which are periodical) provide further variability in cholera outbreaks (Khonje et al. 2012; Rebaudet et al. 2013b). With this in mind, the authors question the applicability of the coastal cholera paradigm in an African context. Inland lakes and rivers act as reservoirs for *Vibrio cholerae*, but the role these systems play in cholera epidemiology remains unclear, and further studies are urgently required to understand these complex dynamics (Rebaudet et al. 2013b).

A worldwide map of suitability for coastal *Vibrio cholerae* survival and proliferation (under current and future climate conditions) was developed by Escobar et al. (2015). They found that increases in pH, sea surface temperature and chlorophyll *a* all helped to create favourable coastal environments for cholera bacteria and that some previously unaffected areas may in future become cholera-harbouring coastlines (Escobar et al. 2015). As such, the African coast-line may become more prominent in driving cholera outbreaks under future climate conditions. A systematic review of the impact of climate change on morbidity and mortality in sub-Saharan Africa found that there was moderate evidence to suggest that the cholera burden will increase as temperatures in Africa rise (Amegah et al. 2016). From a study carried out by Moore et al. (2017), it is clear that climate has a significant bearing on cholera incidence in Africa. The researchers provided strong evidence for a shift in the geographical distribution of cholera occurrence throughout Africa in El Niño years, with the cholera burden shifting away from West Africa, Southern Africa and Madagascar to East Africa in those years (Moore et al. 2017).

3.3.4 Antibiotic Resistance

During the last decade, many reports were published on aspects relating to the antibiotic resistance of *Vibrio cholerae* strains in Africa, showing that the topic is still current on the African research agenda (Pugliese et al. 2009; Scrascia et al. 2009; Islam et al. 2009, 2011; Lamrani Alaoui et al. 2010; Abera et al. 2010; Quilici et al. 2010; Ismail et al. 2011; Kacou-N'douba et al. 2012; Marin et al. 2013;

Akoachere et al. 2013a; Saidi et al. 2014; Smith et al. 2015; Miwanda et al. 2015; Adewale et al. 2016; Eibach et al. 2016; Dengo-Baloi et al. 2017; Sambe-Ba et al. 2017). The susceptibility patterns of *V. cholerae* strains from many African countries were investigated during this period (Fig. 3.3). General statements about the susceptibility of African *Vibrio cholerae* strains to a specific antibiotic or drug class needs to be made with caution, as this is heavily strain-dependent, and may change rapidly over time. Still, the susceptibility of *Vibrio cholerae* to Ciprofloxacin and Doxycycline was often reported in Africa, with resistance to Ampicillin, Co-trimoxazole, Streptomycin, Sulfamethoxazole and Trimethoprim being common. However, reduced susceptibility for fluoroquinolones (including Ciprofloxacin) has been reported in some strains, showing that resistance to even the most effective antibiotics is starting to emerge (Eibach et al. 2016). Susceptibility to Chloramphenicol, Naladixic Acid and Tetracycline varied widely between studies, and a few other drugs (Amoxicillin, Azithromycin, Ceftriaxone, Erythromycin, Furazolidone, Imipenem) were only included in a limited number of studies.

Fig. 3.3 Countries included in peer-reviewed publications on antibiotic resistance/susceptibility profiles for *Vibrio cholerae* (2009–2018)

A few studies also investigated possible mechanisms by which strains acquire resistance genes (Pugliese et al. 2009; Scrascia et al. 2009; Ismail et al. 2011; Marin et al. 2013). It was shown by Pugliese et al. (2009) that the SXT-related integrating conjugative element (ICE) and IncC plasmids played an essential role in shaping antibiotic resistance profiles in Eastern Africa during the late 1990s. They also studied the frequency of resistance transfer (from different donors to a receptor *Vibrio cholerae* strain) and found the frequency under laboratory conditions to be between 1.5×10^{-2} and 1.5×10^{-6}. The presence of the SXT-element and IncC plasmid conveyed resistance was also reported by other authors from Nigeria, South Africa, Senegal and Somalia (Scrascia et al. 2009; Ismail et al. 2011; Marin et al. 2013; Sambe-Ba et al. 2017). While not looking at the specific mechanisms of antibiotic resistance acquisition, Eibach et al. (2016) reported a shift towards more resistant outbreak strains in Ghana from 2011 to 2014. The same concerning trend was seen in the Democratic Republic of Congo (DRC) where a large number of strains' antibiotic susceptibility profiles were investigated; these strains were isolated over 16 years (1997–2012). Moreover, the clonal complex responsible for the 2011/2012 outbreak was only susceptible to a single antimicrobial drug (being resistant to all other antibiotics that were investigated in this study) (Miwanda et al. 2015). Based on their findings, the researchers advise that rapidly emerging multidrug resistance in *Vibrio cholerae* should be monitored more closely.

3.3.5 Water Quality, Sanitation and Hygiene

Insufficient safe water and sanitation coverage are touted by some as the primary driver of persistent cholera in Africa (Mengel 2014). In a study that estimated the effects of improved drinking water and sanitation on cholera, Leidner and Adusumilli (2013) found a statistically significant and negative association between drinking water services and cholera case incidence and severity. They also noted a relatively weak statistical relationship between cholera outbreaks and sanitation services. The study was conducted using data from Africa and Asia (1990–2008) (Leidner and Adusumilli 2013). In another study that looked at the impact of sanitation on cholera outbreaks, Sasaki et al. (2009) investigated the association between drainage networks and cholera cases in Lusaka, Zambia. They found that insufficient drainage networks were statistically associated with cholera incidence and concluded that infrastructure development would help to prevent cholera outbreaks in the long term.

In Douala, Cameroon, the water quality of shallow wells was linked to well characteristics and hygiene behaviour (Akoachere et al. 2013b). The wells were a major source of drinking water in this cholera-endemic area. Inadequacies in the location and construction of wells, as well as subpar hygiene and sanitation practices, were found to be responsible for the poor well water quality observed during the study (Akoachere et al. 2013b). In Freetown, Sierra Leone, the consumption of unsafe water, street-vended water and crabs was found to be a significant risk factor for cholera (Nguyen et al. 2014). As in other studies, the authors called for the implementation of effective household-level water treatment interventions and highlighted the need for water quality and hygiene education (Masauni et al. 2010; Cavallaro et al. 2011; Akoachere et al. 2013b; Nguyen et al. 2014). Two studies that investigated different outbreaks in Uganda identified faecally contaminated drinking water as the source, or driver, of the specific outbreaks (Kwesiga et al. 2017; Oguttu et al. 2017). In another study that was done in the Democratic Republic of the Congo (DRC), a clear association was observed between reduced availability of tap water and a higher incidence of suspected cholera cases (Jeandron et al. 2015). Together, these findings emphasize the importance of clean water provision in cholera-affected areas.

Crustaceans, as a risk factor in *Vibrio* infections, were also investigated in Cote d'Ivoire (Traore et al. 2012). *Vibrio* spp. were isolated from 7.8% of crustacean samples studied; the samples were bought from seven markets (six in Abidjan and one in Dabou). Roughly one-quarter of the *Vibrios* were found to be *Vibrio cholerae*—albeit none carried the genes for cholera toxin (*ctxA and ctxB*). However, the possibility that crustaceans may carry epidemic *V. cholerae* strains cannot be ruled out (Traore et al. 2012). The widespread practice of boiling crustaceans before consumption reduces the risk of infection, but as shown by Scheelbeek et al. (2009), some steps performed during food preparation may re-contaminate the food source, or the preparer's hands. The authors investigated fish preparation in Monrovia, Liberia. They detected *Vibrio cholerae* in the excreta of fish caught in estuarine waters and observed fish preparation steps in 30 households. The results of the study showed that even though frying and boiling were used to prepare the fish (effectively killing the bacteria), contaminated water was often used during fish preparation. Hygiene practices, like hand washing, were also often found to be inadequate (Scheelbeek et al. 2009). The authors believe that fish to hand (and then on to oral) transmission could be responsible for cholera infections.

3.4 Concluding Remarks

Scientific discoveries help society to understand and respond to health challenges like cholera. However, it is imperative that research efforts are directed and focused on addressing the most relevant questions. This is especially true if the studies focus on complex systems like that presented by cholera in Africa, where the multifaceted nature requires in-depth exploration. During the 10-year review period (2009–2018), a large number of journal papers were published on research pertaining to cholera and *Vibrio cholerae* in Africa. The majority of the reviewed publications had an epidemiology theme, with aspects relating to the field of phylogenetics, and those covering oral cholera vaccine (OCV) considerations also eliciting a substantial amount of research. Some other research themes were also addressed in this chapter, with host- and environmental-related aspects, antimicrobial resistance, monitoring, detection and diagnosis covered (amongst others).

In this review, recent literature, spanning a decade of African cholera-related research, was investigated in an attempt to identify research trends, reflect on recent scientific successes and identify possible research gaps.

Epidemiology, evident by the number of publications related to this field, was shown to be a critical research area. Reports on cholera outbreaks and epidemics were published from many African countries, and often the case data was used to identify the origins and drivers of the disease in a specific area. The value of surveillance should not be underestimated, as it will allow countries to implement targeted interventions that will reduce cholera morbidity and mortality in Africa (Mintz and Tauxe 2013). The public health and economic benefits gained from proactive, targeted interventions far outweigh their costs. In our opinion, studies employing an epidemiological approach will remain crucial in the fight against cholera, without which the source, transmission, risk and impact of cholera could not be investigated. These studies help to focus attention on cholera ravaged areas and help to create a sympathetic political will. For instance, the recent cholera outbreak in Zimbabwe led to excellent publications that highlight the plight of disease-stricken communities, calling for humanitarian and political aid (Bateman 2009; Chambers 2009).

Holmgren (2012) made a case for the control of cholera in Africa using vaccination. He cites recent successes in the use of vaccines in combatting endemic and epidemic cholera outbreaks, and how these have helped to change the attitude of authorities towards OCVs. He concludes that 'it is becoming increasingly unconscionable not to include oral cholera vaccine in the public health response to this disease' (Holmgren 2012). There are technical and logistical reasons preventing the widespread use of oral cholera vaccination, but it is the opinion of von Seidlein et al. (2013) that the most significant barrier has been the lack of a loud political voice of the underprivileged populations affected by cholera. Another potential hurdle for OCV utilization is community acceptance. Fortunately, most studies report a high level of anticipated community acceptance. However, a study in Zanzibar indicated that the actual OCV uptake during a vaccination campaign was lower than expected (Schaetti et al. 2012a). Sociocultural determinants impact on vaccine acceptance and uptake and further research in this area may help to clarify and solve potential uptake-associated challenges elsewhere. During the last decade, vaccination has become a viable intervention or prevention strategy, and it was often cited for its ability to decrease morbidity, mortality, and the economic losses experienced during cholera outbreaks. Two recent developments accelerated OCV usage in Africa. Firstly, the creation of the Global Oral Cholera Vaccine Emergency Stockpile (in 2013) helped to increase vaccine availability. Secondly, many researchers provided evidence for the effectiveness of a single-dose OCV application (instead of the recommended two-dose regiment), a modification that allows for better coverage or decreased costs during vaccination campaigns. Sadly, a global shortage of oral cholera vaccines still hampers widespread use of this effective option.

Oral rehydration therapy (ORT) remains the treatment option of choice during cholera epidemics, but antibiotics have, when necessary, been used to reduce the duration of disease, lessen organism shedding and treat severe cases. Even though the specific use of antibiotics during cholera epidemics has been limited, the emergence of multidrug resistant *V. cholerae* strains has been reported from many African countries. The surveillance of antibiotic susceptibility patterns in *V. cholerae* strains continues to be necessary to ensure that antibiotics remain an effective treatment option.

Host-related aspects of cholera were an important research area over the last decade in Africa. The research focussed on the role that humans play in driving or causing cholera outbreaks, and the impacts that cholera outbreaks have on communities and individuals. Human movement and large gatherings of people were often associated with increased cholera risks, as was poor hygiene, food handling practices and living in polluted areas. How people respond to (or take preventative action against) cholera outbreaks depended on their level of awareness or their perceptions regarding the disease. Perceptions were often found to be heavily influenced by sociocultural beliefs. In general, those communities that were better educated and more aware of cholera risks were more resilient (than non-informed settlements) during outbreaks. Informed communities often identified water and sanitation as important drivers of outbreaks, and residents were more likely to visit healthcare facilities during an outbreak. Collectively, the findings highlight the need for studies with a sociocultural focus, as the views of individuals and communities impact on the success of vaccination, education, treatment and prevention strategies.

Accurate and sensitive methods for the diagnosis of cholera are critical in the fight against this re-emerging disease. In the African context, such a diagnostic test needs to be rapid, easy-to-do and cost-effective; something that many of the conventional methods are not. African reports on the use of rapid dipsticks, like Crystal VC, in conjunction with a pre-enrichment step have been encouraging and may be more widely utilized in future.

Studies on co-infections in Africa (with cholera) showed that the risk for cholera increased when an individual was also sick with malaria or HIV/AIDS (Mushayabasa and Bhunu 2012; Okosun and Makinde 2014). However, the wider role that co-infections in Africa play in host susceptibility and lowering of the cholera infectious dose remains unclear, and further research in this area may provide valuable insights.

The importance of an integrated approach towards cholera prevention and mitigation has been recognized by governments and institutions in Africa, and several reports were published on initiatives aimed at the elimination and reduction of the disease. Foremost amongst these initiatives are the *Global Alliance against Cholera* (GAAC) and the *Initiative against Diarrheal and Enteric diseases in Africa and Asia* (IDEA) (Haaser 2014; Hessel 2014). It is believed that an interdisciplinary, inter-sectoral and cross-border approach will provide momentum for the fight against cholera in Africa.

Aquatic reservoirs can contribute to cholera persistence in endemic areas; therefore, efforts to manage and mitigate cholera outbreaks should include environmental monitoring exercises. In this review, an increased interest in *V. cholerae* monitoring was observed over the latter part of the last decade, with many studies investigating different types of environmental samples. Water was the sample type most often tested, and results from various studies suggest that *V. cholerae* environmental strains (lacking cholera toxin genes) are common and widespread in African surface waters. Cholera bacteria that belong to the O1 serogroup (which is associated with cholera in Africa) and those strains that can produce cholera toxin (CT) were infrequently detected but were still present in surface waters sporadically. Two separate studies also report on the presence of O139 serogroup bacteria in African waters. This was an important finding given that serogroup O139 has not been known to cause African outbreaks and has only been detected in Asian territories. *Vibrio cholerae* was also found in some other sample types that include wastewater, excreta from livestock, and aquatic organisms like lake fly larvae, fish and crustaceans. The numbers and virulence of pathogenic *V. cholerae* contained within environmental reservoirs may be influenced by climatic events, and considerable research efforts were directed towards studying the impact of short- and long-term weather changes on cholera outbreaks. Increases in air temperature, sea surface temperature and rainfall were often cited to trigger cholera outbreaks, and in many studies, the research was aimed at developing cholera early warning systems. However, additional factors and drivers may need to be considered as cholera outbreak predictions still lack sufficient accuracy. Climate change scenarios predict an increase in future temperatures and researchers estimated that this could lead to a significant rise in cholera cases. Collectively, the research alludes to the role that aquatic reservoirs play in the African cholera landscape, and it is evident that *V. cholerae* residing in marine, lake and river environments responds to climatic stimuli. Nonetheless, the precise contribution and mechanism of these reservoirs to the cholera burden in Africa remains unclear. What has become clear though is that the cholera paradigm, in which the epidemiology of this waterborne disease is mainly driven by climate-induced variations in coastal cholera reservoirs, is not entirely applicable to Africa (as it is for the Asian Sub-continent) (Rebaudet et al. 2013b). Rebaudet et al. provided a convincing argument for cholera epidemiology driven mostly by inland factors (Rebaudet et al. 2013a, b). They hypothesize that the overriding drivers of cholera in Africa are climate events in association with human behavioural factors, local living conditions and (possibly) inland lakes and rivers.

The exact drivers of cholera outbreaks and the real source of the outbreak strains have been a divisive issue amongst researchers over the last two decades with two different theories proposed. One view is that *V. cholerae* is an autochthonous inhabitant of coastal and freshwater systems, and therefore climate events (for instance heavy rainfalls) may trigger bacterial proliferation in the aquatic reservoirs, thereby leading to outbreaks in nearby communities. This theory was promoted by Colwell et al. 1994 and has gained much support over the years (Colwell and Huq 1994). In this chapter, many publications are covered that provide substantial evidence for the role of environmental reservoirs in cholera outbreaks. Another view is that African cholera outbreaks are primarily driven by imported *V. cholerae* strains and that the seventh pandemic cholera was caused by a single expanded clonal lineage that was imported from Asia into Africa via a series of waves since 1970 (Mutreja et al. 2011; Weill et al. 2017). The latter theory downplays the role that environmental reservoirs have in the cholera paradigm, and highlights travel-associated dissemination by human hosts instead. The differing views were mostly advocated in a mutually exclusive way, meaning that only one theory could explain the cause of cholera outbreaks. However, it is our opinion that both of these processes contribute significantly to cholera outbreaks in Africa. Unfortunately, African researchers had to operate within a research domain that was fractured along two lines globally, that while it is clear that cholera bacteria leverage both 'disease-causing' and 'environmental survival' mechanisms to thrive. We think that the division created by the two opposing theories has stunted cholera research progress in Africa and beyond. Rather than disproving an opposing theory, researchers should accept the role that both factors play in the holistic process and aim to understand the relationship between these two modes of survival.

The field of genotyping and phylogenetics is often not accessible to a broad audience, given its inherent complexity. However, findings from this research area provide valuable insight into the characteristics, origins, evolution and the clonal relationship of *V. cholerae* strains. The information enhances our ability to understand the global spread and the impact of short- and long-term evolution on pathogenic *V. cholerae* strains. Strain to strain variation often accounts for differences in (amongst other) virulence and host immunity (Ssemakalu et al. 2013, 2014), and knowledge herein can help to design effective intervention strategies. Mutreja et al. (2011) underlined the value of phylogenetics in the fight against cholera when they showed how cholera spread globally using a series of overlapping waves in recent history. Our understanding of this model was further expanded by a thorough study done by Weill et al. (2017), where large African outbreaks were attributed to *V. cholerae* strains imported from Asia throughout the last four decades. Continued research efforts will maintain and expand our current understanding of the Asia-Africa cholera interaction, and in future, we may even discover Asian *V. cholerae* strains with an African origin.

Despite considerable progress, cholera in Africa is still characterized by many unknowns. It is foreseen that concerted efforts in cholera prevention and mitigation will, in the long-term, free Africa from the cholera burden. These successes will take time, and focused research efforts will play an important role in overcoming the persistent cholera threat. Foremost, in our opinion, is the need to investigate and understand the interplay between cholera host-related and cholera environmental drivers. Also, the need for a rapid, cost-effective and accurate cholera diagnostic methods remains. This is a much-required tool that will ensure a swift outbreak response in a resource-limited continent such as Africa.

References

Abera B, Bezabih B, Dessie A (2010) Antimicrobial suceptibility of V. cholerae in north west, Ethiopia. Ethiop Med J 48:23–28

Abia ALK, Ubomba-Jaswa E, Genthe B, Momba MNB (2016a) Quantitative microbial risk assessment (QMRA) shows increased public health risk associated with exposure to river water under conditions of riverbed sediment resuspension. Sci Total Environ 566–567:1143–1151. https://doi.org/10.1016/j.scitotenv.2016.05.155

Abia ALK, Ubomba-Jaswa E, Momba MNB (2016b) Competitive survival of Escherichia coli, Vibrio cholerae, Salmonella typhimurium and Shigella dysenteriae in riverbed sediments. Microb Ecol 72:881–889. https://doi.org/10.1007/s00248-016-0784-y

Abia ALK, Ubomba-Jaswa E, Momba MNB (2017) Riverbed sediments as reservoirs of multiple vibrio cholerae virulence-associated genes: a potential trigger for cholera outbreaks in developing countries. J Environ Public Health. https://doi.org/10.1155/2017/5646480

Acquah H, Malm K, Der J et al (2016) Cholera outbreak following a marriage ceremony in Medinya, Western Ghana. Pan Afr Med J 25:3. https://doi.org/10.11604/pamj.supp.2016.25.1.6167

Adagbada AO, Adesida SA, Nwaokorie FO et al (2012) Cholera epidemiology in Nigeria: an overview. Pan Afr Med J:12

Adewale AK, Pazhani GP, Abiodun IB et al (2016) Unique clones of Vibrio cholerae O1 El Tor with Haitian Type ctxB allele implicated in the recent cholera epidemics from Nigeria, Africa. PLoS One 11:e0159794. https://doi.org/10.1371/journal.pone.0159794

Adjei EY, Malm KL, Mensah KN et al (2017) Evaluation of cholera surveillance system in Osu Klottey District, Accra, Ghana (2011–2013). Pan Afr Med J 28:224. https://doi.org/10.11604/pamj.2017.28.224.10737

Ahmed HA, El Bayomi RM, Hussein MA et al (2018) Molecular characterization, antibiotic resistance pattern and biofilm formation of Vibrio parahaemolyticus and V. cholerae isolated from crustaceans and humans. Int J Food Microbiol 274:31–37. https://doi.org/10.1016/j.ijfoodmicro.2018.03.013

Ahmed S, Bardhan PK, Iqbal A et al (2011) The 2008 cholera epidemic in Zimbabwe: experience of the icddr,b team in the field. J Health Popul Nutr 29:541–546

Akoachere J-FT, Masalla T, Njom H (2013a) Multi-drug resistant toxigenic Vibrio cholerae O1 is persistent in water sources in New Bell-Douala, Cameroon. BMC Infect Dis 13:366. https://doi.org/10.1186/1471-2334-13-366

Akoachere J-FT, Omam L-A, Massalla T (2013b) Assessment of the relationship between bacteriological quality of dug-wells, hygiene behaviour and well characteristics in two cholera endemic localities in Douala, Cameroon. BMC Public Health 13:692. https://doi.org/10.1186/1471-2458-13-692

Amegah AK, Rezza G, Jaakkola JJK (2016) Temperature-related morbidity and mortality in sub-Saharan Africa: a systematic review of the empirical evidence. Environ Int 91:133–149. https://doi.org/10.1016/j.envint.2016.02.027

Awalime DK, Davies-Teye BBK, Vanotoo LA et al (2017) Economic evaluation of 2014 cholera outbreak in Ghana: a household cost analysis. Health Econ Rev 7:45. https://doi.org/10.1186/s13561-017-0182-2

Azman AS, Luquero FJ, Rodrigues A et al (2012) Urban cholera transmission hotspots and their implications for reactive vaccination: evidence from Bissau city, Guinea Bissau. PLoS Negl Trop Dis 6:e1901. https://doi.org/10.1371/journal.pntd.0001901

Azman AS, Parker LA, Rumunu J et al (2016a) Effectiveness of one dose of oral cholera vaccine in response to an outbreak: a case-cohort study. Lancet Glob Health 4:e856–e863. https://doi.org/10.1016/S2214-109X(16)30211-X

Azman AS, Rumunu J, Abubakar A et al (2016b) Population-level effect of cholera vaccine on displaced populations, South Sudan, 2014. Emerg Infect Dis 22:1067–1070. https://doi.org/10.3201/eid2206.151592

Bartels SA, Greenough PG, Tamar M, VanRooyen MJ (2010) Investigation of a cholera outbreak in Ethiopia's Oromiya region. Disaster Med Public Health Prep 4:312–317. https://doi.org/10.1001/dmp.2010.44

Bateman C (2009) Cholera – getting the basics right. South African Med J 99:138–142

Bekolo CE, van Loenhout JAF, Rodriguez-Llanes JM et al (2016) A retrospective analysis of oral cholera vaccine use, disease severity and deaths during an outbreak in South Sudan. Bull World Health Organ 94:667–674. https://doi.org/10.2471/BLT.15.166892

Bhuiyan NA, Nusrin S, Ansaruzzaman M et al (2012) Genetic characterization of Vibrio cholerae O1 strains isolated in Zambia during 1996–2004 possessing the unique VSP-II region of El Tor variant. Epidemiol Infect 140:510–518

Binagwaho A, Nyatanyi T, Nutt CT, Wagner CM (2012) Disease outbreaks: support for a cholera vaccine stockpile. Nature 487:39. https://doi.org/10.1038/487039c

Bockemuhl J, Schröter G (1975) The El Tor cholera epidemic in Togo (West Africa) 1970-1972. Tropenmed Parasitol 26:312–322

Bompangue D (2014) Planning for the elimination of cholera: An example of integrated national plan. 16th Int Congr Infect Dis Abstr 21, Supple:71. doi:https://doi.org/10.1016/j.ijid.2014.03.572

Bwire G, Debes AK, Orach CG et al (2018a) Environmental surveillance of Vibrio cholerae O1/O139 in the five African Great Lakes and other major surface water sources in Uganda. Front Microbiol. https://doi.org/10.3389/fmicb.2018.01560

Bwire G, Malimbo M, Kagirita A et al (2015) Nosocomial cholera outbreak in a mental hospital: challenges and lessons learnt from Butabika National Referral Mental Hospital, Uganda. Am J Trop Med Hyg 93:534–538. https://doi.org/10.4269/ajtmh.14-0730

Bwire G, Malimbo M, Makumbi I et al (2013a) Cholera surveillance in Uganda: an analysis of notifications for the years 2007–2011. J Infect Dis 208:S78–S85

Bwire G, Malimbo M, Maskery B et al (2013b) The burden of cholera in Uganda. PLoS Negl Trop Dis 7:e2545. https://doi.org/10.1371/journal.pntd.0002545

Bwire G, Munier A, Ouedraogo I et al (2017a) Epidemiology of cholera outbreaks and socio-economic characteristics of the communities in the fishing villages of Uganda: 2011–2015. PLoS Negl Trop Dis 11:e0005407. https://doi.org/10.1371/journal.pntd.0005407

Bwire G, Mwesawina M, Baluku Y et al (2016) Cross-border cholera outbreaks in sub-Saharan Africa, the mystery behind the silent illness: what needs to be done? PLoS One 11:e0156674. https://doi.org/10.1371/journal.pone.0156674

Bwire G, Orach CG, Abdallah D et al (2017b) Alkaline peptone water enrichment with a dip-stick test to quickly detect and monitor cholera outbreaks. BMC Infect Dis 17:726. https://doi.org/10.1186/s12879-017-2824-8

Bwire G, Sack DA, Almeida M et al (2018b) Molecular characterization of vibrio cholerae responsible for cholera epidemics in Uganda by PCR, MLVA and WGS. PLoS Negl Trop Dis. https://doi.org/10.1371/journal.pntd.0006492

Cartwright EJ, Patel MK, Mbopi-Keou FX et al (2013) Recurrent epidemic cholera with high mortality in Cameroon: persistent challenges 40 years into the seventh pandemic. Epidemiol Infect 141:2083–2093. https://doi.org/10.1017/S0950268812002932

Cavallaro EC, Harris JR, da Goia MS et al (2011) Evaluation of pot-chlorination of wells during a cholera outbreak, Bissau, Guinea-Bissau, 2008. J Water Health 9:394–402

Ceccarelli D, Spagnoletti M, Bacciu D et al (2011) New V. cholerae atypical El Tor variant emerged during the 2006 epidemic outbreak in Angola. BMC Microbiol 11:130

Chambers K (2009) Zimbabwe's battle against cholera. Lancet 373:993–994. https://doi.org/10.1016/S0140-6736(09)60591-2

Choi SY, Lee JH, Jeon YS et al (2010a) Multilocus variable-number tandem repeat analysis of vibrio cholerae O1 El Tor strains harbouring classical toxin B. J Med Microbiol 59:763–769. https://doi.org/10.1099/jmm.0.017939-0

Choi SY, Lee JH, Kim EJ et al (2010b) Classical RS1 and environmental RS1 elements in Vibrio cholerae O1 El Tor strains harbouring a tandem repeat of CTX prophage: revisiting Mozambique in 2005. J Med Microbiol 59:302–308. https://doi.org/10.1099/jmm.0.017053-0

Ciglenecki I, Sakoba K, Luquero FJ et al (2013) Feasibility of mass vaccination campaign with oral cholera vaccines in response to an outbreak in Guinea. PLoS Med 10:e1001512. https://doi.org/10.1371/journal.pmed.1001512

Colwell RR, Huq A (1994) Environmental reservoir of Vibrio cholerae the causative agent of Choleraa. Ann N Y Acad Sci 740:44–54. https://doi.org/10.1111/j.1749-6632.1994.tb19852.x

Cowman G, Otipo S, Njeru I et al (2017) Factors associated with cholera in Kenya, 2008-2013. Pan Afr Med J 28:101. https://doi.org/10.11604/pamj.2017.28.101.12806

Croucher NJ, Harris SR, Fraser C et al (2011) Rapid pneumococcal evolution in response to clinical interventions. Science 331:430–434. https://doi.org/10.1126/science.1198545

Cummings MJ, Wamala JF, Eyura M et al (2012) A cholera outbreak among semi-nomadic pastoralists in northeastern Uganda: epidemiology and interventions. Epidemiol Infect 140:1376–1385. https://doi.org/10.1017/S0950268811001956

Dalusi L, Lyimo TJ, Lugomela C et al (2015) Toxigenic Vibrio cholerae identified in estuaries of Tanzania using PCR techniques. FEMS Microbiol Lett. https://doi.org/10.1093/femsle/fnv009

Dangbé E, Irépran D, Perasso A, Békollé D (2018) Mathematical modelling and numerical simulations of the influence of hygiene and seasons on the spread of cholera. Math Biosci 296:60–70. https://doi.org/10.1016/j.mbs.2017.12.004

Date K, Person B, Nygren B et al (2013) Evaluation of a rapid cholera response activity – Nyanza Province, Kenya, 2008. J Infect Dis 208(Suppl):S62–S68. https://doi.org/10.1093/infdis/jit198

Debes AK, Ateudjieu J, Guenou E et al (2016a) Clinical and environmental surveillance for vibrio cholerae in resource constrained areas: application during a 1-year surveillance in the far north region of Cameroon. Am J Trop Med Hyg 94:537–543. https://doi.org/10.4269/ajtmh.15-0496

Debes AK, Ateudjieu J, Guenou E et al (2016b) Evaluation in Cameroon of a novel, simplified methodology to assist molecular microbiological analysis of V. cholerae in resource-limited settings. PLoS Negl Trop Dis 10:e0004307. https://doi.org/10.1371/journal.pntd.0004307

Deen J, von Seidlein L, Luquero FJ et al (2016) The scenario approach for countries considering the addition of oral cholera vaccination in cholera preparedness and control plans. Lancet Infect Dis 16:125–129. https://doi.org/10.1016/S1473-3099(15)00298-4

Démolis R, Botão C, Heyerdahl LW et al (2018) A rapid qualitative assessment of oral cholera vaccine anticipated acceptability in a context of resistance towards cholera intervention in Nampula, Mozambique. Vaccine 36:6497–6505. https://doi.org/10.1016/j.vaccine.2017.10.087

Dengo-Baloi LC, Sema-Baltazar CA, Manhique LV et al (2017) Antibiotics resistance in El tor vibrio cholerae 01 isolated during cholera outbreaks in Mozambique from 2012 to 2015. PLoS One 12:e0181496. https://doi.org/10.1371/journal.pone.0181496

Desai SN, Akalu Z, Teshome S, et al (2014) A randomized, double-blind, controlled trial to evaluate the safety and immunogenicity of killed oral cholera vaccine (Shanchol®) in healthy individuals in Ethiopia. 16th Int Congr Infect Dis Abstr 21, Suppl:431–432. doi: https://doi.org/10.1016/j.ijid.2014.03.1310

de Magny GC, Thiaw W, Kumar V, Manga NM, Diop BM, Gueye L, Kamara M, Roche B, Murtugudde R, Colwell RR (2012) Cholera outbreak in Senegal in 2005: was climate a factor? PLoS One. 7(8): e44577

Djomassi LD, Gessner BD, Andze GO, Mballa GA (2013) National surveillance data on the epidemiology of cholera in Cameroon. J Infect Dis 208(Suppl):S92–S97. https://doi.org/10.1093/infdis/jit197

Djouma FN, Ateudjieu J, Ram M et al (2016) Factors associated with fatal outcomes following cholera-like syndrome in far north region of Cameroon: a community-based survey. Am J Trop Med Hyg 95:1287–1291. https://doi.org/10.4269/ajtmh.16-0300

du Preez M, van Der Merwe MR, Cumbana A, le Roux W (2010) A survey of vibrio cholerae O1 and O139 in estuarine. Water SA 36:615–620

Eibach D, Herrera-León S, Gil H et al (2016) Molecular epidemiology and antibiotic susceptibility of vibrio cholerae associated with a large cholera outbreak in Ghana in 2014. PLoS Negl Trop Dis 10:e0004751. https://doi.org/10.1371/journal.pntd.0004751

Ekra KD, Attoh-Touré H, Bénié BVJ et al (2009) Five years of cholera surveillance in Côte d'Ivoire during social and political crisis, 2001 to 2005. Bull Soc Pathol Exot 102:107–109

Emikpe BO, Morenikeji OA, Jarikre TA (2016) Zoo animals' disease pattern in a university zoological garden, Ibadan, Nigeria. Asian Pacific J Trop Dis 6:85–89. https://doi.org/10.1016/S2222-1808(15)60991-4

Escobar LE, Ryan SJ, Stewart-Ibarra AM et al (2015) A global map of suitability for coastal vibrio cholerae under current and future climate conditions. Acta Trop 149:202–211. https://doi.org/10.1016/j.actatropica.2015.05.028

Essoya LD, Gessner BD, Kossi B et al (2013) National surveillance data on the epidemiology of cholera in Togo. J Infect Dis 208:S115–S119

Faruque SM, Asadulghani, Rahman MM et al (2000) Sunlight-induced propagation of the lysogenic phage encoding cholera toxin. Infect Immun 68:4795–4801

Faruque SM, Islam MJ, Ahmad QS et al (2006) An improved technique for isolation of environmental Vibrio cholerae with epidemic potential: monitoring the emergence of a multiple-antibiotic-resistant epidemic strain in Bangladesh. J Infect Dis 193:1029–1036. https://doi.org/10.1086/500953

Faruque SM, Tam VC, Chowdhury N et al (2007) Genomic analysis of the Mozambique strain of vibrio cholerae O1 reveals the origin of El Tor strains carrying classical CTX prophage. Proc Natl Acad Sci U S A 104:5151–5156. https://doi.org/10.1073/pnas.0700365104

Feikin DR, Tabu CW, Gichuki J (2010) Does water hyacinth on East African lakes promote cholera outbreaks?. The American journal of tropical medicine and hygiene, 83(2), 370–373

Finger F, Genolet T, Mari L et al (2016) Mobile phone data highlights the role of mass gatherings in the spreading of cholera outbreaks. Proc Natl Acad Sci 113:6421–6426. https://doi.org/10.1073/pnas.1522305113

Gabutti G, Aquilina M, Cova M et al (2012) Prevention of fecal-orally transmitted diseases in travelers through an oral anticholeric vaccine (WC/rBS). J Prev Med Hyg 53:199–203

Garrine M, Mandomando I, Vubil D et al (2017) Minimal genetic change in vibrio cholerae in Mozambique over time: multilocus variable number tandem repeat analysis and whole genome sequencing. PLoS Negl Trop Dis 11:e0005671. https://doi.org/10.1371/journal.pntd.0005671

Gauzere BA, Aubry P (2012) Cholera epidemics on Reunion Island during the 19th century. Med Sante Trop 22:131–136. https://doi.org/10.1684/mst.2012.0044

Gbary AR, Dossou JP, Sossou RA et al (2011a) Epidemiologic and medico-clinical aspects of the cholera outbreak in the Littoral department of Benin in 2008. Med Trop (Mars) 71:157–161

Gbary AR, Sossou RA, Dossou JP et al (2011b) The determinants of the low case fatality rate of the cholera epidemic in the Littoral department of Benin in 2008. Sante Publique 23:345–358

Gdoura M, Sellami H, Nasfi H et al (2016) Molecular detection of the three major pathogenic vibrio species from seafood products and sediments in Tunisia using real-time PCR. J Food Prot 79:2086–2094. https://doi.org/10.4315/0362-028X.JFP-16-205

George G, Rotich J, Kigen H et al (2016) Notes from the field: ongoing cholera outbreak – Kenya, 2014–2016. MMWR Morb Mortal Wkly Rep 65:68–69. https://doi.org/10.15585/mmwr.mm6503a7

Grout L, Martinez-Pino I, Ciglenecki I et al (2015) Pregnancy outcomes after a mass vaccination campaign with an Oral cholera vaccine in Guinea: a retrospective cohort study. PLoS Negl Trop Dis 9:e0004274. https://doi.org/10.1371/journal.pntd.0004274

Guerra J, Mayana B, Djibo A et al (2012) Evaluation and use of surveillance system data toward the identification of high-risk areas for potential cholera vaccination: a case study from Niger. BMC Res Notes 5:231. https://doi.org/10.1186/1756-0500-5-231

Gujral L, Sema C, Rebaudet S et al (2013) Cholera epidemiology in mozambique using national surveillance data. J Infect Dis 208:S107–S114

Haaser F (2014) Guidance for effective elimination of cholera epidemics in a sustainable manner in the Democratic Republic of Congo and other high risk countries. 16th Int Congr Infect Dis Abstr 21, Supple:70. doi: https://doi.org/10.1016/j.ijid.2014.03.571

Halder K, Das B, Nair GB, Bhadra RK (2010) Molecular evidence favouring step-wise evolution of Mozambique Vibrio cholerae O1 El Tor hybrid strain. Microbiology 156:99–107. https://doi.org/10.1099/mic.0.032458-0

Harris JR, Cavallaro EC, de Nobrega AA et al (2009) Field evaluation of crystal VC rapid dipstick test for cholera during a cholera outbreak in Guinea-Bissau. Tropical Med Int Health 14:1117–1121. https://doi.org/10.1111/j.1365-3156.2009.02335.x

Harris SR, Feil EJ, Holden MTG et al (2010) Evolution of MRSA during hospital transmission and intercontinental spread. Science 327:469–474. https://doi.org/10.1126/science.1182395

Hessel L (2014) The initiative against diarrheal and enteric diseases in Africa and Asia: The role of field actors to successfully address the the fight against cholera. 16th Int Congr Infect Dis Abstr 21, Supple:112. doi: https://doi.org/10.1016/j.ijid.2014.03.659

Heyerdahl LW, Ngwira B, Demolis R et al (2018) Innovative vaccine delivery strategies in response to a cholera outbreak in the challenging context of Lake Chilwa. A rapid qualitative assessment. Vaccine 36:6491–6496. https://doi.org/10.1016/j.vaccine.2017.10.108

Holmgren J (2012) A case for control of cholera in Africa by vaccination. Lancet Infect Dis 12:818–819. https://doi.org/10.1016/S1473-3099(12)70204-9

Hounmanou YMG, Mdegela RH, Dougnon TV et al (2016) Toxigenic vibrio cholerae O1 in vegetables and fish raised in wastewater irrigated fields and stabilization ponds during a non-cholera outbreak period in Morogoro, Tanzania: an environmental health study. BMC Res Notes 9:466. https://doi.org/10.1186/s13104-016-2283-0

Hove-Musekwa SD, Nyabadza F, Chiyaka C et al (2011) Modelling and analysis of the effects of malnutrition in the spread of cholera. Math Comput Model 53:1583–1595. https://doi.org/10.1016/j.mcm.2010.11.060

I.V.I. (International Vaccine Institute) (2015) Oral Cholera Vaccine

Ilboudo PG, Huang XX, Ngwira B et al (2017) Cost-of-illness of cholera to households and health facilities in rural Malawi. PLoS One 12:e0185041. https://doi.org/10.1371/journal.pone.0185041

Ilboudo PG, Le Gargasson J-B (2017) Delivery cost analysis of a reactive mass cholera vaccination campaign: a case study of Shanchol vaccine use in Lake Chilwa, Malawi. BMC Infect Dis 17:779. https://doi.org/10.1186/s12879-017-2885-8

Islam MS, Mahmud ZH, Ansaruzzaman M et al (2011) Phenotypic, genotypic, and antibiotic sensitivity patterns of strains isolated from the cholera epidemic in Zimbabwe. J Clin Microbiol 49:2325–2327. https://doi.org/10.1128/JCM.00432-11

Islam MS, Midzi SM, Charimari L et al (2009) Susceptibility to fluoroquinolones of vibrio cholerae O1 isolated from diarrheal patients in Zimbabwe. JAMA 302:2321–2322. https://doi.org/10.1001/jama.2009.1750

Ismail H, Smith AM, Archer BN et al (2012) Case of imported Vibrio cholerae O1 from India to South Africa. J Infect Dev Ctries 6:897–900. https://doi.org/10.3855/jidc.2448

Ismail H, Smith AM, Sooka A, Keddy KH (2011) Genetic characterization of multidrug-resistant, extended-spectrum-ß- lactamase-producing Vibrio cholerae O1 outbreak strains, Mpumalanga, South Africa, 2008. J Clin Microbiol 49:2976–2979

Ismail H, Smith AM, Tau NP et al (2013) Cholera outbreak in South Africa, 2008–2009: laboratory analysis of Vibrio cholerae O1 strains. J Infect Dis 208:S39–S45. https://doi.org/10.1093/infdis/jit200

Iyer AS, Bouhenia M, Rumunu J et al (2016) Immune responses to an Oral cholera vaccine in internally displaced persons in South Sudan. Sci Rep 6:35742

Jeandron A, Saidi JM, Kapama A et al (2015) Water supply interruptions and suspected cholera incidence: a time-series regression in the Democratic Republic of the Congo. PLoS Med 12:e1001893. https://doi.org/10.1371/journal.pmed.1001893

Jeuland M, Cook J, Poulos C et al (2009) Cost-effectiveness of new-generation oral cholera vaccines: a multisite analysis. Value Health 12:899–908. https://doi.org/10.1111/j.1524-4733.2009.00562.x

Jeuland M, Lucas M, Clemens J, Whittington D (2010) Estimating the private benefits of vaccination against cholera in Beira, Mozambique: a travel cost approach. J Dev Econ 91:310–322. https://doi.org/10.1016/j.jdeveco.2009.06.007

Jutla A, Aldaach H, Billian H et al (2015) Satellite based assessment of Hydroclimatic conditions related to cholera in Zimbabwe. PLoS One 10:e0137828. https://doi.org/10.1371/journal.pone.0137828

Kaas RS, Ngandjio A, Nzouankeu A et al (2016) The Lake Chad Basin, an isolated and persistent reservoir of Vibrio cholerae O1: a genomic insight into the outbreak in Cameroon, 2010. PLoS One 11:e0155691. https://doi.org/10.1371/journal.pone.0155691

Kaboré S, Cecchi P, Mosser T, Toubiana M, Traoré O, Ouattara AS, Traoré AS, Barro N, Colwell RR, Monfort P (2018) Occurrence of Vibrio cholerae in water reservoirs of Burkina Faso. Research in Microbiology, 169(1):1–10

Kachwamba Y, Mohammed AA, Lukupulo H et al (2017) Genetic characterization of vibrio cholerae O1 isolates from outbreaks between 2011 and 2015 in Tanzania. BMC Infect Dis 17:157. https://doi.org/10.1186/s12879-017-2252-9

Kacou-N'douba A, Anne JC, Okpo LS et al (2012) Antimicrobial resistance of Vibrio cholerae O1 isolated during a cholera epidemic in 2011 in dry season in Cote d'Ivoire. J Infect Dev Ctries 6:595–597

Keddy KH, Sooka A, Parsons MB et al (2013) Diagnosis of vibrio cholerae o1 infection in Africa. J Infect Dis 208:S23–S31

Kelvin AA (2011) Outbreak of cholera in the republic of Congo and the Democratic Republic of Congo, and cholera worldwide. J Infect Dev Ctries 5:688–691

Keshav V, Potgieter N, Barnard TG (2010) Detection of Vibrio cholerae O1 in animal stools collected in rural areas of the Limpopo Province. Water SA 36:167–171

Khatib AM, Ali M, von Seidlein L et al (2012) Effectiveness of an oral cholera vaccine in Zanzibar: findings from a mass vaccination campaign and observational cohort study. Lancet Infect Dis 12:837–844. https://doi.org/10.1016/S1473-3099(12)70196-2

Khonje A, Metcalf CA, Diggle E et al (2012) Cholera outbreak in districts around Lake Chilwa, Malawi: lessons learned. Malawi Med J 24:29–33

Kiiru J, Mutreja A, Mohamed AA et al (2013) A study on the geophylogeny of clinical and environmental vibrio cholerae in Kenya. PLoS One 8:–e74829. https://doi.org/10.1371/journal.pone.0074829

Kiiru JN, Saidi SM, Goddeeris BM et al (2009) Molecular characterisation of vibrio cholerae O1 strains carrying an SXT/R391-like element from cholera outbreaks in Kenya: 1994–2007. BMC Microbiol 9:275. https://doi.org/10.1186/1471-2180-9-275

Kim J-H, Mogasale V, Burgess C, Wierzba TF (2016) Impact of oral cholera vaccines in cholera-endemic countries: a mathematical modeling study. Vaccine 34:2113–2120. https://doi.org/10.1016/j.vaccine.2016.03.004

Kim SY, Choi Y, Mason PR et al (2011) Potential impact of reactive vaccination in controlling cholera outbreaks: an exploratory analysis using a Zimbabwean experience. S Afr Med J 101:659–664

Kwesiga B, Pande G, Ario AR et al (2017) A prolonged, community-wide cholera outbreak associated with drinking water contaminated by sewage in Kasese District, western Uganda. BMC Public Health 18:30. https://doi.org/10.1186/s12889-017-4589-9

Kyelem CG, Bougouma A, Thiombiano RS et al (2011) Cholera epidemic in Burkina Faso in 2005: epidemiologic and diagnostic aspects. Pan Afr Med J 8:1

Lamrani Alaoui H, Oufdou K, Mezrioui NE (2010) Determination of several potential virulence factors in non-o1 Vibrio cholerae, Pseudomonas aeruginosa, faecal coliforms and streptococci isolated from Marrakesh groundwater. Water Sci Technol 61:1895–1905. https://doi.org/10.2166/wst.2010.263

Langa JP, Sema C, De Deus N et al (2015) Epidemic waves of cholera in the last two decades in Mozambique. J Infect Dev Ctries 9:635–641. https://doi.org/10.3855/jidc.6943

le Roux WJ, van Blerk GN (2011) The use of a high resolution melt real-time polymerase chain reaction (PCR) assay for the environmental monitoring of vibrio cholerae. Afr J Microbiol Res 5:3520–3526. https://doi.org/10.5897/AJMR11.695

Leckebusch GC, Abdussalam AF (2015) Climate and socioeconomic influences on interannual variability of cholera in Nigeria. Health Place 34:107–117. https://doi.org/10.1016/j.healthplace.2015.04.006

Leidner AJ, Adusumilli NC (2013) Estimating effects of improved drinking water and sanitation on cholera. J Water Health 11:671–683. https://doi.org/10.2166/wh.2013.238

Ley B, Khatib AM, Thriemer K et al (2012) Evaluation of a rapid dipstick (crystal VC) for the diagnosis of cholera in Zanzibar and a comparison with previous studies. PLoS One 7:e36930. https://doi.org/10.1371/journal.pone.0036930

Loharikar A, Briere E, Ope M et al (2013) A national cholera epidemic with high case fatality rates – Kenya 2009. J Infect Dis 208(Suppl):S69–S77. https://doi.org/10.1093/infdis/jit220

Lotfi NM, El-Shatoury SA, Hanora A, Saleh Ahmed R (2016) Isolating non-O1/non-O39 vibrio cholerae from Chironomus transvaalensis larvae and exuviae collected from polluted areas in Lake Manzala, Egypt. J Asia Pac Entomol 19:545–549. https://doi.org/10.1016/j.aspen.2016.05.007

Luque Fernandez MA, Schomaker M, Mason PR et al (2012) Elevation and cholera: an epidemiological spatial analysis of the cholera epidemic in Harare, Zimbabwe, 2008-2009. BMC Public Health 12:442. https://doi.org/10.1186/1471-2458-12-442

Luquero FJ, Grout L, Ciglenecki I et al (2013) First outbreak response using an oral cholera vaccine in Africa: vaccine coverage, acceptability and surveillance of adverse events, Guinea, 2012. PLoS Negl Trop Dis 7:e2465. https://doi.org/10.1371/journal.pntd.0002465

Luquero FJ, Grout L, Ciglenecki I et al (2014) Use of Vibrio cholerae vaccine in an outbreak in Guinea. N Engl J Med 370:2111–2120

Machado A, Bordalo AA (2016) Detection and quantification of vibrio cholerae, Vibrio parahaemolyticus, and Vibrio vulnificus in coastal waters of Guinea-Bissau (West Africa). EcoHealth 13:339–349. https://doi.org/10.1007/s10393-016-1104-1

Madoroba E, Momba MNB (2010) Prevalence of vibrio cholerae in rivers of Mpumalanga province, South Africa as revealed by polyphasic characterization. Afr J Biotechnol 9:7295–7301. https://doi.org/10.5897/AJB10.321

Mahamud AS, Ahmed JA, Nyoka R et al (2012) Epidemic cholera in Kakuma refugee camp, Kenya, 2009: the importance of sanitation and soap. J Infect Dev Ctries 6:234–241

Makoutode M, Diallo F, Mongbo V et al (2010) Assessment of the quality of response to the 2008 cholera outbreak in Contonou (Benin). Sante Publique 22:425–435

Maponga BA, Chirundu D, Gombe NT et al (2015) Cholera: a comparison of the 2008-9 and 2010 outbreaks in Kadoma City, Zimbabwe. Pan Afr Med J 20:221. https://doi.org/10.11604/pamj.2015.20.221.5197

Mari L, Bertuzzo E, Righetto L et al (2012) Modelling cholera epidemics: the role of waterways, human mobility and sanitation. J R Soc Interface 9:376–388. https://doi.org/10.1098/rsif.2011.0304

Marin MA, Thompson CC, Freitas FS et al (2013) Cholera outbreaks in Nigeria are associated with multidrug resistant atypical El Tor and non-O1/non-O139 Vibrio cholerae. PLoS Negl Trop Dis 7:e2049. https://doi.org/10.1371/journal.pntd.0002049

Martinez-Pino I, Luquero FJ, Sakoba K et al (2013) Use of a cholera rapid diagnostic test during a mass vaccination campaign in response to an epidemic in Guinea, 2012. PLoS Negl Trop Dis 7:e2366. https://doi.org/10.1371/journal.pntd.0002366

Masauni S, Mohammed M, Leyna GH, et al (2010) Controlling persistent cholera outbreaks in Africa: lessons from the recent Cholera Outbreak, West District Unguja Zanzibar, Tanzania, 2009. 14th Int Congr Infect Dis Abstr 14, Supple: e28. doi: https://doi.org/10.1016/j.ijid.2010.02.1549

Mason PR (2009) Zimbabwe experiences the worst epidemic of cholera in Africa. J Infect Dev Ctries 3:148–151

Mayega RW, Musenero M, Nabukenya I et al (2013) A descriptive overview of the burden, distribution and characteristics of epidemics in Uganda. East Afr J Public Health 10:397–402

Mbopi-Keou F-X, Dempouo Djomassi L, Ondobo Andze G, et al (2012) The dynamics of cholera epidemics in Cameroon since 2010: facts and perspectives. 15th Int Congr Infect Dis Abstr 16, Supple: e137. doi: https://doi.org/10.1016/j.ijid.2012.05.311

McGuigan KG, Conroy RM, Mosler H-J et al (2012) Solar water disinfection (SODIS): a review from bench-top to roof-top. J Hazard Mater 235–236:29–46. https://doi.org/10.1016/j.jhazmat.2012.07.053

Mengel MA (2014) Cholera in Africa: new momentum in fighting an old problem. Trans R Soc Trop Med Hyg 108:391–392

Mengel MA, Delrieu I, Heyerdahl L, Gessner BD (2014) Cholera outbreaks in Africa. Curr Top Microbiol Immunol 379:117–144. https://doi.org/10.1007/82_2014_369

Merten S, Schaetti C, Manianga C et al (2013a) Local perceptions of cholera and anticipated vaccine acceptance in Katanga province, Democratic Republic of Congo. BMC Public Health 13:60. https://doi.org/10.1186/1471-2458-13-60

Merten S, Schaetti C, Manianga C et al (2013b) Sociocultural determinants of anticipated vaccine acceptance for acute watery diarrhea in early childhood in Katanga Province, Democratic Republic of Congo. Am J Trop Med Hyg 89:419–425. https://doi.org/10.4269/ajtmh.12-0643

Mgbeahuruike EE, Yrjönen T, Vuorela H, Holm Y (2017) Bioactive compounds from medicinal plants: focus on piper species. S Afr J Bot 112:54–69. https://doi.org/10.1016/j.sajb.2017.05.007

Mintz ED, Tauxe RV (2013) Cholera in Africa: a closer look and a time for action. J Infect Dis 208(Suppl):S4–S7. https://doi.org/10.1093/infdis/jit205

Miwanda B, Moore S, Muyembe J-J et al (2015) Antimicrobial drug resistance of Vibrio cholerae, Democratic Republic of the Congo. Emerg Infect Dis 21:847–851. https://doi.org/10.3201/eid2105.141233

Mohamed AA, Oundo J, Kariuki SM et al (2012) Molecular epidemiology of geographically dispersed Vibrio cholerae, Kenya, January 2009-May 2010. Emerg Infect Dis 18:925–931. https://doi.org/10.3201/eid1806.111774

Moore S, Miwanda B, Sadji AY et al (2015) Relationship between distinct African cholera epidemics revealed via MLVA Haplotyping of 337 Vibrio cholerae isolates. PLoS Negl Trop Dis 9:e0003817. https://doi.org/10.1371/journal.pntd.0003817

Moore SM, Azman AS, Zaitchik BF et al (2017) El Niño and the shifting geography of cholera in Africa. Proc Natl Acad Sci U S A 114:4436–4441. https://doi.org/10.1073/pnas.1617218114

Moro PL, Sukumaran L (2017) Cholera vaccination: pregnant women excluded no more. Lancet Infect Dis 17:469–470. https://doi.org/10.1016/S1473-3099(17)30055-5

Morof D, Cookson ST, Laver S et al (2013) Community mortality from cholera: urban and rural districts in Zimbabwe. Am J Trop Med Hyg 88:645–650. https://doi.org/10.4269/ajtmh.11-0696

Msyamboza KP, M'bang'ombe M, Hausi H et al (2016) Feasibility and acceptability of oral cholera vaccine mass vaccination campaign in response to an outbreak and floods in Malawi. Pan Afr Med J 23:203. https://doi.org/10.11604/pamj.2016.23.203.8346

Mukandavire Z, Liao S, Wang J et al (2011) Estimating the reproductive numbers for the 2008–2009 cholera outbreaks in Zimbabwe. Proc Natl Acad Sci U S A 108:8767–8772. https://doi.org/10.1073/pnas.1019712108

Mukandavire Z, Morris JG (2015) Modeling the epidemiology of cholera to prevent disease transmission in developing countries. Microbiol Spectr. https://doi.org/10.1128/microbiolspec.VE-0011-2014

Musekiwa A, Volmink J (2011) Oral rehydration salt solution for treating cholera: = 310 mOsm/L solutions. Cochrane Database Syst Rev (12.):CD003:CD003754). https://doi.org/10.1002/14651858.CD003754.pub3

Mushayabasa S, Bhunu CP (2012) Is HIV infection associated with an increased risk for cholera? Insights from a mathematical model. Biosystems 109:203–213. https://doi.org/10.1016/j.biosystems.2012.05.002

Mutonga D, Langat D, Mwangi D et al (2013) National surveillance data on the epidemiology of cholera in Kenya, 1997–2010. J Infect Dis 208:S55–S61

Mutreja A, Kim DW, Thomson NR et al (2011) Evidence for several waves of global transmission in the seventh cholera pandemic. Nature 477:462–465. https://doi.org/10.1038/nature10392

Muyembe JJ, Bompangue D, Mutombo G et al (2013) Elimination of cholera in the Democratic Republic of the Congo: the new national policy. J Infect Dis 208:S86–S91

Naha A, Chowdhury G, Ghosh-Banerjee J et al (2013) Molecular characterization of high-level-cholera-toxin-producing El Tor variant Vibrio cholerae strains in the Zanzibar Archipelago of Tanzania. J Clin Microbiol 51:1040–1045. https://doi.org/10.1128/JCM.03162-12

Nair GB, Qadri F, Holmgren J et al (2006) Cholera due to altered El tor strains of vibrio cholerae O1 in Bangladesh. J Clin Microbiol 44:4211–4213. https://doi.org/10.1128/JCM.01304-06

Ncube A, Jordaan AJ, Mabela BM (2016) Assessing the knowledge, attitudes and practices regarding cholera preparedness and prevention in Ga-Mampuru village, Limpopo, South Africa. Jamba (Potchefstroom, South Africa) 8:164. https://doi.org/10.4102/jamba.v8i2.164

Nguyen VD, Sreenivasan N, Lam E et al (2014) Cholera epidemic associated with consumption of unsafe drinking water and street-vended water--eastern Freetown, Sierra Leone, 2012. Am J Trop Med Hyg 90:518–523. https://doi.org/10.4269/ajtmh.13-0567

Ngwa MC, Young A, Liang S et al (2017) Cultural influences behind cholera transmission in the Far North Region, Republic of Cameroon: a field experience and implications for operational level planning of interventions. Pan Afr Med J 28:311. https://doi.org/10.11604/pamj.2017.28.311.13860

Njagarah JB (2015) Modelling optimal control of cholera in communities linked by migration. Comput Math Methods Med. https://doi.org/10.1155/2015/898264

Njagarah JBH, Nyabadza F (2014) A metapopulation model for cholera transmission dynamics between communities linked by migration. Appl Math Comput 241:317–331. https://doi.org/10.1016/j.amc.2014.05.036

Nkoko DB, Giraudoux P, Plisnier P, Tinda AM, Piarroux M, Sudre B, Horion S, Tamfum J-JM, Ilunga BK, Piarroux R (2011) Dynamics of Cholera Outbreaks in Great Lakes Region of Africa, 1978–2008. Emerging Infectious Diseases, 17(11), 2026–2034

Noora CL, Issah K, Kenu E et al (2017) Large cholera outbreak in Brong Ahafo region, Ghana. BMC Res Notes 10:389. https://doi.org/10.1186/s13104-017-2728-0

Nsagha DS, Atashili J, Fon PN et al (2015) Assessing the risk factors of cholera epidemic in the Buea Health District of Cameroon. BMC Public Health 15:1128. https://doi.org/10.1186/s12889-015-2485-8

Ntema VM, Potgieter N, Barnard TG (2010) Detection of Vibrio cholerae and Vibrio parahaemolyticus by molecular and culture based methods from source water to household container-stored water at the point-of-use in South African rural communities. Water Sci Technol 61:3091–3101

Nyambedha EO, Sundaram N, Schaetti C et al (2013) Distinguishing social and cultural features of cholera in urban and rural areas of Western Kenya: implications for public health. Glob Public Health 8:534–551. https://doi.org/10.1080/17441692.2013.787107

Oguttu DW, Okullo A, Bwire G et al (2017) Cholera outbreak caused by drinking lake water contaminated with human faeces in Kaiso Village, Hoima District, Western Uganda, October 2015. Infect Dis Poverty 6:146. https://doi.org/10.1186/s40249-017-0359-2

Ohene S-A, Klenyuie W, Sarpeh M (2016) Assessment of the response to cholera outbreaks in two districts in Ghana. Infect Dis Poverty 5:99. https://doi.org/10.1186/s40249-016-0192-z

Ohene-Adjei K, Kenu E, Bandoh DA et al (2017) Epidemiological link of a major cholera outbreak in Greater Accra region of Ghana, 2014. BMC Public Health 17:801. https://doi.org/10.1186/s12889-017-4803-9

Okosun KO, Makinde OD (2014) A co-infection model of malaria and cholera diseases with optimal control. Math Biosci 258:19–32. doi: S0025-5564(14)00178-3 [pii]

Oladele DA, Oyedeji KS, Niemogha MT et al (2012) An assessment of the emergency response among health workers involved in the 2010 cholera outbreak in northern Nigeria. J Infect Public Health 5:346–353. https://doi.org/10.1016/j.jiph.2012.06.004

Olanrewaju OE, Adepoju KA (2017) Geospatial assessment of cholera in a rapidly urbanizing environment. J Environ Public Health. https://doi.org/10.1155/2017/6847376

Ontweka LN, Deng LO, Rauzier J et al (2016) Cholera rapid test with enrichment step has diagnostic performance equivalent to culture. PLoS One 11:e0168257. https://doi.org/10.1371/journal.pone.0168257

Onyango D, Karambu S, Abade A et al (2013) High case fatality cholera outbreak in Western Kenya, August 2010. Pan Afr Med J 15:109

Opare J, Ohuabunwo C, Afari E et al (2012) Outbreak of cholera in the East Akim municipality of Ghana following unhygienic practices by small-scale gold miners, November 2010. Ghana Med J 46:116–123

Osei FB, Duker A., Stein A (2012) Evaluating Spatial and Space-Time Clustering of Cholera in Ashanti-Region-Ghana. In: Cholera Ed. Gowder, S.J.T. https://doi.org/10.5772/36316 ISBN: 978-953-51-0415-5

Oyedeji KS, Niemogha MT, Nwaokorie FO et al (2013) Molecular characterization of the circulating strains of Vibrio cholerae during 2010 cholera outbreak in Nigeria. J Health Popul Nutr 31:178–184

Parker LA, Rumunu J, Jamet C et al (2017) Adapting to the global shortage of cholera vaccines: targeted single dose cholera vaccine in response to an outbreak in South Sudan. Lancet Infect Dis 17:e123–e127. https://doi.org/10.1016/S1473-3099(16)30472-8

Paz S (2009) Impact of temperature variability on cholera incidence in Southeastern Africa, 1971–2006. EcoHealth 6:340–345

Penrose K, de Castro MC, Werema J, Ryan ET (2010) Informal urban settlements and cholera risk in Dar Es Salaam, Tanzania. PLoS Negl Trop Dis 4:e631. https://doi.org/10.1371/journal.pntd.0000631

Peprah D, Palmer JJ, Rubin GJ et al (2016) Perceptions of oral cholera vaccine and reasons for full, partial and non-acceptance during a humanitarian crisis in South Sudan. Vaccine 34:3823–3827. https://doi.org/10.1016/j.vaccine.2016.05.038

Poncin M, Zulu G, Voute C et al (2018) Implementation research: reactive mass vaccination with single-dose oral cholera vaccine, Zambia. Bull World Health Organ 96:86–93. https://doi.org/10.2471/BLT.16.189241

Pugliese N, Maimone F, Scrascia M et al (2009) SXT-related integrating conjugative element and IncC plasmids in Vibrio cholerae O1 strains in eastern Africa. J Antimicrob Chemother 63:438–442

Pugliese-Garcia M, Heyerdahl LW, Mwamba C et al (2018) Factors influencing vaccine acceptance and hesitancy in three informal settlements in Lusaka, Zambia. Vaccine 36:5617–5624. https://doi.org/10.1016/j.vaccine.2018.07.042

Quilici ML, Massenet D, Gake B et al (2010) Vibrio cholerae O1 variant with reduced susceptibility to ciprofloxacin, Western Africa. Emerg Infect Dis 16:1804–1805. https://doi.org/10.3201/eid1611.100568

Quinones M, Davis BM, Waldor MK (2006a) Activation of the Vibrio cholerae SOS response is not required for intestinal cholera toxin production or colonization. Infect Immun 74:927–930. https://doi.org/10.1128/IAI.74.2.927-930.2006

Quinones M, Kimsey HH, Ross W et al (2006b) LexA represses CTXphi transcription by blocking access of the alpha C-terminal domain of RNA polymerase to promoter DNA. J Biol Chem 281:39407–39412. https://doi.org/10.1074/jbc.M609694200

Rebaudet S, Mengel MA, Koivogui L et al (2014) Deciphering the origin of the 2012 cholera epidemic in Guinea by integrating epidemiological and molecular analyses. PLoS Negl Trop Dis 8:e2898

Rebaudet S, Sudre B, Faucher B, Piarroux R (2013a) Cholera in coastal Africa: a systematic review of its heterogeneous environmental determinants. J Infect Dis 208:S98–S106

Rebaudet S, Sudre B, Faucher B, Piarroux R (2013b) Environmental determinants of cholera outbreaks in inland africa: a systematic review of main transmission foci and propagation routes. J Infect Dis 208:S46–S54

Reimer AR, van Domselaar G, Stroika S et al (2011) Comparative genomics of Vibrio Cholerae from Haiti, Asia, and Africa. Emerg Infect Dis 17:2113–2121

Reyburn R, Deen JL, Grais RF et al (2011a) The case for reactive mass oral cholera vaccinations. PLoS Negl Trop Dis 5:e952. https://doi.org/10.1371/journal.pntd.0000952

Reyburn R, Kim DR, Emch M et al (2011b) Climate variability and the outbreaks of cholera in Zanzibar, East Africa: a time series analysis. Am J Trop Med Hyg 84:862–869. https://doi.org/10.4269/ajtmh.2011.10-0277

Roy M, Zinck RD, Bouma MJ, Pascual M (2014) Epidemic cholera spreads like wildfire. Sci Rep 4:1–7. https://doi.org/10.1038/srep03710

Saidi SM, Chowdhury N, Awasthi SP et al (2014) Prevalence of vibrio cholerae O1 El tor variant in a cholera-endemic zone of Kenya. J Med Microbiol 63:415–420. https://doi.org/10.1099/jmm.0.068999-0

Saidou K, Cecchi P, Mosser T et al (2017) Occurrence of Vibrio cholerae in water reservoirs of Burkina Faso. Res Microbiol. https://doi.org/10.1016/j.resmic.2017.08.004

Sambe-Ba B, Diallo MH, Seck A et al (2017) Identification of atypical El TorV. Cholerae O1 Ogawa hosting SXT element in Senegal, Africa. Front Microbiol 8:748. https://doi.org/10.3389/fmicb.2017.00748

Sasaki S, Suzuki H, Fujino Y et al (2009) Impact of drainage networks on cholera outbreaks in Lusaka, Zambia. Am J Public Health 99:1982–1987. https://doi.org/10.2105/AJPH.2008.151076

Sauvageot D, Njanpop-Lafourcade B-M, Akilimali L et al (2016) Cholera incidence and mortality in sub-Saharan African sites during multi-country surveillance. PLoS Negl Trop Dis 10:e0004679. https://doi.org/10.1371/journal.pntd.0004679

Sauvageot D, Saussier C, Gobeze A et al (2017) Oral cholera vaccine coverage in hard-to-reach fishermen communities after two mass campaigns, Malawi, 2016. Vaccine 35:5194–5200. https://doi.org/10.1016/j.vaccine.2017.07.104

Schaetti C, Ali SM, Hutubessy R et al (2012a) Social and cultural determinants of oral cholera vaccine uptake in Zanzibar. Hum Vaccin Immunother 8:1223–1229. https://doi.org/10.4161/hv.20901

Schaetti C, Chaignat CL, Hutubessy R et al (2011) Social and cultural determinants of anticipated acceptance of an oral cholera vaccine prior to a mass vaccination campaign in Zanzibar. Hum Vaccin 7:1299–1308. https://doi.org/10.4161/hv.7.12.18012

Schaetti C, Khatib AM, Ali SM et al (2010) Social and cultural features of cholera and shigellosis in peri-urban and rural communities of Zanzibar. BMC Infect Dis 10:339. https://doi.org/10.1186/1471-2334-10-339

Schaetti C, Sundaram N, Merten S et al (2013) Comparing sociocultural features of cholera in three endemic African settings. BMC Med 11:206. https://doi.org/10.1186/1741-7015-11-206

Schaetti C, Weiss MG, Ali SM et al (2012b) Costs of illness due to cholera, costs of immunization and cost-effectiveness of an oral cholera mass vaccination campaign in Zanzibar. PLoS Negl Trop Dis 6:e1844. https://doi.org/10.1371/journal.pntd.0001844

Scheelbeek P, Treglown S, Reid T, Maes P (2009) Household fish preparation hygiene and cholera transmission in Monrovia, Liberia. J Infect Dev Ctries 3:727–731

Scrascia M, Pugliese N, Maimone F et al (2009) Clonal relationship among vibrio cholerae O1 El tor strains isolated in Somalia. Int J Med Microbiol 299:203–207. https://doi.org/10.1016/j.ijmm.2008.07.003

Shikanga OT, Mutonga D, Abade M et al (2009) High mortality in a cholera outbreak in western Kenya after post-election violence in 2008. Am J Trop Med Hyg 81:1085–1090. https://doi.org/10.4269/ajtmh.2009.09-0400

Smith AM, Keddy KH, Ismail H et al (2012) Possible laboratory contamination leads to incorrect reporting of Vibrio cholerae O1 and initiates an outbreak response. J Clin Microbiol 50:480–482. https://doi.org/10.1128/JCM.05785-11

Smith AM, Njanpop-Lafourcade B-M, Mengel MA et al (2015) Comparative characterization of Vibrio cholerae O1 from five sub-Saharan African countries using various phenotypic and genotypic techniques. PLoS One 10:e0142989. https://doi.org/10.1371/journal.pone.0142989

Songsore J (2017) The complex interplay between everyday risks and disaster risks: the case of the 2014 cholera pandemic and 2015 flood disaster in Accra, Ghana. Int J Disaster Risk Reduct 26:43–50. https://doi.org/10.1016/j.ijdrr.2017.09.043

Ssemakalu CC, Ubomba-Jaswa E, Motaung KS, Pillay M (2014) Influence of solar water disinfection on immunity against cholera – a review. J Water Health 12:393–398. https://doi.org/10.2166/wh.2014.158

Ssemakalu CC, Woulter LR, Pillay M (2013) Impact of solar irradiation on cholera toxin secretion by different strains of Vibrio cholerae. S Afr J Sci 109

Sule IB, Yahaya M, Aisha AA et al (2017) Descriptive epidemiology of a cholera outbreak in Kaduna state, Northwest Nigeria, 2014. Pan Afr Med J 27:172. https://doi.org/10.11604/pamj.2017.27.172.11925

Sundaram N, Schaetti C, Chaignat CL et al (2013) Socio-cultural determinants of anticipated acceptance of an oral cholera vaccine in Western Kenya. Epidemiol Infect 141:639–650. https://doi.org/10.1017/S0950268812000829

Sundaram N, Schaetti C, Merten S et al (2016) Sociocultural determinants of anticipated oral cholera vaccine acceptance in three African settings: a meta-analytic approach. BMC Public Health 16:36. https://doi.org/10.1186/s12889-016-2710-0

Tebou PLF, Tamokou J-d-D, Ngnokam D et al (2017) Flavonoids from Maytenus buchananii as potential cholera chemotherapeutic agents. S Afr J Bot 109:58–65. https://doi.org/10.1016/j.sajb.2016.12.019

Teklehaimanot GZ, Genthe B, Kamika I, Momba MNB (2015) Prevalence of enteropathogenic bacteria in treated effluents and receiving water bodies and their potential health risks. Sci Total Environ 518–519:441–449. https://doi.org/10.1016/j.scitotenv.2015.03.019

Thompson CC, Freitas FS, Marin MA et al (2011) Vibrio cholerae O1 lineages driving cholera outbreaks during seventh cholera pandemic in Ghana. Infect Genet Evol 11:1951–1956. https://doi.org/10.1016/j.meegid.2011.08.020

Trærup S, Ortiz R, Markandya A (2011) The costs of climate change: a study of cholera in Tanzania. Int J Environ Res Public Health 8:4386–4405. https://doi.org/10.3390/ijerph8124386

Traore SG, Bonfoh B, Krabi R et al (2012) Risk of Vibrio transmission linked to the consumption of crustaceans in coastal towns of Cote d'Ivoire. J Food Prot 75:1004–1011. https://doi.org/10.4315/0362-028X.JFP-11-472

Ujjiga TTA, Wamala JF, Mogga JJH et al (2015) Risk factors for sustained cholera transmission, Juba County, South Sudan, 2014. Emerg Infect Dis 21:1849–1852. https://doi.org/10.3201/eid2110.142051

Valia R, Taviani E, Spagnoletti M et al (2013) Vibrio cholerae O1 epidemic variants in Angola: a retrospective study between 1992 and 2006. Front Microbiol 4:354

Valneva (2019) https://www.valneva.ca/en/ [Accessed November 29, 2019]

Vezzulli L, Grande C, Tassistro G et al (2017) Whole-genome enrichment provides deep insights into Vibrio cholerae Metagenome from an African River. Microb Ecol 73:734–738. https://doi.org/10.1007/s00248-016-0902-x

von Seidlein L, Jiddawi M, Grais RF et al (2013) The value of and challenges for cholera vaccines in Africa. J Infect Dis 208(Suppl):8–14. https://doi.org/10.1093/infdis/jit194

W.H.O. (2011) Cholera annual report, 2010. Wkly Epidemiol Rec 86(31):325–340

W.H.O. (2012) Cholera annual report, 2011. Wkly Epidemiol Rec 87(31–32):289–304

W.H.O. (2013) Cholera annual report, 2012. Wkly Epidemiol Rec 88(31):321–336

W.H.O. (2014) Cholera annual report, 2013. Wkly Epidemiol Rec 89(31):345–356

W.H.O. (World Health Organization) (2014) Weekly epidemiological record Relevé épidémiologique hebdomadaire. 205–220

Weill F-X, Domman D, Njamkepo E et al (2017) Genomic history of the seventh pandemic of cholera in Africa. Science 358:785–789. https://doi.org/10.1126/science.aad5901

Yirenya-Tawiah DR, Darkwa A, Dzodzomenyo M (2018) Environmental surveillance for vibrio cholerae in selected households' water storage systems in Accra metropolitan area (AMA) prior to the 2014 cholera outbreak in Accra, Ghana. Environ Sci Pollut Res 25:28335–28343. https://doi.org/10.1007/s11356-018-2860-y

Chapter 4
Reservoirs of *Cryptosporidium* and *Giardia* in Africa

Lisa M. Schaefer, W. J. le Roux, and Akebe Luther King Abia

4.1 Introduction

Cryptosporidium and *Giardia* are the most commonly identified waterborne protozoan pathogens. Although they are responsible for waterborne outbreaks globally (Baldursson and Karanis 2011), the highest burden of disease from *Cryptosporidium* and *Giardia* is in developing countries, including those in Africa (Feng and Xiao 2011; Bouzid et al. 2018). These pathogens may be transmitted by contaminated drinking water, recreational water, and treated and untreated wastewater (Ongerth and Karanis 2018). Other transmission routes include contaminated food, usually from infected food handlers or food irrigated with contaminated water (Carmena 2010). In Africa, *Cryptosporidium* and *Giardia* infections are prevalent among communities that lack access to clean potable water supplies. Over 300 million people in sub-Saharan Africa are without access to safe drinking water (WHO 2015). Surface waters, often polluted due to uncontrolled sewage discharge, poorly managed wastewater treatment plants (Dungeni and Momba 2010), and inadequate waste disposal facilities from settlements located close to rivers (Uneke and Uneke 2007), thus form alternative, if not the only water source available within many countries on the continent. The sub-Saharan African population lacking access to sanitation has increased since 1990, influenced by rapid population growth and slow progress in attaining sanitation targets (WHO 2015). Unsafe water supplies and inadequate sanitation and hygiene conditions increase the transmission of diarrheal diseases such as *Cryptosporidium* and *Giardia*. In addition to the challenges that

L. M. Schaefer (✉) · W. J. le Roux
Water Centre, Council for Scientific and Industrial Research (CSIR), Pretoria, South Africa
e-mail: lschaefer@csir.co.za

A. L. K. Abia
Antimicrobial Research Unit, College of Health Sciences, University of KwaZulu-Natal, Durban, South Africa

© Springer Nature Switzerland AG 2020
A. L. K. Abia, G. R. Lanza (eds.), *Current Microbiological Research in Africa*,
https://doi.org/10.1007/978-3-030-35296-7_4

115

Africa faces regarding unsafe water supplies, the prevalence of HIV and malnutrition also predisposes communities to the diarrheal diseases caused by these parasites.

4.2 Cryptosporidiosis and Giardiasis: Clinical Features, Transmission, and Diagnosis

Cryptosporidiosis is a diarrheal illness characterized by short-term diarrhea that resolves spontaneously in immunocompetent individuals or persists as a life-threatening illness in the immunocompromised (Current and Garcia 1991). Giardiasis is characterized by acute or chronic diarrhea, with a severity that varies between individuals. *Giardia* infections tend to be self-limiting in immunocompetent individuals (Cotton et al. 2011). Chronic *Giardia* infections in children are associated with malnutrition, stunting, and reduced cognitive function (Berkman et al. 2002). Some hosts infected with *Giardia* or *Cryptosporidium* may remain asymptomatic (Hunter and Thompson 2005; Cotton et al. 2011). Transmission of *Cryptosporidium* or *Giardia* typically occurs following ingestion of the infectious oocysts or cysts through contaminated food or water or directly through the fecal-oral route. *Cryptosporidium hominis* is transmitted only between humans, but the zoonotic transmission of *Cryptosporidium parvum* occurs by direct contact with infected cattle or indirect transmission through drinking water (Hunter and Thompson 2005). It is also suggested that *Cryptosporidium* may be transmitted via respiratory secretions (Sponseller et al. 2014). *Giardia duodenalis* (synonymous with *G. lamblia* and *G. intestinalis*) of assemblages A and B is the only species known to infect humans, and these infections do not appear to occur by zoonotic transmission (Hunter and Thompson 2005; Xiao and Fayer 2008), although a number of genetic variants within assemblages A and B of *G. duodenalis* in animal species have been shown to have zoonotic potentials (Ryan and Cacciò 2013). *Giardia* has also been reported to have the potential to be sexually transmitted (Escobedo et al. 2018).

Diagnosis of *Cryptosporidium* and *Giardia* is usually by examining stool samples microscopically, which requires technical competence. In sub-Saharan Africa, microbiological methods for clinical investigation of diarrheal diseases are usually restricted to identifying conventional enteric bacteria such as *Escherichia coli*, *Salmonella,* and *Shigella*. Isolates are often not fully characterized due to a lack of resources and expertise (Opintan et al. 2010).

4.3 Cryptosporidiosis and Giardiasis: Disease Burden

The protozoan parasites *Cryptosporidium* and *Giardia* are significant causes of diarrheal disease (Chalmers and Davies 2010; Feng and Xiao 2011), with *Cryptosporidium* being the most frequently isolated protozoan pathogen globally (WHO 2009). *Cryptosporidium* and *Giardia* infections are widespread in adult and immunocompetent populations in Africa, due to the unhygienic and improper disposal of wastewater and the use of polluted surface waters as a primary source of potable water (Putignani and Menichella 2010).

It is estimated that 2.9 million diarrheal episodes caused by *Cryptosporidium* occur annually in children aged <24 months in sub-Saharan Africa (Sow et al. 2016). The Global Enteric Multicenter Study (GEMS) has shown that during the first 5 years of life, *Cryptosporidium* is the second leading contributor to moderate-to-severe diarrheal disease in sub-Saharan Africa, with the first being rotavirus (Kotloff et al. 2013). Infection with this pathogen is responsible for the excessive mortality in children aged 12–23 months (Kotloff et al. 2013). Cryptosporidiosis is often reported as a cause of life-threatening illness in individuals with HIV/AIDS and is associated with diarrhea and malnutrition in young children in sub-Saharan Africa. *Cryptosporidium* is also a significant pathogen regardless of HIV prevalence (Mølbak et al. 1993; Kotloff et al. 2013). The disease burden in both developed and developing nations may be largely underestimated due to the number of self-limiting and asymptomatic infections. The lack of efficient diagnosis of the causative agents of diarrhea, as well as the use of microscopy for routine clinical diagnosis (with low specificity and sensitivity), may also contribute to an underestimation of *Cryptosporidium* infection (Bouzid et al. 2018).

Giardia duodenalis infections occur worldwide (Feng and Xiao 2011), but *Giardia* is notably prevalent in areas with poor sanitary conditions and limited water treatment facilities. Although the GEMS reported that *Giardia* was not significantly positively associated with moderate-to-severe diarrhea in sub-Saharan Africa in children under 5 (Kotloff et al. 2013), high infection rates of *Giardia* have been reported in developing countries including those in Africa (Feng and Xiao 2011). *Giardia duodenalis*, the species infecting only humans, is estimated to cause 2.8×10^8 cases of intestinal disease per annum worldwide (Lane and Lloyd 2002). In the developing world, *Giardia* has been found in stool samples of approximately 15% of children aged 0–1 year (McCormick 2014). A systematic review and meta-analysis of endemic pediatric giardiasis revealed that responses to *Giardia lamblia* infections differed between industrialized populations and developing populations (Muhsen and Levine 2012). In industrialized countries, *G. lamblia* can cause both acute and persistent diarrhea in adults and children. *Giardia* does not generally cause acute pediatric diarrhea among infants and children in developing countries but is associated with persistent diarrhea. This may be due to the age of initial exposure and the frequency of re-exposure to *Giardia* infections that may offer protection against symptomatic disease (Muhsen and Levine 2012).

The WHO included *Cryptosporidium* and *Giardia* in the Neglected Disease Initiative in 2004. The initiative includes diseases that are a global burden, have a common link with poverty, and impair development and socioeconomic improvements (Savioli et al. 2006). Effective control measures for *Cryptosporidium* are desperately needed in sub-Saharan Africa due to pervasive conditions of poor sanitation and hygiene, the limited availability of antiretrovirals, and the high prevalence of cryptosporidiosis in children (independent of HIV infection) (Mor and Tzipori 2008). The urgent need for effective intervention on the African continent was also emphasized by Aldeyarbi et al. (2016) when reviewing the epidemiology and transmission dynamics of *Cryptosporidium* in Africa. *Giardia* is considered to have a significant public health impact due to its high prevalence and propensity to cause major outbreaks as well as the effects it causes on growth and cognitive functions in infected children (Feng and Xiao 2011).

4.4 Occurrences and Prevalence of *Cryptosporidium* and *Giardia* in Africa

Cryptosporidium and *Giardia* are prevalent in domestic and wild animals (Table 4.1), the environment and food crops (Table 4.2), and in humans in Africa, with molecular typing indicating anthroponotic (human to human) and zoonotic or potentially zoonotic potentials for *Giardia* (animal to human) transmission cycles (Squire and Ryan 2017).

4.5 Epidemiology

Despite the high prevalence of these protozoan parasites in developing countries, relatively little is known about the epidemiology of cryptosporidiosis and giardiasis in Africa (Aldeyarbi et al. 2016; Squire and Ryan 2017). Epidemiological data regarding these parasites mostly originate from outbreak investigations (Savioli et al. 2006). A review of global waterborne protozoan parasite outbreaks documented for the period between 2004 and 2010 showed that while 46.7% of the outbreaks occurred in Australia, 30.6% in North America, 16.5% in Europe, and 3.5% from Asia, no outbreaks were reported in Africa (Baldursson and Karanis 2011). This is unlikely as the highest prevalence of protozoan parasite infections is suggested to occur in developing counties, and more outbreaks would be expected. The higher rate of reported waterborne protozoan parasitic outbreaks in developed nations was ascribed to better technological capabilities and the establishment of surveillance systems (Baldursson and Karanis 2011). This highlights that the lack of capabilities and data reporting play a significant role in influencing the apparent global distribution pattern of diseases due to these protozoa, and in particular, that from the African continent.

Young children are especially vulnerable to infection from *Cryptosporidium*, independent of their HIV status and cryptosporidiosis peaks in children in sub-Saharan Africa aged 6–12 months and becomes less clinically significant with age (Mor and Tzipori 2008). Breastfeeding may give some protection by providing antibodies or preventing exposure to contaminated water (Mor and Tzipori 2008). A lower prevalence of *Giardia* infections was found in breastfed children aged 7–12 months (Tellevik et al. 2015), compared to those that had been weaned.

Table 4.1 Some environmental reservoirs of *Cryptosporidium* and *Giardia* in Africa

Country	Environmental sources/reservoirs	Method of detection	References
Cameroon	Lake water, wastewater, rivers	Ziehl–Neelsen method and Lugol iodine coloration	Ajeagah et al. (2007), Gideon et al. (2007)
Côte d'Ivoire	Urban wastewater and lagoon water	Sodium acetate formalin (SAF) technique	Yapo et al. (2014)
Egypt	Swimming pool, leafy vegetables	Microscopy using Lugol's iodine and modified Ziehl–Neelsen	Abd El-Salam (2012), Eraky et al. (2014)
Ethiopia	Drinking water sources	U.S. EPA method 1623	Atnafu et al. (2012)
Ethiopia	Fruits and vegetables	Sedimentation concentration	Tefera et al. (2014)
Ghana	Sachet water	Not mentioned (outsourced)	Ndur et al. (2015)
Kenya	River and surface water	Filtration and immunomagnetic bead separation (IMS)-immunoantibody staining method, calcium carbonate flocculation (CCF) and sucrose floatation method, polymerase chain reaction coupled with the restriction fragment length polymorphism (PCR-RFLP);	Kato et al. (2003), Muchiri et al. (2009)
Morocco	Coriander, carrots, radish, mint, potatoes, irrigation water	Not described	Amahmid et al. (1999)
Nigeria	Surface waters	Filtration, backwashing, concentration and modified Ziehl–Neelsen staining technique	Uneke and Uneke (2007)
Tanzania	Surface waters	Microscopy using Lugol's iodine and modified Ziehl–Neelsen	Kusiluka et al. (2005),
Tunisia	Raw and treated wastewater; sludge	Microscopy, PCR, sequencing	Khouja et al. (2010), Ben Ayed et al. (2012)
South Africa	Harvested rainwater, treated wastewater effluent, irrigation water, vegetables	PCR, modified Ziehl–Neelsen acid-fast technique	Dungeni and Momba (2010), Duhain (2012), Dobrowsky et al. (2014)
Uganda	Natural and communal piped tap water	Ziehl–Neelsen stain and *Giardia* cysts by zinc sulfate floatation technique, PCR, sequencing	Sente et al. (2016)
Zimbabwe	Surface waters, wells, springs, taps	Zinc sulfate floatation technique, microscopy	Dalu et al. (2011), Mtapuri-Zinyowera et al. (2014)

Table 4.2 Some animal reservoirs of *Cryptosporidium* and *Giardia* in Africa

Country	Animal sources/ reservoirs	Method of detection	References
Algeria	Lamb, goat, cattle, chicken, Turkey, horse, donkey	Nested-PCR, PCR-RFLP, sequencing, modified Ziehl–Neelsen acid-fast technique, ELISA	Smith and Nichols (2010), Baroudi et al. (2013, 2017, 2018), Laatamna et al. (2013, 2015), Abbas et al. (2015), Ouakli et al. (2018)
Cameroon	Domestic Guinea pig	Sedimentation test, ELISA	Meutchieye et al. (2017)
Central African republic	Gorillas	Nested-PCR, PCR-RFLP, sequencing	Sak et al. (2013)
Egypt	Calves, ruminant animals, dairy cattle	Nested-PCR, Ziehl–Neelsen method, copro-antigen *RIDA®QUICK* test, and real-time PCR	Amer et al. (2013), Helmy et al. (2014), Ghoneim et al. (2017)
Ghana	Calves	Modified Ziehl–Neelsen (MZN) staining technique	Mensah et al. (2018)
Ethiopia	Dairy cattle	Modified Ziehl–Neelsen (MZN) microscopy, nested PCR, PCR-RFLP, sequencing	Wegayehu et al. (2016); Manyazewal et al. (2018)
South Africa	Calves	Ziehl–Neelsen (MZN) staining technique, PCR, sequencing	Samra et al. (2016)
Sudan	Calves	Modified Ziehl–Neelsen (MZN) microscopy, nested PCR, PCR-RFLP, sequencing	Taha et al. (2017)
Tanzania	Baboons, chimpanzees, goats, cattle	PCR-RFLP, sequencing	Kusiluka et al. (2005), Parsons et al. (2015)
Zambia	Pigs	Merifluor® *Cryptosporidium/Giardia* immunofluorescence assay	Siwila and Mwape (2012)

In the African context, the most critical epidemiological scenarios that drive and exacerbate infections from *Cryptosporidium* and *Giardia* include access to safe water, the prevalence of HIV infections, and malnutrition. These scenarios are discussed in the following section.

4.5.1 Waterborne Cryptosporidiosis and Giardiasis

Cryptosporidium and *Giardia* may be transmitted to humans through the consumption of water contaminated with the infectious oocysts of *Cryptosporidium* or the cysts of *Giardia*. The presence of these parasites in water sources is caused by human and animal fecal contamination. *Cryptosporidium* and *Giardia* are ubiquitous in aquatic environments, and both pathogens have been isolated from rivers, lakes, treated drinking water, recreational waters, and groundwater, as indicated in Table 4.1 (Carmena 2010). Contamination of municipal drinking water systems and recreational waters serve as significant routes of exposure in the developed world (Karanis et al. 2007). In the developing communities of sub-Saharan Africa, inadequate water treatment, poor hygiene practices, and drinking untreated or poorly treated water predisposes communities to protozoan infections (Sente et al. 2016). The problem is exacerbated during the rainy season when human and animal fecal matter is washed into water bodies used by communities for domestic purposes (Sente et al. 2016). In Africa, water sources are shared by humans, domesticated animals, and wild animals, which may shed oocysts into the water, increasing the risk of surface water contamination with *Cryptosporidium* (Aldeyarbi et al. 2016). While the zoonotic potential of *Giardia* has been established, more research is required to confirm zoonotic transmission (Ryan and Cacciò 2013).

Cryptosporidium and *Giardia* possess numerous characteristics that facilitate their waterborne transmission. The oocysts and cysts are environmentally robust, which allows them to survive for long periods and persist in the aquatic environment. *Cryptosporidium* oocysts can survive for 6 months and *Giardia* cysts can survive for 2–3 months in surface waters (Smith et al. 2006). Zoonotic transmission of *Cryptosporidium* increases the likelihood of environmental contamination and waterborne transmission (Karanis et al. 2007). Humans are also significant contributors to the contamination of surface waters, and infected persons can shed up to 1×10^{10} oocysts during symptomatic *Cryptosporidium* infections and 1.44×10^9 cysts per day by humans infected with *Giardia* (Karanis et al. 2007). Both parasites have low infectious doses with as low as nine oocysts for *C. parvum* (Okhuysen et al. 1999) and a median infectious dose of 25–100 *Giardia* cysts (Smith et al. 2006). The small size of the oocysts (4–6 µm) and cysts (8–12 × 7–10 µm length x width) also enables penetration of the physical barriers of water treatment (Karanis et al. 2007). The (oo)cysts of *Cryptosporidium* and *Giardia* are resistant to many water treatment disinfectants. *Cryptosporidium* spp. are extremely chlorine-tolerant (Betancourt and Rose 2004), and in conventionally treated drinking water, *Cryptosporidium* is considered to be more of a public health threat due to the oocysts being more resistant to chemical disinfection (Smith et al. 2006).

4.5.2 Cryptosporidiosis and Giardiasis in Immunocompromised Groups (HIV/AIDS)

Cryptosporidium is associated with self-limiting diarrhea in immunocompetent individuals, or the infection may be asymptomatic. The infection in immunocompromised individuals with HIV/AIDS may cause severe and fatal diarrhea (O'Connor et al. 2011). *Cryptosporidium* is an opportunistic pathogen in the immunocompromised (Assefa et al. 2009), and is responsible for morbidity and mortality of individuals with HIV/AIDS in developing countries (O'Connor et al. 2011). Studies in Ugandan and Tanzanian children indicated that *Cryptosporidium* infection was found more often in HIV-positive children than those who were HIV-negative (Tumwine et al. 2005; Tellevik et al. 2015). *Cryptosporidium* infections are also significantly higher among HIV-positive adults, especially in those with lower CD4 T-cell counts (Assefa et al. 2009). Reports have shown that immunocompromised individuals with HIV/AIDS are more likely to be infected with *Giardia* (Sanyaolu et al. 2011; Boaitey et al. 2012; Adamu et al. 2013), but *Giardia* is not considered to be an opportunistic pathogen in the immunocompromised (Stark et al. 2009).

Regarding HIV prevalence, in 2017, the African region was the worst affected region worldwide, with an HIV prevalence in adults aged 15–49 of 4.1% compared to the global prevalence of 0.8% (WHO 2018). The HIV epidemic thus serves as a significant contributor to the occurrence of cryptosporidiosis and giardiasis in Africa (Squire and Ryan 2017). In 2017, just 59% of adults (>15 years old) and 52% of children (0–14 years old) with HIV could access antiretroviral therapy (UNAIDS 2018). Chronic diarrhea may be reduced significantly in HIV-positive individuals receiving antiretroviral therapy (Elfstrand and Florén 2010; Adamu et al. 2013). The absence of antiretroviral treatment, together with the lack of effective treatment specific for cryptosporidiosis in immunocompromised individuals, also intensifies this disease in vulnerable populations (O'Connor et al. 2011).

4.5.3 Cryptosporidiosis and Giardiasis in Malnourished Young Children

The United Nations International Children's Emergency Fund (UNICEF) list infectious diseases such as diarrhea as a critical determinant of stunting. In 2011 one in three children in Africa under the age of five were stunted, hampering cognitive development and hindering learning ability later in life (UNICEF 2013). As causative agents of diarrhea, *Cryptosporidium* and *Giardia* adversely affect growth and nutritional status, especially during early childhood.

There is a significantly higher prevalence of cryptosporidiosis among malnourished children (Hunter and Nichols 2002; Mor and Tzipori 2008; Bouzid et al. 2017). These children are more at risk of death and prolonged illness (Hunter and Nichols 2002). Sub-Saharan Africa has the highest prevalence of undernourishment globally with over 220 million hungry people in 2014–2016. Africa as a whole, and specifically sub-Saharan Africa, has failed to reach the millennium development goal of reducing hunger by half by 2015. The undernourished population even increased by 44 million between 1990–1992 and 2014–2016 (FAO 2015). The relationship between malnutrition and cryptosporidiosis is complicated, and studies are conflicting (Hunter and Nichols 2002; Bouzid et al. 2017). Malnutrition may be a risk factor for *Cryptosporidium* infection due to impaired cell-mediated immunity. It is also possible that *Cryptosporidium* infection impairs nutrient absorption, causing weight loss and growth stunting (Mor and Tzipori 2008).

There is an increased prevalence of *Giardia* in children with chronic diarrhea and malnutrition (Gendrel et al. 2003; Botero-Garcés et al. 2009). *Giardia* is associated with malnutrition and stunting (weight to height) in early childhood and has been linked to reduced cognitive function at a later age (Berkman et al. 2002; Simsek et al. 2004; Koruk et al. 2010). This may be due to a loss of intestinal surface area, and reduced absorption or digestion of nutrients caused by giardiasis (Halliez and Buret 2013). In addition to growth retardation, *Giardia* is associated with malabsorption of micronutrients, iron deficiency and anemia (Ertan et al. 2002; Simsek et al. 2004) as well as zinc deficiencies in school children (Ertan et al. 2002; Quihui et al. 2010).

4.6   Environmental Monitoring and Detection

Methods for the isolation and identification of *Cryptosporidium* and *Giardia* have been reviewed (Adeyemo et al. 2018; Ahmed and Karanis 2018). Monitoring for *Cryptosporidium* and *Giardia* has been used for risk assessment purposes, examining water treatment efficacy, and waterborne outbreak investigations (Betancourt and Rose 2004). Monitoring is valuable due to the recalcitrant nature of the protozoan (oo)cysts and because the usual indicators for water quality are not necessarily indicative of water safety for protozoa (Rose et al. 2002; Betancourt and Rose 2004). In the USA and the UK, national occurrence monitoring programs for *Cryptosporidium* oocysts in water have been undertaken (Rose et al. 2002). In South Africa, *Cryptosporidium* and *Giardia* monitoring in drinking water supply systems is not widespread, and only a few studies have mainly focused on the final treated water (Sigudu et al. 2014).

The detection of these protozoa in water is complicated by the fact that oocyst and cyst concentrations in water are characteristically low concerning the limit of detection (Ongerth and Karanis 2018). The recovery efficiencies vary widely, and as a result the analysis of water samples should be at a minimum every month using sufficient sample volumes (Ongerth and Karanis 2018). Analysis of waterborne oocysts/cysts involves concentration, purification, and detection. The USEPA has validated and approved Method 1623 (ISO15553) for the simultaneous detection of waterborne *Cryptosporidium* and *Giardia* (USEPA 2012). This method involves filtration, immunomagnetic separation of the cysts and oocysts, and an immunofluorescence assay for the determination of protozoan concentrations. Confirmation is obtained through staining using DAPI (4,6-diamidino-phenylindole) and differential interference contrast microscopy (USEPA 2012). Due to improved recoveries of cysts and oocysts with the USEEPA method 1623, 10 L sample volumes have generally been used for water monitoring; however, Ongerth and Karanis (2018) are of the opinion that using 10 L sample volumes may have resulted in a misleading impression that *Cryptosporidium* and *Giardia* are widespread, but intermittent in surface water. Analysis of samples using sufficient sample volumes of 50 L has shown that *Cryptosporidium* and *Giardia* are practically universal in surface water (Ongerth and Karanis 2018).

The detection of *Cryptosporidium* and *Giardia* in the environment is influenced by the type of water sampled, and recoveries tend to be lower in more turbid samples. Turbidity due to the presence of inorganic and organic debris could influence the concentration, separation, and examination of samples for the detection of *Cryptosporidium* oocysts and *Giardia* cysts. Chemicals added to finished waters during treatment, as well as naturally occurring debris, may also cause interference. False negatives may occur due to losses of (oo)cysts during concentration, purification, and detection (Grundlingh and de Wet 2004). When samples are examined by fluorescence microscopy for the detection of (oo)cysts, interference may occur due to organisms/debris that autofluoresce or algae and yeast cells that demonstrate non-specific fluorescence, resulting in false positives (Grundlingh and de Wet 2004). This approved conventional approach is time-consuming and requires a skilled technician in fluorescence microscopy. This method is also unable to distinguish different species of *Cryptosporidium* and species of *Giardia*.

Molecular assays like the polymerase chain reaction (PCR) can differentiate *Cryptosporidium* and *Giardia* spp. that are infective to humans from those that are not infective. PCR has also been shown to increase detection sensitivity by 10^3–10^4 fold compared to immunofluorescence microscopy (Lowery et al. 2000). Many PCR-based detection methods have been described, highlighting the need for an optimized standardized method for the detection of *Cryptosporidium* and *Giardia* (Guillot and Loret 2010). Some of the methods described include using PCR-RFLP to detect and discriminate the human pathogenic species *C. parvum* and *C. hominis* in water samples (Xiao et al. 2004; Ochiai et al. 2005) and *G. duodenalis* from other *Giardia* species in wastewater samples (Sulaiman et al. 2004). A multiplex PCR assay for the simultaneous detection of *Cryptosporidium* and *Giardia* has been demonstrated (Rochelle et al. 1997), while a real-time multiplex PCR assay for simultaneous detection and quantification of *G. lamblia* and *C. parvum* in environmental water samples and sewage has been developed (Guy et al. 2003). An immune-capture-based method for the detection of *C. parvum* and *G. intestinalis* in natural surface waters has been reported (Rimhanen-Finne et al. 2002). The detection of *Cryptosporidium* and *Giardia* with molecular-based techniques is also influenced by the method used for extracting DNA from cysts and oocysts. The presence of PCR-inhibitory substances in water samples and sewage may be detrimental to PCR assays (Stinear et al. 1996; Guy et al. 2003).

Both conventional and molecular detection methods require specialized equipment and technical skills for analysis, and these methods may also not be affordable for developing countries in Africa. The cost of the transport of large sample volumes to laboratories for analysis is also a constraint for water monitoring. In Africa, the accessibility of laboratory testing and the quality of available services is problematic. A survey of online registers in 2009 of leading accreditation providers showed that only 340 laboratories in sub-Saharan Africa were accredited. The majority of the accredited laboratories were in South Africa, with 8.2% of the laboratories in sub-Saharan Africa. Of the accredited laboratories in South Africa, fewer than 10% were public sector laboratories (Gershy-Damet et al. 2010). This can be attributed to a lack of trained laboratory experts, weak quality management systems, and the high cost of participation in international accreditation schemes (Gershy-Damet et al. 2010). In January 2019, only three laboratories were listed on the South African National Accreditation System (SANAS) as being capable of testing for *Cryptosporidium* and *Giardia* in potable and environmental waters (SANAS 2019). As environmental monitoring for waterborne protozoan pathogens is imperative for disease prevention, standardized and cost-effective methods for reliable detection and quantification of *Cryptosporidium* and *Giardia*, from a range of water environments is required. Although the detection methods mentioned above have been employed in some of these African countries (Table 4.1), many other countries cannot afford the cost of the instruments. Also, molecular detection is usually outsourced to developed countries. When these are available, they are mostly used for research purposes, making it challenging to apply molecular detection of *Cryptosporidium* and *Giardia* in the environmental setting. These shortcomings could lead to the underestimation of reservoirs for these pathogens with a consequent underestimation of the potentially associated disease burden.

4.7 Recent Advances in *Cryptosporidium* and *Giardia* Detection and Characterization

The world of microbiology has been revolutionized by the development of sophisticated tools and techniques such as next-generation sequencing, whole-genome sequencing, and digital PCR. These new approaches have allowed for a deeper understanding of microbial communities, genetic make-up, and enumeration at previously unimaginably minute scales. Although highly used for bacteria, these techniques have now widely found their way into the identification and characterization of protozoa such as *Cryptosporidium* and *Giardia*.

The popularity of qPCR increased in the last decades due to its ability to directly quantify microbial species, including *Cryptosporidium* and *Giardia*. However, this technique is limited by the need to construct a standard curve before enumeration, a shortcoming that has been eliminated with the more advanced digital droplet PCR (Gutiérrez-Aguirre et al. 2015). This novel technology, fuelled by recent advances in microfluidics, helps in the absolute quantification of nucleic acid in samples without the need for large sample volumes and calibration (Quan et al. 2018; Pomari et al. 2019). Despite the remarkable success of this technique, it has been demonstrated to be twice as expensive (consumables and labor) than qPCR for the identification and quantification of *Cryptosporidium* and *Giardia* (Yang et al. 2014). This would, therefore, be a significant challenge for the application of the technique in most African settings, where financial resources are highly limited.

Although microscopy is the most widely used technique for the identification of *Cryptosporidium* and *Giardia*, with immunofluorescence remaining the gold standard (Wang et al. 2018), this technique cannot differentiate between strains of the organisms as well as fully characterize specific strains involved in outbreaks, for example. This has been overcome using whole-genome sequencing (WGS). Several authors have used WGS to elucidate the complete genome of *Cryptosporidium* and *Giardia* from different sample types (Guo et al. 2015a, b; Hadfield et al. 2015; Xiao and Feng 2017; Gilchrist et al. 2018). One of the biggest challenges with WGS is the need to isolate enough DNA of extremely high quality (Guo et al. 2015b). This can be a significant challenge when dealing with environmental sample since these parasites are usually sparsely distributed in the environment. Also, WGS has not been widely applied to microbial species in Africa, probably due to its cost. Such instrumentation is only present in some countries such as South Africa. Moreover, many studies using this technique have done so in collaboration with other countries in Europe and America, where the instruments and the technical knowhow are readily available.

4.8   The Impact of Climate Change and Seasonality

The threat of waterborne diseases may be increased by climate change, which will impair water availability, accessibility, and affect demand (Aldeyarbi et al. 2016). Africa has been identified among the most vulnerable regions to climate change as it experiences more climate-sensitive economies than other continents (Bain et al. 2013). It is anticipated that the world's fresh water supply is likely to be affected due to increasingly variable rainfall patterns (Bain et al. 2013). The microbial load of *Cryptosporidium* and *Giardia* in surface water and drinking water reservoirs has been shown to increase after heavy rainfall due to runoff (Kistemann et al. 2002). Temperature and precipitation are significant predictors of the incidence of cryptosporidiosis, particularly in the tropical climates (Jagai et al. 2009). Many studies have shown a correlation between infection with *Cryptosporidium* and the rainy season in sub-Saharan Africa (Mahin and Peletz 2009). No association between site-specific mean temperature or rainfall was found for *Giardia* incidence in a multi-site birth-cohort study (Rogawski et al. 2017). A study on the prevalence of *Cryptosporidium* and *Giardia* in KwaZulu-Natal in South Africa failed to show a correlation with climatic factors such as rainfall, season, or year. This could indicate that hygiene, potable water supply, sanitation, and education may be more significant for *Cryptosporidium* and *Giardia* prevalence in developing countries (Jarmey-Swan et al. 2001). As studies on parasitic seasonality are often contradictory, it is clear that the incidence of *Cryptosporidium* and *Giardia* associated with environmental factors such as humidity, temperature, seasonal variation, and other geographic factors requires further investigation (Guillot and Loret 2010).

4.9 Treatment of Cryptosporidiosis and Giardiasis in Africa

As cryptosporidiosis is self-limiting in immunocompetent patients, but not in immunocompromised individuals, treatment would aim to reduce the duration of diarrhea, prevent complications, eliminate the organism from the host, and reduce mortality (Abubakar et al. 2007). Only one drug, nitazoxanide, is available for use against *Cryptosporidium*, and this drug is not recommended for infants under 1 year of age. Nitazoxanide has been found to reduce the load of parasites and may be useful in immunocompetent individuals (Abubakar et al. 2007). The drug is not effective for cryptosporidiosis in immunocompromised individuals with HIV (Amadi et al. 2009; Abubakar et al. 2007). The number of people in Africa infected with HIV in 2017 was more than 25.7 million (WHO 2017). Nitazoxanide is, therefore, ineffective against the most critical target population in Africa (Squire and Ryan 2017). The only option for most immunocompromised individuals is the use of fluid and electrolyte replacement (Abubakar et al. 2007). There is, therefore, an urgent need for high-quality, new effective drugs for the treatment of cryptosporidiosis, especially in immunocompromised persons (Abubakar et al. 2007; Miyamoto and Eckmann 2015). It is of interest that in HIV-positive patients infected by *Cryptosporidium*, highly active antiretroviral therapy with the inclusion of protease inhibitors appeared to control chronic diarrhea by partially restoring immune functions (Carr et al. 1998; Miao et al. 2000). Eradication of *Cryptosporidium* was only observed after 6 months of treatment (Miao et al. 2000). HIV protease inhibitors may also act as anti-parasitic drugs (Mele et al. 2003; Hommer et al. 2003). The drugs indinavir, saquinavir, and ritonavir have been reported to have anti-*Cryptosporidium* effects in vitro and in vivo. Indinavir has been shown to directly interfere with the cycle of *C. parvum*, resulting in a reduction in oocyst shedding and the number of intracellular parasites (Mele et al. 2003). There is a need to identify potential drug targets throughout the parasite's life cycle to ensure adequate drug design (Miyamoto and Eckmann 2015).

The most commonly utilized drugs for the treatment of giardiasis are the nitro-imidazole class of agents, including metronidazole (Gardner and Hill 2001). Resistance to these drugs has been reported (Ansell et al. 2015). Several other classes of drugs that have good efficacy for the treatment of giardiasis exist, but the dosing regimens are not optimal, and emerging resistance to these drugs is problematic (Miyamoto and Eckmann 2015). The drugs may have adverse effects or be contraindicated in certain situations (Gardner and Hill 2001). Improvements in potency and dosing as well as preventing new forms of drug resistance are priorities for drug development for *Giardia* infections (Miyamoto and Eckmann 2015).

A challenge for new drug development for both parasitic infections is that cryptosporidiosis and giardiasis are considered neglected diseases with low funding priority and limited commercial interest (Miyamoto and Eckmann 2015). Despite there being a large target population in Africa, the treatment of *Cryptosporidium* and *Giardia* has a small market and pharmaceutical companies are often cautious about investing in developing new therapeutics for developing countries (Squire and Ryan 2017).

There are no vaccines currently available for *Cryptosporidium* and *Giardia*. Effective vaccines targeting protozoa have been challenging to develop due to the complexity of these microorganisms. Reduced clinical effects to *Cryptosporidium* may occur after primary infection due to acquired immunity when exposure occurs regularly. In developing countries where primary exposure to cryptosporidiosis occurs at a young age, a therapy given at disease onset to help overcome the primary infection may be more practical and cost-effective (Mor and Tzipori 2008).

4.10   Conclusions and Current and Future Challenges

The establishment of surveillance systems in developing nations would be a first step in controlling parasitic protozoan infections and improving the health of populations (Efstratiou et al. 2017). It is suggested that international collaboration against waterborne protozoan pathogens be initiated and a standardized reporting system be developed (Efstratiou et al. 2017). Seasonal patterns and distributions of *Cryptosporidium* and *Giardia* species in Africa need to be evaluated (Guillot and Loret 2010; Aldeyarbi et al. 2016). This is particularly relevant given that adverse climatic events driven by climate change occur more frequently globally, which may result in large waterborne infectious disease outbreaks (Efstratiou et al. 2017). Active surveillance systems require reliable and affordable diagnostic tools for the detection of protozoan parasites (Efstratiou et al. 2017). Improved clinical diagnostic methods for *Cryptosporidium* and *Giardia* would be especially beneficial in asymptomatic patients (Opintan et al. 2010; Halliez and Buret 2013). Molecular characterization of *Cryptosporidium* and *Giardia* should also be a priority as molecular techniques can provide a genetic characterization of the parasites isolated from water, which may assist in determining the source of contamination (Thompson 2004; Efstratiou et al. 2017).

New drug development is needed for children with diarrhea and malnutrition in developing countries as the present drugs for *Cryptosporidium* are mostly ineffective due to a host-parasite interface that is poorly understood (Mor and Tzipori 2008). As no effective therapy is available in the immunocompromised, *Cryptosporidium* will continue to serve as a prominent threat to HIV-positive individuals in developing countries that lack resources where antiretroviral therapy is not available or affordable (O'Connor et al. 2011). Even though cryptosporidiosis is a serious cause of global diarrhea illness, there has been a lack of development of effective therapies due to a perceived limited market in developed countries (Mor and Tzipori 2008). The sequencing of the genomes of *C. parvum* (Abrahamsen et al. 2004) and *C. hominis* (Xu et al. 2004) are developments that may lead to the identification of new molecular targets for drug development (Mor and Tzipori 2008). The Center for Disease Control and Prevention in the United States has placed cryptosporidiosis among the category B biothreat pathogens, and it is hoped that this will offer an additional incentive for drug development (Mor and Tzipori 2008).

Giardia will continue to be of public health importance due to the high prevalence of giardiasis in young children in developing countries as well as its association with malnutrition and stunting (Halliez and Buret 2013). Malnutrition undermines the resilience of vulnerable populations, reducing their ability to grow economically and their ability to cope and adapt to climate change, placing significant risk on African populations (Bain et al. 2013).

Effective control strategies are needed to reduce the detrimental effects of protozoan infections on human societies (Halliez and Buret 2013). Improved water, sanitation, and hygiene are critical in preventing diarrhea morbidity and mortality caused by protozoan parasites in developing countries (Mahin and Peletz 2009; Omarova et al. 2018), and this cannot be achieved unless there is a commitment from African governments (Squire and Ryan 2017). Community programs must be initiated for public health education with a focus on water safety measures, good sanitary practices, and personal hygiene in order to decrease the infection risks caused by these parasites (Anim-Baidoo et al. 2016; Squire and Ryan 2017).

References

Abbas AM, Bouali F, Audin D (2015) An evaluation of the prevalence of Giardia spp., Cryptosporidium spp and Eimeria spp. in dairy cattle on sixteen farms in three regions at North-Eastern Algeria. Microscope 400:800

Abd El-Salam MM (2012) Assessment of water quality of some swimming pools: a case study in Alexandria, Egypt. Environ Monit Assess 184(12):7395–7406

Abrahamsen MS, Templeton TJ, Enomoto S et al (2004) Complete genome sequence of the apicomplexan, *Cryptosporidium parvum*. Sci 304:441–445

Abubakar II, Aliyu SH, Arugam C et al (2007) Prevention and treatment of cryptosporidiosis in immunocompromised patients. Cochrane Database Syst Rev 24(1):CD004932

Adamu H, Wegayehu T, Petros B (2013) High prevalence of diarrhoegenic intestinal parasite infections among non-ART HIV patients in Fitche Hospital, Ethiopia. PLoS One 8(8):e72634

Adeyemo FE, Singh G, Reddy P, Stenström TA (2018) Methods for the detection of cryptosporidium and Giardia: from microscopy to nucleic acid based tools in clinical and environmental regimes. Acta Trop 184:15–28

Ahmed SA, Karanis P (2018) An overview of methods/techniques for the detection of cryptosporidium in food samples. Parasitol Res 117:629–653

Ajeagah G, Njine T, Foto S et al (2007) Enumeration of Cryptosporidium spp and Giardia spp (oo)cysts in a tropical eutrophic lake: the municipal lake of Yaounde. Int J Environ Sci Technol 4:223–232

Aldeyarbi HM, Abu El-Ezz NMT, Karanis P (2016) Cryptosporidium and cryptosporidiosis: the African perspective. Environ Sci Pollut Res 23(14):13811–13821

Amadi B, Mwiya M, Sianongo S et al (2009) High dose prolonged treatment with nitazoxanide is not effective for cryptosporidiosis in HIV positive Zambian children: a randomised controlled trial. BMC Infect Dis 9:195

Amahmid O, Asmama S, Bouhoum K (1999) The effect of waste water reuse in irrigation on the contamination level of food crops by Giardia cysts and Ascaris eggs. Int J Food Microbiol 49:19–26

Amer S, Zidan S, Adamu H et al (2013) Experimental parasitology prevalence and characterization of Cryptosporidium spp. in dairy cattle in Nile River delta provinces, Egypt. Exp Parasitol 135:518–523. https://doi.org/10.1016/j.exppara.2013.09.002

Anim-Baidoo I, Narh CA, Oddei D et al (2016) *Giardia lamblia* infections in children in Ghana. Pan Afr Med J 24:217

Ansell BRE, McConville MJ, Ma'ayeh SY et al (2015) Drug resistance in *Giardia duodenalis*. Biotechnol Adv 33:888–901

Assefa S, Erko B, Medhin G et al (2009) Intestinal parasitic infections in relation to HIV/AIDS status, diarrhea and CD4 T-cell count. BMC Infect Dis 9:155

Atnafu T, Kassa H, Keil C et al (2012) Presence, viability and determinants of *Cryptosporidium* oocysts and *Giardia* cysts in the Addis Ababa water supply and distribution system. Water Qual Exp Health 4(1):55–65

Bain LE, Awah PK, Geraldine N et al (2013) Malnutrition in sub-Saharan Africa: burden, causes and prospects. Pan Afr Med J 15:120

Baldursson S, Karanis P (2011) Waterborne transmission of protozoan parasites: review of world-wide outbreaks – an update 2004–2010. Water Res 45:6603–6614

Baroudi D, Hakem A, Adamu H et al (2018) Zoonotic cryptosporidium species and subtypes in lambs and goat kids in Algeria. Parasit Vectors 11:1–8. https://doi.org/10.1186/s13071-018-3172-2

Baroudi D, Khelef D, Goucem R et al (2013) Common occurrence of zoonotic pathogen Cryptosporidium meleagridis in broiler chickens and turkeys in Algeria. Vet Parasitol 196:334–340

Baroudi D, Khelef D, Hakem A et al (2017) Molecular characterization of zoonotic pathogens Cryptosporidium spp., Giardia duodenalis and Enterocytozoon bieneusi in calves in Algeria. Vet Parasitol Reg Stud Reports 8:66–69

Ben Ayed L, Yang W, Widmer G et al (2012) Survey and genetic characterization of wastewater in Tunisia for Cryptosporidium spp., Giardia duodenalis, Enterocytozoon bieneusi, Cyclospora cayetanensis and Eimeria spp. J Water Health 10:431–444. https://doi.org/10.2166/wh.2012.204

Berkman DS, Lescano AG, Gilman RH et al (2002) Effects of stunting, diarrhoeal disease, and parasitic infection during infancy on cognition in late childhood: a follow-up study. Lancet 359:564–571

Betancourt WQ, Rose JB (2004) Drinking water treatment processes for removal of Cryptosporidium and Giardia. Vet Parasitol 126:219–234

Boaitey YA, Nkrumah B, Idriss A, Tay SCK (2012) Gastrointestinal and urinary tract pathogenic infections among HIV seropositive patients at the Komfo Anoke Teaching Hospital in Ghana. BMC Res Notes 5:454

Botero-Garcés JH, García-Montoya GM, Grisales-Patiño D et al (2009) Giardia intestinalis and nutritional status in children participating in the complementary nutrition program, Antioquia, Colombia, May to October 2006. Rev Inst Med Trop Sao Paulo 51:155–162

Bouzid M, Hunter PR, Chalmers RM, Tyler KM (2017) *Cryptosporidium* pathogenicity and virulence. Clin Microbiol Rev 26(1):115–134

Bouzid M, Kintz E, Hunter P (2018) Risk factors for *Cryptosporidium* infection in low and middle-income countries: a systematic review and meta-analysis. PLoS Negl Trop Dis 12(6):e0006553

Carmena D (2010) Waterborne transmission of Cryptosporidium and Giardia: detection, surveillance and implications for public health. In: Mendez-Vilas A (ed) Current research, technology and education topics in applied microbiology and microbial biotechnology, 2nd edn. Formatex Research Center, Spain

Carr A, Marriott D, Field A et al (1998) Treatment of HIV-1-associated microsporidiosis and cryptosporidiosis with combination antiretroviral therapy. Lancet 351(9098):256–261

Chalmers RM, Davies AP (2010) Minireview: clinical cryptosporidiosis. Exp Parasitol 124:138–146

Cotton JA, Beatty JK, Buret A (2011) Host-parasite interactions and pathophysiology in *Giardia* infections. Int J Parasitol 41:925–933

Current WL, Garcia LS (1991) Cryptosporidiosis. Clin Microbiol Rev 4(3):325–358

Dalu T, Barson M, Nhiwatiwa T (2011) Impact of intestinal microorganisms and protozoan parasites on drinking water quality in Harare, Zimbabwe. J Water Sanit Hyg Dev 1:153–163

Dobrowsky PH, De Kwaadsteniet M, Cloete TE, Khan W (2014) Distribution of indigenous bacterial pathogens and potential pathogens associated with roof-harvested rainwater. Appl Environ Microbiol 80:2307–2316

Duhain GLMC (2012) Occurrence of *Cryptosporidium* spp. in South African irrigation waters and survival of *Cryptosporidium parvum* during vegetable processing. Dissertation (MSc), University of Pretoria, South Africa

Dungeni M, Momba MNB (2010) The abundance of Cryptosporidium and Giardia spp. in treated effluents produced by four wastewater treatment plants in the Gauteng Province of South Africa. Water SA 36(4):425–432

Efstratiou A, Ongerth JE, Karanis P (2017) Waterborne transmission of protozoan parasites: review of worldwide outbreaks – an update 2011–2016. Water Res 114:14–22

Elfstrand L, Florén C-H (2010) Management of chronic diarrhea in HIV-infected patients: current treatment options, challenges and future directions. HIV AIDS (Auckl) 2:219–224

Eraky MA, Rashed SM, Nasr MES et al (2014) Parasitic contamination of commonly consumed fresh leafy vegetables in Benha, Egypt. J Parasitol Res. https://doi.org/10.1155/2014/613960

Ertan P, Yereli K, Kurt Ö et al (2002) Serological levels of zinc, copper and iron elements among *Giardia lamblia* infected children in Turkey. Pediatr Int 44:286–288

Escobedo AA, Acosta-Ballester G, Almirall P et al (2018) Potential sexual transmission of *Giardia* in an endemic region: a case series. Infez Med 26:145–149

FAO (2015) The state of food insecurity in the world. Meeting the 2015 international hunger targets: taking stock of uneven progress. Food and Agriculture Organization of the United Nations. Available online at: http://www.fao.org/3/a-i4646e.pdf. Accessed: 31 Mar 2019

Feng Y, Xiao L (2011) Zoonotic potential and molecular epidemiology of giardia species and giardiasis. Clin Microbiol Rev 24(1):110–140

Gardner TB, Hill DR (2001) Treatment of giardiasis. Clin Microbiol Rev 14(1):114–128

Gendrel D, Treluyer JM, Richard-lenoble D (2003) Parasitic diarrhea in normal and malnourished children. Fundam Clin Pharmacol 17:189–197

Gershy-Damet G-M, Rotz P, Cross D et al (2010) The World Health Organization African region laboratory accreditation process: improving the quality of laboratory systems in the African region. Am J Clin Pathol 134:393–400

Ghoneim NH, Hassanain MA, Hamza DA et al (2017) Prevalence and molecular epidemiology of cryptosporidium infection in calves and hospitalized children in Egypt. Res J Parasitol 12:19–26

Gideon AA, Njiné T, Nola M et al (2007) Évaluation de l'abondance des formes de résistance de deux protozoaires pathogènes (Giardia sp et Cryptosporidium sp) dans deux biotopes aquatiques de Yaoundé (Cameroun). Cah d'études Rech Francoph. Sante 17:167–172

Gilchrist CA, Cotton JA, Burkey C et al (2018) Genetic diversity of cryptosporidium hominis in a bangladeshi community as revealed by whole-genome sequencing. J Infect Dis 218:259–264

Grundlingh M, de Wet CME (2004) The search for Cryptosporidium oocysts and Giardia cysts in source water used for purification. Water SA 30(5):33–36

Guillot E, Loret J-P (2010) *Waterborne pathogens*: review for the drinking water industry. GWRC report series. IWA Publishing, London, UK

Guo Y, Li N, Lysén C et al (2015b) Isolation and enrichment of cryptosporidium DNA and verification of DNA purity for whole-genome sequencing. J Clin Microbiol 53:641–647

Guo Y, Tang K, Rowe LA et al (2015a) Comparative genomic analysis reveals occurrence of genetic recombination in virulent Cryptosporidium hominis subtypes and telomeric gene duplications in Cryptosporidium parvum. BMC Genomics. https://doi.org/10.1186/s12864-015-1517-1

Gutiérrez-Aguirre I, Racki N, Dreo T, Ravnikar M (2015) Droplet digital PCR for absolute quantification of pathogens. Methods Mol Biol 1302:331–347. https://doi.org/10.1007/978-1-4939-2620-6_24

Guy RA, Payment P, Krull UJ, Horgen PA (2003) Real-time PCR for quantification of Giardia and Cryptosporidium in environmental water samples and sewage. Appl Environ Microbiol 69(9):5178–5185

Hadfield SJ, Pachebat JA, Swain MT et al (2015) Generation of whole genome sequences of new Cryptosporidium hominis and Cryptosporidium parvum isolates directly from stool samples. BMC Genomics. https://doi.org/10.1186/s12864-015-1805-9

Halliez MCM, Buret AG (2013) Extra-intestinal and long term consequences of Giardia duodenalis infections. World J Gastroenterol 19(47):8974–8985

Helmy YA, Klotz C, Wilking H et al (2014) Epidemiology of Giardia duodenalis infection in ruminant livestock and children in the Ismailia province of Egypt: insights by genetic characterization. Parasit Vectors 7:1–11. https://doi.org/10.1186/1756-3305-7-321

Hommer V, Eichholz J, Petry F (2003) Effect of antiretroviral protease inhibitors alone, and in combination with paromomycin, on the excystation, invasion and in vitro development of *Cryptosporidium parvum*. J Antimicrob Chemother 52:359–364

Hunter PR, Nichols G (2002) Epidemiology and clinical features of *Cryptosporidium* infection in immunocompromised patients. Clin Microbiol Rev 15(1):145–154

Hunter PR, Thompson RCA (2005) The zoonotic transmission of *Cryptosporidium* and *Giardia*. Int J Parasitol 35:1181–1190

Jagai JS, Castronova DA, Monchak J, Naumova EN (2009) Seasonality of cryptosporidium: a meta-analysis approach. Environ Res 109(4):465–478

Jarmey-Swan C, Bailey IW, Howgrave-Graham AR (2001) Ubiquity of the water-borne pathogens *Cryptosporidium* and *Giardia* in KwaZulu-Natal populations. Water SA 27(1):57–64

Karanis P, Kourenti C, Smith H (2007) Waterborne transmission of protozoan parasites: a worldwide review of outbreaks and lessons learnt. J Water Health 5:1–38

Kato S, Ascolillo L, Egas J et al (2003) Waterborne *Cryptosporidium* oocyst identification and genotyping: use of GIS for ecosystem studies in Kenya and Ecuador. J Eukaryot Microbiol 50(1):548–549

Khouja LB, Cama V, Xiao L (2010) Parasitic contamination in wastewater and sludge samples in Tunisia using three different detection techniques. Parasitol Res 107(1):109–116

Kistemann T, Claßen T, Koch C et al (2002) Microbial load of drinking water reservoir tributaries during extreme rainfall and runoff. Appl Environ Microbiol 68(5):2188–2197

Koruk I, Simsek Z, Koruk ST et al (2010) Intestinal parasites, nutritional status and physchomotor development delay in migratory farm worker's children. Child Care Health Dev 36:888–894

Kotloff KL, Nataro JP, Blackwelder W et al (2013) Burden and aetiology of diarrhoeal disease in infants and young children in developing countries (the Global Enteric Multicentre Study, GEMS): a prospective, case-control study. Lancet 382(9888):209–222

Kusiluka LJM, Karimuribo ED, Mdegela RH et al (2005) Prevalence and impact of water-borne zoonotic pathogens in water, cattle and humans in selected villages in Dodoma rural and Bagamoyo districts, Tanzania. Phys Chem Earth, Parts A/B/C 30:818–825

Laatamna AE, Wagnerová P, Sak B et al (2013) Equine cryptosporidial infection associated with cryptosporidium hedgehog genotype in Algeria. Vet Parasitol 197:350–353. https://doi.org/10.1016/j.vetpar.2013.04.041

Laatamna AE, Wagnerová P, Sak B et al (2015) Microsporidia and cryptosporidium in horses and donkeys in Algeria: detection of a novel Cryptosporidium hominis subtype family (Ik) in a horse. Vet Parasitol 208:135–142

Lane S, Lloyd D (2002) Current trends in research into the waterborne parasite Giardia. Crit Rev Microbiol 28:123–147

Lowery CJ, Moore JE, Millar BC et al (2000) Detection and speciation of cryptosporidium spp. in environmental water samples by immunomagnetic separation, PCR and endonuclease restriction. J Med Microbiol 49:779–785

Mahin T, Peletz R (2009) Cryptosporidium contamination of water in Africa: the impact on mortality rates for children with HIV/AIDS. Shaw RJ (Ed). In: Water, sanitation and hygiene – Sustainable development and multisectoral approaches: Proceedings of the 34th WEDC International Conference, Addis Ababa, Ethiopia, 18–22 May 2009

Manyazewal A, Francesca S, Pal M et al (2018) Veterinary parasitology: regional studies and reports prevalence, risk factors and molecular characterization of cryptosporidium infection in cattle in Addis Ababa and its environs, Ethiopia. Vet Parasitol Reg Stud Reports 13:79–84

McCormick BJJ (2014) Frequent symptomatic or asymptomatic infections may have long-term consequences on growth and cognitive development. In: Heitdt PJ, Lang D, Riddle RI (eds) . Old Herborn University Seminar Monographs, Institute for Microbiology and Biochemistry, Herborn, Germany

Mele R, Gomez Morales MA, Tosini F, Pozio E (2003) Indinavor reduces Cryptosporidium parvum infection in both in vitro and in vivo models. Int J Parasitol 33:757–764

Mensah GT, Bosompem KMP, Ayeh-kumi F et al (2018) Epidemiology of cryptosporidium Sp upstream the water treatment plants in Kpong and Weija, Ghana. Ghana J Sci 58:5–11

Meutchieye F, Kouam MK, Miegoué E et al (2017) A survey for potentially zoonotic gastrointestinal parasites in domestic cavies in Cameroon (Central Africa). BMC Vet Res 13:1–5

Miao YM, Awad-El-Kariem FM, Franzen C (2000) Eradication of cryptosporidia and microsporidia following successful antiretroviral therapy. J Acquir Immune Defic Syndr 25(2):124–129

Miyamoto Y, Eckmann L (2015) Drug development against the major diarrhea-causing parasites of the small intestine, *Cryptosporidium* and *Giardia*. Front Microbiol 6:1208

Mølbak K, Hojlyng N, Gottschau A et al (1993) Cryptosporidiosis in infancy and childhood mortality in Guinea Bissau, West Africa. BMJ 307:417–420

Mor SM, Tzipori S (2008) Cryptosporidiosis in children in sub-Saharan Africa: a lingering challenge. Clin Infect Dis 47(7):915–921

Mtapuri-Zinyowera S, Ruhanya V, Midzi N et al (2014) Human parasitic protozoa in drinking water sources in rural Zimbabwe and their link to HIV infection. Germs 4:86–91

Muchiri JM, Ascolillo L, Mugambi M et al (2009) Seasonality of cryptosporidium oocyst detection in surface waters of Meru, Kenya as determined by two isolation methods followed by PCR. J Water Health 7(1):67–75

Muhsen K, Levine MM (2012) A systematic review and meta-analysis of the association between Giardia lamblia and endemic pediatric diarrhea in developing countries. Clin Infect Dis 55(S4):S271–S273

Ndur S, Kuma J, Buah W, Galley J (2015) Quality of sachet water produced at Tarkwa, Ghana. Ghana Min J 15(1):22–34

O'Connor RM, Shaffie R, Kang G, Ward H (2011) Cryptosporidiosis in patients with HIV/AIDS. AIDS 25:549–560

Ochiai Y, Takada C, Hosaka M (2005) Detection and discrimination of *Cryptosporidium parvum* and *C. hominis* in water samples by immunomagnetic separation-PCR. Appl Environ Microbiol 71(2):898–903

Okhuysen PC, Chappell CL, Crabb JH et al (1999) Virulence of three distinct *Cryptosporidium parvum* isolates for healthy adults. J Infect Dis 180:1275–1281

Omarova A, Tussupova K, Berndtsson R et al (2018) Protozoan parasites in drinking water: a system approach for improved water, sanitation and hygiene in developing countries. Int J Env Res Public Health 15:495

Ongerth JE, Karanis P (2018) *Cryptosporidium* and *Giardia* in water – key features and basic principles for monitoring and data analysis. PRO 2:691

Opintan JA, Newman MJ, Ayeh-Kuni PF (2010) Pediatric diarrhea in southern Ghana: etiology and association with intestinal inflammation and malnutrition. Am J Trop Med Hyg 83(4):936–943

Ouakli N, Belkhiri A, de Lucio A et al (2018) Cryptosporidium-associated diarrhoea in neonatal calves in Algeria. Vet Parasitol Reg Stud Reports 12:78–84

Parsons MB, Travis D, Lonsdorf EV et al (2015) Epidemiology and molecular characterization of Cryptosporidium spp. in humans, wild primates, and domesticated animals in the Greater Gombe Ecosystem, Tanzania. 1–13. doi: https://doi.org/10.1371/journal.pntd.0003529

Pomari E, Piubelli C, Perandin F, Bisoffi Z (2019) Digital PCR: a new technology for diagnosis of parasitic infections. Clin Microbiol Infect. https://doi.org/10.1016/j.cmi.2019.06.009

Putignani L, Menichella D (2010) Global distribution, public health and clinical impact of the protozoan pathogen cryptosporidium. Interdiscip Perspect Infect Dis 2010:753512

Quan P, Sauzade M, Brouzes E (2018) dPCR: a technology review. Sensors. 18(4):1271. https://doi.org/10.3390/s18041271

Quihui L, Morales GG, Méndez RO et al (2010) Could giardiasis be a risk factor for low zinc status in school children from northwestern Mexico? A cross-sectional study with longitudinal follow-up. BMC Public Health 10:85–91

Rimhanen-Finne R, Hörman A, Ronkainen P, Hänninen M-L (2002) An IC-PCR method for the detection of *Cryptosporidium* and *Giardia* in natural surface waters in Finland. J Microbiol Method 50:299–303

Rochelle PA, De Leon R, Stewart MH, Wolfe RL (1997) Comparison of primers and optimisation of PCR conditions for detection of *Cryptosporidium parvum* and *Giardia lamblia* in water. Appl Environ Microbiol 63(1):106–114

Rogawski ET, Bartelt LA, Platts-Mills JA et al (2017) Determinants and impact of *Giardia* infection in the first 2 years of life in the MAL-ED birth cohort. J Pediatr Infect Dis Soc 6(2):153–160

Rose JB, Huffman DE, Gennaccaro A (2002) Risk and control of waterborne cryptosporidiosis. FEMS Microbiol Rev 26:113–123

Ryan U, Cacciò SM (2013) Zoonotic potential of *Giardia*. Int J Parasitol 43:943–956

Sak B, Petrzelkova KJ, Kvetonova D et al (2013) Long-term monitoring of Microsporidia, Cryptosporidium and Giardia infections in Western Lowland gorillas (Gorilla gorilla gorilla) at different stages of habituation in Dzanga Sangha protected areas, Central African Republic. PLoS One. https://doi.org/10.1371/journal.pone.0071840

Samra NA, Jori F, Cacciò SM et al (2016) Cryptosporidium genotypes in children and calves living at the wildlife or livestock interface of the Kruger National Park, South Africa. Onderstepoort J Vet Res 83(1):a1024. https://doi.org/10.4102/ojvr.v83i1.1024.

SANAS (2019) The South African National Accreditation System. Available online at https://www.sanas.co.za/. Accessed: 16 Jan 2019

Sanyaolu AO, Oyibo WA, Fagbenro-Beyioku F et al (2011) Comparative study of entero-parasitic infections among HIV sero-positive and sero-negative patients in Lagos, Nigeria. Acta Trop 120:268–272

Savioli L, Smith H, Thompson A (2006) *Giardia* and *Cryptosporidium* join the 'neglected disease initiative'. Trend Parasitol 22(5):203–208

Sente C, Erume J, Naigaga I et al (2016) Prevalence of pathogenic free-living amoeba and other protozoa in natural and communal piped tap water from queen Elizabeth protected area, Uganda. Infect Dis Poverty 5:68

Sigudu MV, du Preez HH, Retief F (2014) Application of a basic monitoring strategy for *Cryptosporidium* and *Giardia* in drinking water. Water SA 40(2):297–312

Simsek Z, Zeyrek FY, Kurcer MA (2004) Effect of *Giardia* infection on growth and psychomotor development of children aged 0-5 years. J Trop Pediatr 50(2):90–93

Siwila J, Mwape KE (2012) Prevalence of *Cryptosporidium* spp. and *Giardia duodenalis* in pigs in Lusaka, Zambia. Onderstepoort J Vet Res 79:1–5. https://doi.org/10.4102/ojvr.v79i1.404

Smith HV, Caccio SM, Tait A et al (2006) Tools for investigating the environmental transmission of *Cryptosporidium* and *Giardia* infections in humans. Trends Parasitol 22(4):160–167

Smith HV, Nichols RAB (2010) Cryptosporidium: detection in water and food. Exp Parasitol 124:61–79

Sow SO, Muhsen K, Nasrin D et al (2016) The burden of cryptosporidium diarrheal disease among children <24 months of age in moderate/high mortality regions of sub-Saharan Africa and South Asia, utilizing data from the Global Enteric Muliticenter Study (GEMS). PLOS Negl Trop Dis 10(5):1–20

Sponseller JK, Griffiths JK, Tzipori S (2014) The evolution of respiratory cryptosporidiosis: evidence for transmission by inhalation. Clin Microbiol Rev 27(3):575–586

Squire AA, Ryan U (2017) *Cryptosporidium* and *Giardia* in Africa: current and future challenges. Parasit Vectors 10:195

Stark D, Barratt JLN, van Hal S et al (2009) Clinical significance of enteric protozoa in the immunosuppressed human population. Clin Microbiol Rev 22(4):634–650

Stinear T, Matusan A, Hines K, Sandery M (1996) Detection of a single viable *Cryptosporidium parvum* oocyst in environmental water concentrates by reverse transcription-PCR. Appl Environ Microbiol 62(9):3385–3390

Sulaiman IM, Jiang J, Singh A, Xiao L (2004) Distribution of *Giardia duodenalis* genotypes and subgenotypes in raw urban wastewater in Milwaukee, Wisconsin. Appl Environ Microbiol 70(6):3776–3780

Taha S, Elmalik K, Bangoura B, Lendner M (2017) Molecular characterization of bovine cryptosporidium isolated from diarrheic calves in the Sudan. Parasitol Res. https://doi.org/10.1007/s00436-017-5606-8

Tefera T, Biruksew A, Mekonnen Z, Eshetu T (2014) Parasitic contamination of fruits and vegetables collected from selected Local Markets of Jimma Town, Southwest Ethiopia. Int Sch Res Notices 2014:1–7. https://doi.org/10.1155/2014/382715

Tellevik MG, Moyo SJ, Blomberg B et al (2015) Prevalence of *Cryptosporidium parvum/hominis, Entamoeba histolytica* and *Giardia lamblia* among young children with and without diarrhea in Dar Es Salaam, Tanzania. PLoS Negl Trop Dis 9(10):e0004125

Thompson RCA (2004) The zoonotic significance and molecular epidemiology of *Giardia* and giardiasis. Vet Parasitol 126:15–35

Tumwine JK, Kekitiinwa A, Bakeera-Kitaka S et al (2005) Cryptosporidiosis and microsporidiosis in Ugandan children with persistent diarrhea with and without concurrent infection with the human immunodeficiency virus. Am J Trop Med Hyg 73(5):921–925

UNAIDS (2018) Global HIV and AIDS statistics – 2018 fact sheet. Available online at http://www.unaids.org/sites/default/files/media_asset/UNAIDS_FactSheet_en.pdf Accessed: 21 Mar 2019

Uneke CJ, Uneke BI (2007) Occurrence of *Cryptosporidium* species in surface water in South-Eastern Nigeria: the public health implication. Internet J Health 2:1–6

UNICEF (2013) Child Malnutrition in Africa. United Nations International Children's Emergency Fund Brochure. Available online at https://data.unicef.org/wp-content/uploads/2015/12/Africa_Brochure_2013_158.pdf. Accessed: 22 Mar 2019

USEPA (2012) Method 1623.1: cryptosporidium and giardia in water by filtration/IMS/FA. Environmental Protection Agency. Office of Water, Washington, DC, USA. EPA-816-R-12-001

Wang RJ, Li JQ, Chen YC et al (2018) Widespread occurrence of cryptosporidium infections in patients with HIV/AIDS: epidemiology, clinical feature, diagnosis, and therapy. Acta Trop 187:257–263. https://doi.org/10.1016/j.actatropica.2018.08.018

Wegayehu T, Karim MR, Anberber M et al (2016) Prevalence and genetic characterization of cryptosporidium species in dairy calves in Central Ethiopia. PLoS One 11:1–11. https://doi.org/10.1371/journal.pone.0154647

WHO (2009) Diarrhoea: why children are still dying and what can be done. World Health Organization/United Nations International Children's Emergency Fund (WHO/UNICEF), Geneva/New York. Available online at http://apps.who.int/iris/bitstream/10665/44174/1/9789241598415_eng.pdf. Accessed: 19 Feb 2018

WHO (2015) Water sanitation and hygiene. Key Facts from JMP 2015 Report. World Health Organisation. Available online at https://www.who.int/water_sanitation_health/publications/JMP-2015-keyfacts-en-rev.pdf?ua=1. Accessed: 31 Mar 2019

WHO (2017) HIV/AIDS data and statistics. World Health Organisation. Available online at https://www.who.int/hiv/data/en/. Accessed: 21 Feb 2019

WHO (2018) Global Health Observatory (GHO) data. World Health Organization Available online at, Geneva. http://www.who.int/gho/hiv/en/. Accessed: 14 Nov 2018

Xiao L, Fayer R (2008) Molecular characterisation of species and genotypes of *Cryptosporidium* and *Giardia* and assessment of zoonotic transmission. Int J Parasitol 38:1239–1255

Xiao L, Feng Y (2017) Molecular epidemiologic tools for waterborne pathogens. Food and Waterborne Parasitology 8–9:14–32

Xiao L, Lal AA, Jiang J (2004) Detection and differentiation of *Cryptosporidium* oocysts in water by PCR-RFLP. Methods Mol Biol 268:163–176

Xu P, Widner G, Wang Y et al (2004) The genome of *Cryptosporidium hominis*. Nature 431:1107–1112

Yang R, Paparini A, Monis P, Ryan U (2014) Comparison of next-generation droplet digital PCR (ddPCR) with quantitative PCR (qPCR) for enumeration of cryptosporidium oocysts in faecal samples. Int J Parasitol 44:1105–1113

Yapo R, Koné B, Bonfoh B et al (2014) Quantitative microbial risk assessment related to urban wastewater and lagoon water reuse in Abidjan, Côte d'Ivoire. J Water Health 12(2):301–309

Chapter 5
Microbiological Air Quality in Different Indoor and Outdoor Settings in Africa and Beyond: Challenges and Prospects

Cecilia Oluseyi Osunmakinde, Ramganesh Selvarajan, Henry J. O. Ogola, Timothy Sibanda, and Titus Msagati

5.1 Introduction

Clean air is a fundamental requirement for life. In the last two decades, there has now been a pronounced consciousness about the air quality of both the indoor and outdoor systems, as the microorganisms prevailing in an environment symbolize the diverse human activities (Pasquarella et al. 2000; Barberán et al. 2015a), as well as their functions in different ecosystems (Adams et al. 2016). However, in densely populated areas, airborne microbial contaminants, especially the emerging pathogens and contaminants, can have numerous adverse effects on human health and well-being, including inflammation, toxic effects (allergic responses) and infections (Schmidt et al. 2012; Peter 2014; Wei et al. 2017). The term '*bioaerosol*' refers to airborne biological particles (living and non-living), volatile organic compounds (VOC) formed from the dispersal and excretions of particles released from diverse ecosystems into the atmosphere (Peter 2014; Saramanda et al. 2016). Bioaerosols are generally ubiquitous in the environment and due to their small particle size (<2.5 μm) are easily dispersed in the air. Those aerosols created at the

C. O. Osunmakinde · T. Msagati
Nanotechnology and Water Sustainability Research Unit, College of Science, Engineering and Technology, University of South Africa, Florida, South Africa

R. Selvarajan (✉)
Department of Environmental Sciences, College of Agricultural and Environmental Science, University of South Africa, Florida, South Africa

H. J. O. Ogola
Department of Environmental Sciences, College of Agricultural and Environmental Science, University of South Africa, Florida, South Africa

Centre for Research Innovation and Technology, Jaramogi Oginga Odinga University of Science and Technology, Bondo, Kenya

T. Sibanda
Department of Biological Sciences, University of Namibia, Windhoek, Namibia

© Springer Nature Switzerland AG 2020
A. L. K. Abia, G. R. Lanza (eds.), *Current Microbiological Research in Africa*,
https://doi.org/10.1007/978-3-030-35296-7_5

137

surface of aquatic systems are known to concentrate and carry microbes through the liquid-air interface (Brodie et al. 2007). The sizes of the bioaerosol particulate vary and are in the range of 0.3–100 mm in diameter (Stetzenbach et al. 2004). Bioaerosol includes viruses, fungi, bacteria, pollen and their by-products such as volatile organic compounds (VOCs), mycotoxins and endotoxins (Bernstein et al. 2004; Stetzenbach et al. 2004; Karwowska 2005; Kalwasińska et al. 2012; Saramanda et al. 2016).

Air quality is a global issue, particularly in developing countries, and the declining air quality is a deepening environmental unease. Poor air quality threatens human health and contributes to environmental damage (Peter 2014). The atmospheric air breathed daily contains 78.09% nitrogen, 20.95% oxygen, 0.93% argon, 0.04% carbon dioxide, and other gases in small amounts. Over the last century, a wide range of pollutants is being introduced daily into the environment through various means and sources such as the increased burning of fossil fuels and industrial emissions. Air pollution is broadly divided into the outdoor (indirect contact) or indoor pollution (direct contact); likewise, it can either be natural or human-made pollution (Fig. 5.1).

Natural air pollution arises from the release of by-product from plants or biomass of the ocean, gases released from volcanic eruptions, resuspension of dust and industrial, agricultural and household activities. The human-induced air pollution denotes the natural discharge in the atmosphere, which can give rise to primary and secondary pollutants (outdoor air) (Barnes et al. 1999). The problem associated with air pollution is related to the adverse effects it has on humans and animals (Ghorani-Azam et al. 2016). Also, a wide range of harmful effects that could arise from air pollution might lead to problems in the water bodies, agricultural production, soil and atmospheric air.

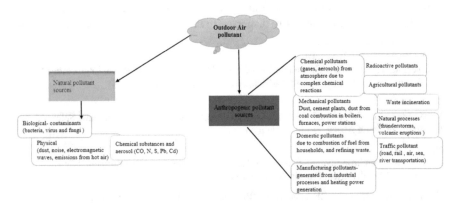

Fig. 5.1 Overview of different sources and causes of outdoor air pollution

5.2   Outdoor Air Pollutants

Outdoor air pollution, which is created by both human activities and naturally occurring events, is among the most significant environmental risk factors for mortality, responsible for 6.4 million deaths in 2015 globally (11% of global deaths) (Cohen et al. 2017). Different pollutants are introduced into the ambient air through the mixture of liquid droplets, solid particles and gases from a range of sources such as heating appliances, manufacturing industries, burning of fossil fuels by motor vehicles (cars, buses and trucks) and power plants (including coal-fired and natural gas plants) (Leung 2015). The most common outdoor air pollutants are oxides of nitrogen, sulphur, ammonia, volatile organic compounds (VOCs), ozone, carbon monoxide and particulate matter (PM) of different particle size (Leung 2015; Hoek 2017). The particulate matter (PM_{10} and $PM_{2.5}$) consists of a complex mixture of solid and liquid particles that are suspended in the air such as ammonia, sodium chloride, nitrates, sulphate, black carbon, mineral dust, water and ultrafine particles (Leung 2015; Hoek 2017). Other pollution sources include smoke from bushfires, volcanic eruptions, windblown dust, whirlwind and emissions from vegetation (pollen and mould spores). However, outdoor air pollutants are broadly divided into the persistent organic pollutants (e.g. pesticides, polychlorinated biphenyls) (Schecter et al. 2006), gaseous pollutants (oxides of N and S, CO, ozone, heavy metals and particulate matter (Bernstein et al. 2004). A summary of the different sources and causes of outdoor pollution is shown in Fig. 5.1. Also, Table 5.1 gives a summary of representative outdoor air contaminants and the adverse health effects they pose.

5.3   Indoor Air Pollutants

In contrast to outdoor air pollutants, indoor air pollutants denote diverse unsafe chemical, physical and biological processes that are emitted from materials that are used for the construction of buildings, house decorations and furniture. These anthropogenic activities introduce certain toxic and hazardous substances that decrease indoor air quality and pose a threat to human health. Several air pollutants have been recognized to exist indoors with adverse human health effects that may be due to exposures to these biological (e.g., aerosols, mites, viruses and bacteria) and chemical (e.g., benzene) particles, and those arising from physical activities (e.g., fine particles). The sources of indoor pollutants include dirt, dust, cooking activities (nitrogen oxides), cleaning (sweeping and vacuuming), furniture burning, tobacco smoke, wood burning heaters, burning of incense, candles, moulds, heating and cooling systems, and ventilation systems (Nazaroff 2004; Zhang and Zhu 2012; Tham 2016; Emmerich et al. 2017). Figure 5.2 shows a schematic diagram of various sources of indoor air pollution.

Indoor air quality in premises is gradually becoming a major concern globally due to increased awareness on its impact on the overall performance, productivity and energy consumption in buildings. Long-term exposure to indoor air pollutants

Table 5.1 Outdoor air pollutants, sources and health effects

Pollutant	Source	Health effect	References
Carbon monoxide (CO)	A colourless, odourless gas by-product of incomplete combustion. Common sources of *CO* include tobacco smoke, emission from powered tools, heaters, cooking equipment fossil fuel-operated space heaters, defective central heating furnaces and automobile exhaust	• reduces oxygen delivery to organs and tissues. • moderate to high levels of CO can cause headaches and fatigue, nausea, dizziness, vomiting and loss of consciousness	Leung (2015), Wei et al. (2017)
Oxides of nitrogen (NO)	• Emitted as NO, which rapidly reacts with ozone or radicals in the atmosphere forming NO_2. • By-product of mobile and stationary exhausts through combustion sources	• Resistance of the lungs to bacterial infection. • Sore throat, cough, nasal congestion, bronchitis, pneumonia	Peden et al. (2004), Mabahwi et al. (2014), Leung (2015), Tham (2016)
Sulphur dioxide (SO_2)	Industrial plants – Combustion and refining of coal, oil and metal-containing ores, gasoline and volcanoes	• Breathing problems, respiratory illness, changes in the lung's functioning, asthma and cardiovascular disease	Peden et al. (2004), Leung (2015)
Polycyclic aromatic hydrocarbons (PAHs), benzene	Incomplete burning of fossil fuels Benzene compounds	Carcinogenic	European Commission (2001), Tsai (2016), Wei et al. (2017)
Heavy metals (Pb, Ag, Mn, Cr, M, cd, as, se) present in coals and wastes	Soil, drinking and recreational, agricultural food products, paints and automobiles	Higher concentration can lead to a reduction in lungs functioning, retardation and brain damage; learning disabilities; heart disease; damage to the nerve system, hearing loss, lung disease	Järup (2003), Peden et al. (2004)
Volatile organic compounds (VOCs), e.g., gasoline	Highly reactive and toxic organics produced by nature and man, and they are highly volatile	Break down of the respiratory system; cancer	Peden et al. (2004), Tsai (2016)
Ozone	• Secondary to aero chemical reaction to nitrogen oxides and VOCs, because of photochemical reactions • Vehicle exhaust and some other chemicals commonly used in industry mix in strong sunlight	Difficulties in breathing, especially for people with asthma and other respiratory diseases Ozone depletion	Leung (2015), Zhou et al. (2015), Wei et al. (2017) Özden et al. 2008

(continued)

Table 5.1   (continued)

Pollutant	Source	Health effect	References
Particulates matter (PM$_{2.5}$-PM$_{10}$) Ultrafine particulate matter (< 0.1 um diameter)	Factories, power plants, refuse incinerators, motor vehicles, construction activity, forest fires, wood smoke, natural windblown dust	Pulmonary inflammation, especially for people with heart and lung diseases	Bernstein et al. (2004), Pöschl (2005), Leung (2015), Wei et al. (2017)
Chemical pollutant (NH$_3$ and non-methane contaminants)	NH$_3$ is the most abundant alkaline gas in the atmosphere, is produced naturally from decomposition of organic matter, plants, animals and wastes, fertilizers	At high concentrations, or in moist areas, ammonia leads to irritation of the throat and respiratory tract	Peden et al. (2004), Tsai (2016), Wei et al. (2017)

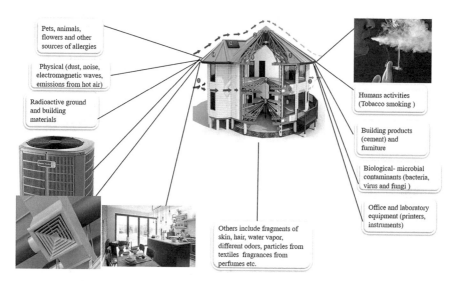

Fig. 5.2   Different sources and causes of indoor air pollution

may lead to serious respiratory and health problems (Cohen et al. 2017). The rise in different health complications has been linked to the ability of the outdoor particles to infiltrate and penetrate the building through various routes such as windows, doors, leakages from buildings and some ventilation systems. However, the impact of indoor air pollution depends on the magnitude of exposure, when one takes into account, exposure intensity, time spent indoors (which is often far more than time spent outdoors) and the number of individuals exposed. This has resulted in indoor air pollution to be ranked as one of the leading causes of global mortality (Rajagopalan and Brook 2012; Cohen et al. 2017). Table 5.2 shows the different sources of indoor air pollutants and associated health risks.

Table 5.2 Indoor air pollutant, sources and adverse health effects

Pollutant	Source	Health effect	References
Building materials, (asbestos-containing product)	Manufacturing, designing and installation of asbestos-based products	Lung cancer, malignant, mesothelioma, asbestosis and skin irritation	Curtis et al. (2006), M. et al. 2013, Vainio (2015), Tham (2016), Wei et al. (2017)
Biological Contaminants (bacteria, fungi, pollen and animal dander)	Human occupants, organic dust, various materials stored in the buildings, and the air inflowing from the ventilation and air conditioning systems	Allergies, asthma, hay fever, pneumonia, and sick building syndrome	Newson et al. (2000), Bernstein et al. (2004), Kalwasińska et al. (2012), Leung (2015), Tham (2016)
Tobacco smoking	From cigarettes, cigar or pipes. These produce some toxic chemicals and gases	• Chronic respiratory symptoms; reduces the level of lung function in children; implicated as a cause of lung cancer in adults, tuberculosis • Heart diseases and general irritation of the eyes, throat, nose and lungs	Tham (2016), Wei et al. (2017)
Heaters, insulation and moist materials. Air conditioning systems	Building materials and ventilating systems	Can lead to dampness, growth of moulds, presence of microbes in buildings can cause infections, allergic or hypersensitivity reactions	Bernstein et al. (2008), Tham (2016)
Chemical pollutants (benzene, formaldehyde, adhesives, cleaners, solvents, combustion by-products and emissions from floor or wall coverings).	• Furniture and decoration materials: All kinds of artificial board, solvents, paints, adhesives, paint, UF foam, synthetic fibre • Formaldehyde is a potent fungicide, which is widely used to preserve sample in scientific research and hospitals	Formaldehyde can cause chronic respiratory disease, colon cancer and brain tumour. Abnormalities, leukaemia, memory and mental of young people descend	Leung (2015), Wei et al. (2017)

(continued)

Table 5.2 (continued)

Pollutant	Source	Health effect	References
Using diverse electronic products. Artificial lights	Ultrasound, radiation, electrical equipment, microwave, TV, electric blanket, telephone and other processes can produce electric radiation, noise, vibration which speeds up the electromagnetic field	Noise and vibration, fatigue, poor concentration, dizziness, headache with nausea, ringing in ears and pounding hearteye problems (burning, dry, gritty eye)	
Particulate matter (PM 2.5–10) Ultrafine particles	Cigarette smoke, degradation of VOCs, cooking, heating systems and resuspension of house-settled dust. Ineffective house cleaning and improper ventilation c	Lung and bladder cancer, heart diseases, nervous systems and respiratory illness	Leung (2015), Wei et al. (2017)
SO_2, NO_2, NH_3 (wood smoke, gas stoves)	Atmospheric formation process begins with the emission of SO_2 from the burning of coal and oil in stationary and mobile sources. Fuel-burning heating systems (wood, oil and natural gas), NH_3 is emitted mainly from agriculture through the spreading and disposal of animal wastes and the use of nitrogenous fertilizers NO_2 indoor fuel-burning stoves (wood, kerosene, natural gas, propane)	High doses of NO_2 exposure cause severe pulmonary oedema, diffuse lung tissue inflammatory injury Marks et al. (2010)	Julvez et al. (2009), Marks et al. (2010), Tham (2016), Leung (2015)
Ozone	Most of the formation contributors are oxidant species, such as NO_2 and non-methane volatile organic compounds	Particularly asthmatic symptoms and respiratory illnesses Depletion of the ozone layer, which protects humans from harmful ultraviolet radiation from the sun	Kim et al. (2013), Leung (2015), Wei et al. (2017)
VOCs (household products, gases)	• Paints, paint strippers • Wood preservatives • Aerosol sprays, disinfectants, • Moth repellents, air fresheners, stored fuels, dry-cleaned clothing and • Pesticide	• Irritation of the eyes, nose, throat. Headaches, loss of coordination and nausea. Damage to the liver, kidney, central nervous system, and allergic skin reaction	Wei et al. (2017)

Air pollution remains a significant challenge in Africa in both indoor and outdoor settings. According to the WHO, approximately 600,000 deaths every year across the continent are associated with air pollution, with 23% of global deaths (12.6 million) linked to environmental factors (WHO 2018). Understanding of the microbial identities, distribution and abundance remains in its infancy in African settings. This review, supported by the standing scientific literature, discusses the existing knowledge on microbiological air quality in both outdoor and indoor settings while highlighting the few identified studies from Africa on this subject. Also, the sampling methods, monitoring techniques, challenges and future perspectives of microbial air quality research is discussed to stimulate the knowledge around this field of interest for improving studies, analysis and simulations.

5.4 Microbiological Quality in Outdoor (Ambient) Settings

In the last two decades, studies geared towards comprehensive monitoring of atmospheric aerosol concentrations have increased, not only for environmental management but also for the assessment of the health impacts of air pollution. Most of these studies have focused on occupational and indoor environments because people spend more than 90% of their time indoors (Ashmore and Dimitroulopoulou 2009; Wichmann et al. 2010; Zhai et al. 2018). However, human beings are in constant exposure to outdoor bioaerosols that constitute a major source of microorganisms indoors. Therefore, understanding the sources and dynamics of outdoor microbial quality and its impact on public health has become imperative.

Soil and water environments and the atmosphere are the main biospheres harbouring microorganisms, which include bacteria, fungi and viruses (Lee et al. 2010). These microorganisms can persist in the biosphere as individual cells or can be associated with other particles, such as aerosols, dust particles, leaf fragments, spores and other bio-pollutants (Maron et al. 2006). Notably, the aerial dispersal is a natural facet of the life cycle of many microorganisms, required for reproduction and the colonization of new sites (Kuske 2006). A recent study by Yamamoto et al. (2015) found that soil dust was enriched with skin-associated yeasts, those of the genera *Rhodotorula, Candida, Cryptococcus, Malassezia* and *Trichosporon*. Another study reported that garden plants contribute minimally to certain airborne fungi upon agitation such as during watering or strong air currents, producing elevated levels of airborne microbes such as *Cladosporium, Penicillium, Alternaria, Epicoccum* and *Pithomyces* spp. (Burge et al. 1982). Similarly, the aerosol dispersal of bacterial pathogens such as those in the genera *Escherichia, Salmonella, Legionella, Neisseria, Bacillus, Francisella, Burkholderia, Clostridium, Brucella* and *Yersinia* pose important health and ecological issues (Kuske 2006).

The structure and diversity of airborne microbial communities have been reported to be generally dependent on intrinsic environmental factors of natural habitats such as nutrient concentration, relative humidity (RH), temperature and UV intensity (Gandolfi et al. 2015). For example, high humidity and temperature provide a favourable environment for the proliferation of airborne microorganisms (Reanprayoon and Yoonaiwong 2012). However, some lipid-enveloped viral particles such as measles and varicella-zoster virus (VZV) are sensitive to changes in temperature, relative humidity and UV radiation (Tang et al. 2006). Similarly, bacterial pathogens, such as *Escherichia coli* and *Klebsiella pneumonia* that tend to behave like enveloped viruses, are less stable at high RH (Tang et al. 2006). In contrast, an airborne pathogen, *Salmonella senftenberg*, has been reported to survive even at high humidity conditions (Doyle and Mazzota 2000). Nkhebenyane et al. (2012) also showed that increased RH (100%) was associated with increased *Bacillus cereus* bioaerosols. Some other pathogenic microbes like *Mycobacterium tuberculosis*, a hardy organism with a thick cell wall and responsible for causing tuberculosis, can withstand extreme environmental conditions in bioaerosols and is of a tremendous public health concern (Tang et al. 2006).

In addition to the contribution from the natural environment, increasing concentrations of bioaerosols from anthropogenic activities and their impact on local air quality is becoming a growing public health concern worldwide (Tarwater et al. 2010; Yassin and Almouqatea 2010; Kaarakainen et al. 2011; Bowers et al. 2013). Consequently, there have been increased interest and a number of studies investigating outdoor microbial air quality from different types of man-made sources such as waste processing operations (Kiviranta et al. 1999; Recer et al. 2001; Prazmo et al. 2003; Taha et al. 2006; Fracchia et al. 2006; Fischer et al. 2008; Sanchez-Monedero et al. 2008), agricultural environments, industrial activities, including environmental phenomena such as dust storms (Zhu et al. 2003; Pearce et al. 2009; Tarwater et al. 2010; Yassin and Almouqatea 2010; Bowers et al. 2011; Kaarakainen et al. 2011; Adams et al. 2013; Hanson et al. 2016a; Kumari et al. 2016; Leung and Lee 2016; Lymperopoulou et al. 2016; Sommer and Moustaka-gouni 2017; Zhen et al. 2017; Du et al. 2018; Liu et al. 2018; Schlatter et al. 2018; Zhai et al. 2018; Wei et al. 2019). In these studies, researchers have been focusing on bioaerosols concentration, biodiversity, sources and environmental impact factors. Findings indicate that microbial diversity and concentrations vary among different types of outdoor environments, depending on the local environment, seasons and meteorological factors.

In highly polluted atmospheric environments, microorganisms in bioaerosols are closely correlated with the concentration of air pollutants (Waters et al. 2016). In urban areas of Beijing experiencing severe smog events, bioaerosols are characterized by higher concentrations of air pollutants, including $PM_{2.5}$, PM_{10}, nitrogen dioxide (NO_2), sulphur dioxide (SO_2) and carbon monoxide (CO), than in rural areas (Chai et al. 2014; She et al. 2017; Liu et al. 2019a). Interestingly, a positive correlation exists between the particulate pollutants ($PM_{2.5}$ and PM_{10}) (Dong et al. 2016; Gou et al. 2016; Liu et al. 2018) and gaseous pollutants (such as NO_2, SO_2 and CO) (Ho et al. 2005) of smog events. Studies using either culture-dependent or next-generation metagenome sequencing have shown that airborne microbes in smog events are composed mainly of 86.1% bacteria, 13.0% eukaryotes, 0.8% archaea and 0.1% viruses in the $PM_{2.5}$ samples, while 80.8% bacteria, 18.3% eukaryotes, 0.8% archaea and 0.1% viruses in the PM_{10} samples (Cao et al. 2014). Dominant bacteria phyla that have been reported include *Proteobacteria* (32.2%),

Cyanobacteria (18.0%), *Actinobacteria* (16.5%), *Firmicutes* (15.5%) and *Bacteroidetes* (11.6%) (Liu et al. 2018). Similarly, studies in other countries indicate that P*roteobacteria*, *Bacteroidetes*, *Cyanobacteria* and *Firmicutes* are the predominant bacterial phyla in atmospheric bioaerosols of polluted environments while most dominant fungal categories were Ascospores, followed by *Cladosporium* and *Aspergillus/Penicillium* (Zhu et al. 2003; Després et al. 2007; Pearce et al. 2009; Fahlgren et al. 2011; Bowers et al. 2013; Barberán et al. 2015b; Hanson et al. 2016a; Lymperopoulou et al. 2016; Sommer and Moustaka-gouni 2017; Schlatter et al. 2018). In these studies, the fungal diversities and concentrations exhibited significant diurnal and seasonal variations, where *Aspergillus, Cladosporium, Ganoderma, Arthrinium/Papularia, Cercospora, Periconia, Alternaria* and *Botrytis* were significantly higher in PM_{10}.

Waste processing facilities that manage large quantities of both organic and organic waste materials are also significant sources of potentially pathogenic microorganisms to the surrounding environment (Kiviranta et al. 1999; Taha et al. 2006; Cyprowski et al. 2008; Fischer et al. 2008). More importantly, the potential spread of microorganisms of faecal origin is of great concern to nearby residents. High concentrations of airborne microbes in various occupational settings have been linked to adverse health effects experienced by the workers in the waste treatment facilities (Vincken and Roels 1984; Bünger et al. 2000) and populations living near such facilities (Recer et al. 2001; Herr et al. 2003; Fracchia et al. 2006; Fischer et al. 2008). During organic waste treatment, the characteristics of bioaerosols depend on many factors including the composition of the sewage, size and capacity of the facility and meteorological factors, but is dominated by enteric bacteria (of genus *Pseudomonas, Staphylococcus* and *Streptococcus*), fungi (*Trichophyton* and *Microsporum*), protozoan cysts and worm eggs, enteroviruses and retroviruses (Prazmo et al. 2003; Fracchia et al. 2006; Cyprowski et al. 2008; Ko et al. 2008; Kumari et al. 2016). Among the fungi, *Aspergillus* and *Penicillium* predominate at municipal solid waste landfill sites and other waste treatment facilities (Bünger et al. 2000; Prazmo et al. 2003; Fracchia et al. 2006; Kumari et al. 2016).

Although most atmospheric bioaerosols are soil-associated and non-pathogenic to humans, the identification of several respiratory microbial allergens and pathogens, whose relative abundance correlates to increased concentrations of PM pollution, is of great public health concern to environmental scientists, health workers and city planners (Cohen et al. 2005; Curtis et al. 2006). The findings that bioaerosols from sewage treatment plants are dominated with drug-resistant staphylococci and streptococci have a tremendous influence on human health and are, therefore, a priority challenge for modern medicine. Also, health problems such as respiratory infections, digestive disorders and skin allergies, attributable to endotoxins and mycotoxins produced by these bacteria and fungi, have been reported (Vincken and Roels 1984; Kiviranta et al. 1999; Bünger et al. 2000; Herr et al. 2003; Cyprowski et al. 2008). For example, *A. fumigatus,* a common fungus isolated in outdoor atmospheric bioaerosols is known to cause invasive infections such as chronic pulmonary aspergillosis and bronchitis in immunocompromised and immunocompetent individuals with underlying lung damage, and allergic disease of the respiratory system (Denning et al. 2014). The *A. fumigatus*-associated allergic bronchopulmonary aspergillosis (ABPA) may affect patients with asthma or cystic fibrosis and constitutes the principal clinical disorder due to *Aspergillus* hypersensitivity (Fairs et al. 2010; Knutsen and Slavin 2011; Denning et al. 2014). These examples illustrate the importance of studying the dynamic characteristics of bioaerosols and the factors that affect them, as it helps researchers to positively identify environments that encourage more proliferation of bacteria or pathogens, and to take active control measures against microbial pollution and airborne diseases.

Indoor and outdoor air pollution sources remain a significant environmental and health issue and a policy challenge in the African continent (WHO 2019). Furthermore, there are no formally regulated standards for bioaerosol levels in most countries, with most of them having very limited air quality monitoring capacity. Most African countries have been experiencing faster-growing economies in the last two decades, coupled with increased urbanization and population growth (UNECA 2017). The associated environmental pollution and the created microenvironments related to increasing urbanization is proving to be a big challenge, where particulate matter (the air pollutant of primary concern for human health), industrial and municipal wastes, and escalating informal settlements have enormous negative impacts on public health (WHO 2019). These microenvironments play a crucial role in bioaerosol generation, propagation, aerosolization, resuspension, controlling diffusion, transportation and intermolecular interactions, and warrant investigation. Therefore, increased funding on research studies targeting increased awareness and providing references for a better understanding of outdoor air quality (OAQ) and public health outcomes in different environmental settings across Africa is urgently needed. Specifically, more data is needed to establish the basis for guidelines for protective zones between the sources of outdoor bioaerosols and residential areas.

5.5 Microbiological Quality in Indoor Settings

Similar to outdoor settings, indoor air quality (IAQ) is a major problem worldwide. In the modern era, people spend 80–90% of their time indoors, and therefore indoor air quality poses a significant potential public health threat; however, IAQ is still receiving relatively little attention from researchers, regulatory officials and environmental analysts. Major health problems like allergies and respiratory diseases have been associated with poor IAQ. Also, even minor ailments like headaches and eye irritation can cause discomfort and distraction, ultimately leading to lower productivity. According to the World Health Organisation (WHO), different respiratory and cardiovascular diseases caused by polluted indoor air can cause premature death. About 7 million deaths are caused by ambient (outdoor) and indoor air pollution, and 3.8 million deaths per year are attributed solely to indoor air pollution (WHO 2018). Therefore, it is crucial to investigate the overall quality of air in indoor settings.

Indoor air is arguably the fastest and most highly efficient means of pathogen spread in a given setting. Airborne pathogens discharged into the air may settle on different environmental surfaces like walls, ventilation systems, fans, and false ceilings, among others, which could then become secondary vehicles for the spread of infectious agents indoors (Prussin and Marr 2015). Table 5.3 summarizes different studies of microbiological air quality in indoor settings.

The microbiome of indoor environments contains a large number of different taxonomic groups. For example, a survey of homes across the United States revealed, on average, approximately 7000 different types (operationally defined as operational taxonomic units (OTUs) based on sequence similarity) of bacteria and 2000 types of fungi per house in the dust on the upper trim of an inside door (Barberán et al. 2015a). Similarly, Hewitt et al. (2012) found that bacterial communities on indoor surfaces in offices were distinguishable from those outside, with bacterial counts being higher on surfaces in the indoor system than outside. A study in a neonatal intensive care unit (NICU) in a hospital also identified approximately 12,000 bacterial OTUs on various surfaces per room having different bacterial genera including *Staphylococcus*, *Corynebacterium*, *Lactococcus*, *Firmicutes* and *Actinobacteria*, and fungi such as *Cladosporium*, *Penicillium* and *Aspergillus* (Barberán et al. 2015a; Prussin and Marr 2015). The multitude of recent studies examining various indoor microbiomes reveals that microbial communities in indoor environments are complex and highly variable. Researchers have also witnessed that microbial communities are vastly different between different types of indoor environments such as schools, houses, working offices and hospitals, such that even different rooms within the same building (for example, living room vs. bathroom) exhibit distinct microbiomes.

Table 5.3 Studies on microbiological indoor air quality in Africa and other countries

Country	Type of indoor system	No of samples/ duration	Method	Abundant microbes	References
India	Hospital- ICU	February–April 2006	Cultivation method	*S. aureus, Micrococci, Klebsiella* sp., *A. flavus K. pneumonia, A. niger, Pseudomonas* sp.	Sudharsanam et al. (2008)
	Hospital - OT	February–April 2006	Cultivation method	*S. aureus, Micrococci, Pseudomonas* sp.	Sudharsanam et al. (2008)
China	Airline cabin	57 samples	Cultivation assisted 16 s amplicon sequencing	*Staphylococcus epidermidis* and *Pseudomonas luteola; Brachybacterium paraconglomeratum* and *Deinococcus daejeonensis*	Liu et al. (2019b)

(continued)

Table 5.3   (continued)

Country	Type of indoor system	No of samples/ duration	Method	Abundant microbes	References
Poland	School	Spring of 2016 and 2017	Cultivation method	*Staphylococcus lentus, Staphylococcus epidermidis, Staphylococcus sciuri, Staphylococcus chromogens, Kocuria rosea, Micrococcus* spp., *Bacillus circulans, Bacillus subtilis, Bacillus mycoides, Bacillus cereus, Bacillus pumilus, Brevibacterium* spp., *Corynebacterium auris, Corynebacterium tuberculostearicum, Corynebacterium propinquum, Pseudomonas* spp.	Bragoszewska et al. (2018)
	University	59 samples	Cultivation method	*Micrococcus* spp., *Bacillus* spp., *Staphylococcus* spp., *Sarcina* spp., *Serratia* spp., *Cladosporium* spp., *Penicillium viridicatum* and *Penicilluim expansum, Aspergillus niger* and *Aspergillus flavus. Cladosporium herbarum, Alternaria alternata, Mucor* spp., *Rhizopus nigricans* and *Epicoccum* spp.	Stryjakowska-Sekulska et al. (2007)
	Office building	7 May to 7 June 2017	Cultivation method	*Macrococcus equipercicus, Micrococcus luteus D, Staphylococcus xylosus, Staphylococcus* spp., *Gemella haemolysans, Corynebacterium tuberculostearicum, Nocardia shimofusensis/ higoensis, Janibacter anophelis/hoylei, Bacillus pseudomycoides, Pseudomonas putida* and *Enterococcus faecium*	Bragoszewska et al. (2018)

Table 5.3 (continued)

Country	Type of indoor system	No of samples/ duration	Method	Abundant microbes	References
Slovenia	Kindergartens	51 samples	Cultivation method	*Staphylococcus* spp. and *Enterobacteria; Cladosporium* spp., *Penicillium* spp., *Aspergillus* spp., *Mucor* spp., *Phoma* spp., *Fusarium* spp., *Monilia* spp., *Scopulariopsis* spp. and *Rhodotorula* spp.	Rejc et al. (2019)
Portugal	Hospital	June 2013 (summer) and February 2014 (winter)	Cultivation method	*Staphylococcus* (including *Staphylococcus aureus, Staphylococcus capitis, Staphylococcus hominis, Staphylococcus epidermidis* and *Staphylococcus warneri*) and *Micrococcus* (including *Micrococcus luteus* and *Micrococcus lylae*); *Neisseria, Brevibacterium casei, Proteus* and *Shigella* sp.; *Penicillium* and *Aspergillus*	Cabo Verde et al. (2015)
South Africa	Hospital (HVAC)	Six-months (period not mentioned)	Cultivation assisted MALDI	*A. Oxydans, B. megaterium, P. chrysogenum Anaerococcus* spp., *A. polychromogene*	Malebo and Shale (2013)
Nigeria	Primary health care	Two months duration	Cultivation assisted 16 s amplicon sequencing	*Bacillus cereus, Proteus mirabilis, Chryseobacterium* sp., *Staphylococcus epidermidis* and *Staphylococcus aureus*	Robinson and Wemedo (2019)
	Offices	50 samples	Cultivation method	*Staphylococcus* spp., *Streptococcus* spp. and *Micrococcus* spp.; *Cladosporium* spp., *Aspergillus* spp., *Penicillium* spp., *Fusarium* spp., and *Candida* spp.	Oluwakemi Omolola (2019)

(continued)

Table 5.3 (continued)

Country	Type of indoor system	No of samples/ duration	Method	Abundant microbes	References
Ethiopia	Public primary schools	51 samples	Cultivation method	*Bacillus* sp., *Staphylococcus aureus* and coagulase-negative *Staphylococcus* (CoNS) species	Andualem et al. (2019)
	Hospital	15 February to 30 April 2017	Cultivation method	Coagulase-negative staphylococci (CoNS), *S. aureus* and *Klebsiella* sp.	Getachew et al. (2018)
Egypt	Hospital both ICU and OT	January – December 2013	Cultivation method	*Bacillus atrophaeus* and *B. subtilis; Staphylococcus aureus, Bacillus atrophaeus, B. pumilus* and *P. glucanolyticus; Aspergillus, Alternaria, Cladosporium, Penicillium, Cladosporium, Fusarium* and *Scopulariopsis*	Osman et al. (2018)
Ghana	Research institute	January to May 2017	Cultivation method	*Staphylococcus aureus* and *Streptococcus* sp.	Abiola et al. (2018)

Poor air quality within school buildings can cause health problems in students and teachers and affect learning and working (Śmiełowska et al. 2017). For instance, Bragoszewska et al. (2018) investigated the bacterial aerosols in three different educational buildings that included preschool, primary school and high school during spring 2016 and 2017. Results of this study demonstrated that the primary school had the highest concentration of bacteria (2205 CFU/m³), while the high school established the lowest bacterial concentration (391 CFU/m³). The authors further confirmed that the Gram-positive cocci *Staphylococcus lentus, Staphylococcus epidermidis, Staphylococcus sciuri, Staphylococcus chromogens* and *Micrococcus* sp. were the most frequently occurring species in the indoor environment (Bragoszewska et al. 2018). Similarly, in Slovenia, the microbial air quality was assessed in two kindergartens using the culture method during four seasons. The study confirmed that indoor quality was most affected during spring seasons and harboured common fungal species including *Cladosporium, Penicillium* and *Aspergillus* along with bacterial members such as staphylococci and *Enterobacteriaceae* (Rejc et al. 2019). However, there are limited studies in Africa. Recently, a study was conducted in Ethiopian primary schools to determine the bacterial load during March 29–April 26, 2018, and it was found that *Staphylococcus aureus*, Coagulase-negative *Staphylococcus* species and *Bacillus* species were recurrent bacterial members (Andualem et al. 2019). Likewise, the university, office buildings, public utility buildings and research institutes are prone to indoor air contamination, and the extent of the contamination results in the presence of harmful microbial pathogens (Table 5.3) that are linked to the spread of infectious diseases.

Notably, indoor air quality in hospitals, primary healthcare settings and clinics plays an important role in infection, especially in infants and pregnant women and patients with reduced or impaired immunity (Śmiełowska et al. 2017). Recently, numerous studies were targeting the quality of indoor in different hospital settings, especially the intensive care units (ICU) and operation theatres (OT). An assessment of the microbial air quality in both ICU and OT in a public hospital in India found that the air contained different bacterial groups. For example, the ICU was frequently dominated by *S. aureus, K. pneumonia*, micrococci, *Pseudomonas* sp., *Klebsiella* sp., *A. flavus* and *A. niger* while the OT carried fewer bacteria members such as *S. aureus, Micrococci* and *Pseudomonas* sp. (Sudharsanam et al. 2008). However, few studies in Africa were carried out in hospital settings to assess the microbial air quality. For instance, Malebo and Shale (2013) determined the bacterial counts and dominant members colonizing the Heating Ventilation Air-Conditioning Systems (HVAC) at a South African hospital and concluded that the genera *Anaerococcus, Arthrobacter, Bacillus, Staphylococcus* and *Streptomyces* were dominant bacterial members identified along with the fungus *Penicillium chrysogenum*. Other recent findings from Nigerian and Ethiopian hospitals demonstrated the same bacterial members including *Bacillus cereus, Proteus mirabilis, Chryseobacterium* sp., *Staphylococcus epidermidis* and *Staphylococcus aureus*, confirming that the hospital settings were highly prone to specific bacteria, especially Gram-positive bacterial members such as *Bacillus* and *Staphylococcus* sp. (Getachew et al. 2018; Robinson and Wemedo 2019) which are considered as highly resistant endospore-forming and emerging antibiotic-resistant pathogens.

The emergence of airborne pathogens and the evolution of drug resistance are a key challenge for the ability to treat and control airborne-related infections. Much attention is to be focused today on airborne pathogenic microorganisms that have developed resistance to specific antibiotic treatments, or entire types or classes of antibiotics (Levetin et al. 2001). A recent study reported that emerging pathogens such as noroviruses and *Clostridium difficile* have also been detected in indoor air, with a strong potential for airborne dissemination (Ijaz et al. 2016). Some pathogenic microbes, for instance, *Legionella*, may be transmitted via aerosols from poorly maintained air conditioners or water distribution systems. Anyone who works in a building is at risk of acquiring the disease, which is often wrongly assumed to be a flu-like illness (Lim 2011). People leave their microbial footprint as a part of the indoor microbiome (Wu et al. 2016). For example, humans carry different types of bacteria and viruses in the respiratory tract and saliva and discharge the microorganisms into the built environment in aerosols during coughing, sneezing, talking, and even just breathing (Prussin and Marr 2015). Higher levels of occupancy and activity will influence the abundance and composition of bacteria and other microbes (which they carry) in indoor systems (Adams et al. 2016). Hospodsky et al. (2012) found that human occupancy in a university classroom increased the total bacterial genome concentration in indoor air compared to unoccupied periods.

Another study from Hewitt et al. (2012) revealed that humans are the primary sources of bacterial contamination in offices, and detected different bacterial genera such as *Streptococcus*, *Corynebacterium*, *Flavimonas*, *Lactobacillus*, and members of the *Burkholderiales*. Furthermore, they found different pathogens in indoor office surface, including *Neisseria*, *Shigella*, *Streptococcus* and *Staphylococcus* sp. Indoor air quality is, therefore, a prominent public health concern in many countries, including Africa that requires a clear understanding of the microbial load inside the indoor system.

5.6   Monitoring Techniques for Microbial Air Quality

Air sampling is a crucial function of any quality control for detecting and measuring chemical and microbial contaminants in both the outdoor and indoor settings. The techniques used for detection and monitoring of air quality are either qualitative or quantitative based on different principles, the extent of use and their cost-effectiveness (Bernstein et al. 2008). There are two main techniques for microbial air sampling, namely the active and passive monitoring sampling technique (Napoli et al. 2012), based on the air volume and the use of a driving force or naturally (Fig. 5.3) (Khan et al. 2018).

Active sampling is a microbial sampling monitoring technique whereby an air sampling device or pump is used to force air into or onto a collection medium, for example, Petri dish with nutrient agar over a specified period (Napoli et al. 2012). The sampled plates are then incubated, and the number of microorganisms present is measured in CFU (colony forming units)/m^3 of air; this approach is often applicable when the microbial load of the air is low.

Active monitoring utilizes microbiological air sampling devices that physically draw a known volume of air over, or through, a microbial particle collection medium either by impaction, impingement or filtration techniques (Mandal and Brandl 2011; Napoli et al. 2012). In an impingement sampler, a liquid medium for particle/bioaerosol collection is used, and the sampled air is drawn using a suction pump through a narrow inlet tube into a small flask containing the collection medium. This technique accelerates the air towards the surface of the collection medium; the flow rate is determined by the inlet tube diameter. Contact of the air with the surface of the liquid changes its direction abruptly, leading to the impingement of any suspended particle into the collection liquid. The collection liquid can then be cultured to enumerate viable microorganisms (Mandal and Brandl 2011).

Unlike the liquid medium, a solid or adhesive medium such as agar is used to culture bioaerosol particles collected with impactor samplers. In this system, the air is drawn into the sampler by a pump through a perforated plate. As the air makes contact with the collection surface, suspended particles are collected by impact as they hit the solid collection agar surface; these are then incubated directly, and observable colonies are enumerated (Mandal and Brandl 2011).

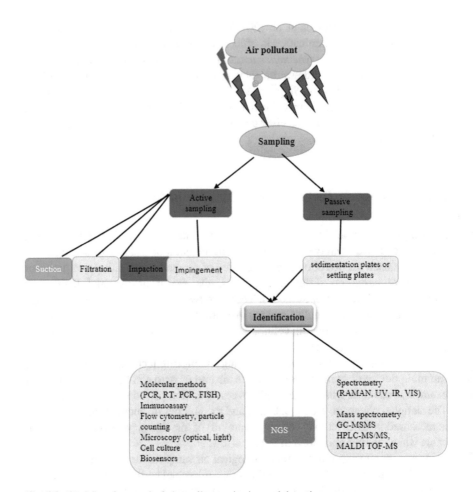

Fig. 5.3 Workflow for a typical air quality monitoring and detection

Filtration sampler systems are commonly used alternatives, whereby a known air volume is drawn through a polycarbonate or cellulose acetate membrane filter using a pump or vacuum line. The filter is then transferred onto the surface of an agar medium before incubation. Alternatively, the filter can be transferred onto gelatine, which is subsequently dissolved and analysed through culture or other rapid methods (Mandal and Brandl 2011).

Passive sampling is a microbial sampling monitoring technique comprising the use of sedimentation plates or settling plates. Petri dishes containing culture media are exposed to air for some time, then incubated to allow visible colonies to develop and then the microbial load is determined, usually expressed in CFU/plate/time or

CFU/m^2/h (Napoli et al. 2012). The method is based on the principle of diffusion through a static air layer or permeation through a membrane without the use of pumps and the devices used are referred to as diffusive samplers. The air streams freely around a filter, membrane or another medium (sorbent), which captures pollutants during the period of passive air sampling. Settle plates used in passive sampling are often restricted in their application since they are only appropriate for monitoring viable biological particles that sediment out of the air and settle onto a surface over the time of exposure. Comparatively, passive sampling is inexpensive, easy to use and requires no special equipment. However, passive air sampling suffers from the main disadvantage that it might not be able to detect smaller suspended particles or droplets in the air, and consequently the data generated might not be quantitative. Also, they are vulnerable to interference and contamination from non-airborne sources, and the agar growth medium in the plates may deteriorate on more prolonged exposure in the environment. Furthermore, settle plates may quickly become overgrown in heavily contaminated conditions making data interpretation difficult.

Despite the inherent shortcomings, both active and passive sampling have been useful for qualitative analysis of bioaerosols and the data they produce have been helpful in detecting underlying trends in airborne contamination for early warning systems in both indoor and outdoor settings under different environments (Yousefi and Rama 1992; Pasquarella et al. 2000; Damialis and Gioulekas 2006; Das and Gupta-Bhattacharya 2008; Filali Ben Sidel et al. 2015; Matinyi et al. 2018). For example, the level of *Legionella* contamination in air samples from the bathrooms of a multicentre health facility in Italy was evaluated using active and passive sampling methods (Montagna et al. 2017). In this study, although the contamination level varied among each sampling technique, *Legionella* air contamination was detected in 36.4% of the sampled health facilities by at least one of the methods. Similarly, the bacterial and fungal contamination levels were measured and monitored in the operating and non-operating theatre using both active and passive sampling in a hospital in Parma (Pasquarella et al. 2000). In Poland, comparison of microbial air quality of an air-conditioned office and a naturally ventilated office space using active sampling indicated the presence of *Staphylococcus xylosus*, *Bacillus species*, *Micrococcus luteus* and *Macrococcus equipercicus* (Brągoszewska et al. 2018). In Ethiopia, the bacterial load of *Staphylococcus aureus* and *Streptococcus pyogenes* was determined in the indoor air samples of a hospital ward using the passive and active sampling methods (Gizaw et al. 2016). A study on exploring the outdoor and indoor fungal and bacterial contamination in different public places in Egypt, using both passive and active sampling methods, reported a high prevalence of fungi such as *Aspergillus niger* and *Penicillium* spp. (Yassin and Almouqatea 2010). These examples illustrate the potential utility of both active and passive sampling techniques in studying and monitoring microbial air quality in Africa.

5.7 Techniques for Detection, Identification and Monitoring of Airborne Microbes and Other Pollutants

The principle of detection and identification of microbial air pollutants involve three key steps: (i) recovery and concentration; (ii) purification and separations, and (iii) assay and characterization for identification. Figure 5.3 illustrates a summary of the different techniques used for microbial air pollutant analysis.

5.7.1 Culture-Based Methods

Culture methods, long regarded as the 'gold standard' in the detection and identification of microorganisms, are based on enrichment of target microbes from a sample in specific liquid growth media, then plating onto selective/differential agar. These methods usually require the confirmation of pure isolates using a diverse set of further analysis such as morphological, biochemical, serological and other tests, including molecular approaches. Compared to other methods, culture-based techniques for enumeration and identification of microbial air pollutants are well established, simple, inexpensive and can be used for both quantitative or qualitative analyses (Zeng et al. 2006; Dungan and Leytem 2009; Hubad and Lapanje 2013). Due to these advantages, the bulk of studies on microbial air quality monitoring in Africa have been based on culture-based methods (Yousefi and Rama 1992; Yassin and Almouqatea 2010; Filali Ben Sidel et al. 2015; Gizaw et al. 2016; Matinyi et al. 2018; Andualem et al. 2019). Despite its more extensive use, culture-based methods still suffer from various disadvantages when applied to microbial air quality monitoring. For instance, these methods still rely on the growth of the target microorganisms in one or more nutrient media, making them labour-intensive and yield results after several days of repeated culture and confirmation steps. Also, since these methods usually use selective media for identifying microorganisms, they tend to be biased as they favour the growth of some organisms while suppressing the growth of others. Even when non-selective media are used, fast-growing organisms still grow at the detriment of slow growers. As such, culture methods cannot be used to have a complete picture of the microbial quality of air.

5.7.2 Culture-Independent Methods

Flow cytometry detection has also been used as an enumeration technique to investigate viable and non-viable cells from bioaerosols based on the principle of optical detection of scattered light and fluorescence allowing identification of microbes based on their morphology (Mandal and Brandl 2011; Ou et al. 2017). Airborne pathogens such as *Escherichia coli*, *Aspergillus fumigatus* and *Penicillium brevicompactum* conidia have been analysed using flow cytometry at different concentrations (Prigione et al. 2004; Ou et al. 2017).

Other enumeration techniques that are used for the microbial analysis of bioaerosols include immunofluorescence microscopy and scanning electron microscopy (Terzieva et al. 1996). For example, the total viral- and bacteria-like particle concentration from nine different locations have previously been assessed in both outdoor and indoor air environment using fluorescent microscopy (Prussin and Marr 2015).

Currently, there is increased use of conventional molecular/genomic-based methods in the detection and characterization of microbial contaminants from indoor and outdoor air environments. In contrast to culture-based methods, some of the molecular methods can be useful for examining diverse microbial contamination of air without the need for cultivation. These techniques range from polymerase chain reaction (PCR), immunological methods (ELISA test), radioimmunoassays, use of biosensors, fluorescence in situ hybridization (FISH), next-generation sequencing (NGS) and post-genomic techniques (such as proteomics, metatranscriptomics and metabolomics). Comparatively, techniques such as PCR, immunological and radioimmunoassays have been widely used to study and monitor microbial air quality worldwide. For example, conventional and quantitative polymerase chain reaction (PCR, RT-PCR, qPCR) have been used widely due to their sensitivity and rapid line application for the detection of a specific microbial air pollutant as well as microbial communities (Zeng et al. 2006; Dungan and Leytem 2009; Hospodsky et al. 2010; Mandal and Brandl 2011). Also, PCR coupled with the use of specific fluorescent dyes or probes for the total quantification of specific organisms has been applied in microbial detection and quantification in outdoor and indoor environments (Dungan and Leytem 2009). A study by Lignell and his co-workers compared the analysed results between the detection of house dust fungi and *Streptomycetes* using a combination of culture and quantitative polymerase chain reaction (qPCR). From their study, the qPCR method had a higher sensitivity in identifying taxonomy and functional niches, as well as improved understanding in quantifying the different level of concentrations of the target species (Lignell et al. 2008). On the other hand, immunological detection techniques such as enzyme-linked immunosorbent assay (ELISA), memory lymphocyte immune stimulation assay (MELISA), and radioimmunoassays have also been used to detect bacterial and viral air pathogens (Ding et al. 2015).

Fluorescence in situ hybridization (FISH), a technique that utilizes specific fluorescent probes that bind to the ribosomal RNA of a microbial cell has also been used to characterize microbial air contaminants (Deloge-Abarkan et al. 2007; Dungan and Leytem 2009; Mandal and Brandl 2011). Biosensors, comprising integrated devices that can provide quantitative or semi-quantitative analytical information through a biochemical receptor system, represent other potential molecular methods that can be applied in microbial air quality monitoring (Badihi-Mossberg et al. 2007). Biosensors are classified based on signal transduction used for detecting pollutants, and they can be electrochemical, optical, mass sensitive or thermal sensors (Nigam and Shukla 2015). The potential application of biosensors to monitor chemical air pollutants such as formaldehyde, SO_2 and PAH in an indoor setting has been reported (Badihi-Mossberg et al. 2007). However, with improving technologies, potential applications of biosensors, specifically those embedded with a biological recognition element such as an immunochemical, enzymatic, non-enzymatic receptor and nucleic acid biosensors, to detect and quantify microbial air contaminants may be possible in future. Despite the increased application of molecular/genomic methods, consideration of the strengths and limitations of each method provides different insights into the nature and behaviour of bacteria. Culture methods are essential in revealing phenotypic characteristics and behaviour, while the latter elucidate the genotypic information. As neither type of information is superior to the other, both methods should be viewed as specialized and complementary tools which are suited to answering different experimental questions related to detection, identification and monitoring of microbial air quality under different environmental conditions and settings.

5.7.3   Use of Next-Generation Sequencing (NGS) Technologies

The bulk of the current understanding of airborne microorganisms comes from culture-based studies, despite the shortcoming that the majority of the environmental microbes are not culturable. However, the recent advancements such as next-generation sequencing (NGS) technologies have seen a rise in publications utilizing high-throughput metagenome sequencing in studying the bioaerosols microbial diversity and concentrations in different environments (Zhu et al. 2003; Dong et al. 2016; Leung and Lee 2016; Sommer and Moustaka-gouni 2017; Abd et al. 2018; Du et al. 2018; Liu et al. 2018, 2019a). In contrast to first-generation sequencing, NGS generally produces voluminous (often millions) short DNA sequence reads (usually between 25 and 400 bp in length) and at a relatively low cost and short time. NGS technology has been widely used in recent years and includes a few major products with different chemistries, including 454 Sequencing (Roche Applied Science), Solexa (now Illumina) Technology (Illumina inc.), SOLiD (Applied Biosystems), Ion Torrent (Life Technologies Corporation) and Nanopure (Oxford Nanopore Technologies). These technologies are being increasingly applied in whole-genome (WGS), transcriptome, epigenome and small RNA sequencing (small RNA-seq), molecular marker and gene discovery, comparative and evolutionary genomics, and association studies, which have all helped decipher and further advance understanding of the distribution, variation and the potential metabolic activities of airborne microorganisms from different indoor and outdoor ecosystems (Table 5.4).

In recent years, several studies have utilized high-throughput sequencing technology to analyse the microbial composition and its functions from airborne settings. For instance, various airborne microorganisms, including double-stranded DNA viruses and sequences of several respiratory pathogens and allergens in the air, were successfully identified after a severe smog event in China (Cao et al. 2014). The study also revealed that microbial abundance increased with increased air

Table 5.4 Some of the studies using the next-generation sequencing approaches to study the environmental microbiome of air

Research target	Country	Methods	Sequencing platform	Conclusions	References
Indoor and outdoor urban environments	USA	WGS pyrosequencing; 16S rDNA metagenome sequencing	454 titanium	WGS and 16S rDNA metagenome sequencing can be used to identify airborne bacteria that are metabolically active, and to differentiate between transient members and those that use air as a habitat.	Yooseph et al. (2013)
Severe smog event	China	Shotgun sequencing	Illumina MiSeq and HiSeq 2000	With sufficient sequencing depth, airborne microbes including bacteria, archaea, fungi, and dsDNA viruses can be identified at the species level. Majority of the inhalable microorganisms in severe smog event are soil-associated and non-pathogenic to human	Cai et al. (2014)
Microbial functioning in clouds	France	Shotgun sequencing; multiple displacement amplification (MDA) of genomic DNA and total RNAs; metatranscriptomics	Illumina MiSeq	Microbes influence the cloud physical and chemical processes, via oxidant capacity, iron speciation and availability, amino acids distribution and carbon and nitrogen fates	Amato et al. (2019)

(continued)

Table 5.4 (continued)

Research target	Country	Methods	Sequencing platform	Conclusions	References
Airborne environment in urban spaces	USA	Shotgun sequencing	Illumina HiSeq 2000	Metagenomic complexity of urban aerosols temporally dependent and genomic analytical techniques can be used for biosurveillance and monitoring of threats to public health	Be et al. (2014)
Antibiotic resistance genes (ARGs) in severe smog event	China	Shotgun sequencing	454 titanium Illumina MiSeq and HiSeq 2000	Tetracycline, β-lactam and aminoglycoside resistance genes had high abundance in airborne PMs. Airborne PMs contained a higher level of ARGs in smog days than in non-smog days	Hu et al. (2018)
Indoor air quality in a library	Italy	I6S rDNA metagenomic sequencing	Illumina MiSeq	Libraries have a low microbial load (IGCM/ $m^3 < 1000$) characterized by different species, including several cellulose metabolizing bacteria. Workers and visitors appeared a relevant source of microbial contamination. Air biodiversity assayed by NGS seems a promising marker for studying IAQ	Valeriani et al. (2017)

(continued)

Table 5.4 (continued)

Research target	Country	Methods	Sequencing platform	Conclusions	References
Bioaerosols Childcare facilities	Korea	16S rRNA (V1–V3 regions) targeted amplicon sequencing	Roche/454 GS junior system	The bacterial community in the indoor air contain diverse bacteria associated with both humans and the outside environment. In contrast, the fungal community was derived mainly from the surrounding outdoor environment and not from human activity	Shin et al. (2015)
Microbiome of the built environment	USA	16S rRNA (V1–V3 regions) targeted sequencing	Illumina MiSeq	Offices have city-specific bacterial communities, which is dependent on different usage patterns with human skin contributing heavily to the composition of built environment surfaces bacterial diversity	Chase et al. (2016)
Hospital air	USA	Shotgun sequencing	llumina HiSeq 2500	Shotgun metagenomic sequencing approach can be used to characterize the resistance determinants of pathogen genomes that are uncharacteristic for an otherwise consistent hospital air microbial metagenomic profile	King et al. (2015)

(continued)

Table 5.4 (continued)

Research target	Country	Methods	Sequencing platform	Conclusions	References
Mycobiome in urban residences	Hong Kong	18S/5.8S rDNA (ITS1) target sequencing	Illumina MiSeq	Occupants exert a weaker influence on surface fungal communities compared to bacterial communities, and local environmental factors, including air currents, appear to be stronger determinants of indoor airborne mycobiome than ventilation strategy, human occupancy, and room type	Tong et al. (2017)
Indoor air bacteria in residences	USA	16S rRNA-based pyrosequencing and quantitative PCR	Roche GS FLX+ system	Indoor air in residences harbours a diverse bacterial community originating from both outdoor and indoor sources and is strongly influenced by household characteristics	Miletto and Lindow (2015)

(continued)

Table 5.4 (continued)

Research target	Country	Methods	Sequencing platform	Conclusions	References
Indoor dust and outdoor air	USA	18S/5.8S rDNA (ITS1) target sequencing 16S rRNA (V1–V3 regions) targeted pyrosequencing	Roche-454 life sciences titanium	Microbiome sequencing is applicable for different types of environmental samples (indoor dust, and low biomass air particulate samples), and offers the potential to study how whole communities of microbes (including unculturable taxa) influence human health	Hanson et al. (2016a, b)
Microbiome and allergens in the air of bedrooms of allergy disease patients	USA	16S rRNA (V1–V3 regions) targeted pyrosequencing	Illumina MiSeq	Microbial communities can be differentiated between rural, suburban, and urban homes and houses that were physically closer to each other have significantly more similar microbiota. It is possible to determine significant links between allergen burden and the microbiota in the air from the same sample and that these links relate to the characteristics of the home and neighbourhoods	Richardson et al. (2019)

(continued)

Table 5.4 (continued)

Research target	Country	Methods	Sequencing platform	Conclusions	References
Viruses in aerosol in animal slaughterhouses	New Zealand	Shotgun sequencing	Illumina HiSeq2000	Carefully designed large multi-region longitudinal studies can be used to elucidate viral causes of cancer caused by exposure to a bioaerosol. Metagenomic data provides a baseline and starting point for such investigations, including the development of diagnostics and targeted approaches for specific agents	Hall et al. (2013)
Office space bacterial diversity	USA	Multiplex pyrosequencing; cell culture	454 life sciences FLX genome sequencer	Comprehensive molecular analysis using cultures and metagenomic analysis of office building microbial diversity shows the potential of studying patterns and origins of indoor bacterial contamination	Hewitt et al. (2012)

pollution. In another study, Yooseph et al. (2013) successfully identified highly diverse microbial communities using metagenomics focussing on the genes involved in metabolism, transport, translation and signal transduction in microorganisms in indoor and outdoor air. Similarly, Tringe et al. (2008) studied the airborne metagenome in an indoor environment and identified genes responsible for adaptive mechanisms involved in resistance to desiccation and oxidative damage.

Concerning airborne viral diversity, NGS plays a significant role in establishing diverse airborne viruses in the near-surface atmosphere over three distinct land locations (residential, forest, and industrial), including those that infect plants and animals (Whon et al. 2012). These studies illustrate that NGS technologies are becoming convenient tools for monitoring airborne viral, bacterial and fungal pathogens and studying their global distribution patterns. With continuous progress and advances in the field, NGS will help identify the airborne microbes, and their genes and metabolic pathways that potentially contribute to atmospheric transformation, meteorological applications, environmental bioremediation, and health, and facilitate the search for ways to incorporate these finding into novel applications.

In addition to the microbiological quality, other airborne contaminants are determined using different analytical methods. For instance, spectrophotometry has been used for the detection of some air pollutants such as particulate matter concentrations (Language et al. 2016), as well as for the detection of SO_2 CO, ozone and Cr (Salem et al. 2009; Homa et al. 2017). Another powerful analytical detection method is the chromatographic separation approach; analysis can also be done using chromatographic mass spectrometry techniques (Thompson 2008). Some of the air pollutants such as volatile organic compounds (VOCs), mycotoxins from biological pollutants and chemical pollutants (e.g. formaldehyde) can be analysed using Gas chromatographic separation (GC-MS/MS) (Badihi-Mossberg et al. 2007) and high performance liquid chromatography (HPLC-MS/MS) (Salem et al. 2009). The main advantage of using chromatographic methods is that it can also identify unknown compounds (Srivastava and Majumdar 2011).

Liquid chromatography-tandem mass spectrometry (LC-MS/MS) is a powerful qualitative and quantitative analytical technique with a wide range of applications. For example, agricultural practices use tremendous amounts of airborne pesticides such as organophosphorus (OP) pesticides chlorpyrifos (CPF), azinphos-methyl (AZM), and their oxygen analogues, chlorpyrifos-oxon (CPF-O) and azinphos-methyl-oxon (AZM-O), which have been associated with higher temperatures, higher levels of ozone, dry weather, interaction with hydroxyl radicals, and photodegradation via ultraviolet light (Armstrong et al. 2014). Identification and quantification of these airborne contaminants require more sensitivity, accuracy and precision when compared to the traditional GC-MS method and, therefore, LC-MS/MS can be considered an appropriate alternative analytical method. Apart from these pollutants, microbial metabolites related to adverse human health effects have been observed in indoor air settings. Studies using LC-MS/MS detected 33 different microbial metabolites, including toxic bacterial metabolites and mycotoxins from indoor environments (Täubel et al. 2011). However, the current information on the natural occurrence of toxic microbial metabolites in indoor and ambient environments is limited. Applying these analytical methods will increase the knowledge of the variety of microbial metabolites present in both indoor and outdoor settings.

5.8   Challenges and Prospects

Despite the global increase in research on outdoor and indoor air quality and their relationship to public health, there are no significant studies done in Africa. Currently, there are numerous studies mainly targeting the chemistry of criteria pollutants and air toxics in Africa (Schwela 2012; Simwela et al. 2018; Katoto et al. 2019). For example, a systematic review of the literature identified 60 articles assessing the association between Air Associated Pollutants (AAP) and health outcomes in sub-Saharan Africa (Katoto et al. 2019). In contrast, only a few studies have focused on microbial air quality, with the majority being culture-based (Malebo and Shale 2013; Abiola et al. 2018; Getachew et al. 2018; Osman et al. 2018; Andualem et al. 2019; Oluwakemi Omolola 2019; Robinson and Wemedo 2019). Due to inherent shortcomings of culture-based techniques, the current state of affairs indicates that there is very little knowledge about the microbial content and diversity in both indoor/outdoor systems in Africa, which demonstrates the need to survey microbiome, using next-generation sequencing technologies. Based on the limitations of the existing molecular methods that target specific viruses, and specific bacterial indicators, new methodologies such as metagenomics are vital for the identification of unique or unlooked-for microbiomes in the aerial ecosystems in the continent. This would provide references for a better understanding of indoor/outdoor air microbiome quality and public health outcomes in different environmental settings across African.

Techniques such as NGS and metagenomics are still emerging techniques for the identification and diversification of microbiomes from different ecosystems in Africa. Although the causes of low research output in Africa are multifaceted, there are two major issues hindering research on modern high-throughput genomics and bioinformatics projects in African institutions: (1) shortage of trained bioinformaticians and (2) infrastructural problems. Generally, investment in science in Africa remains low by international standards with many countries contributing paltry <1% of GDP to research (Djikeng et al. 2012; Beuadry et al. 2018). African countries still do not have the resources to develop their genomic projects on a large scale. For example, it is estimated that there were only 13 high-throughput sequencing infrastructure on the Africa continent by 2015 (Prifti and Zucker 2015), illustrating how the continent lags far behind in terms of NGS equipment. Over the years, Africa has also witnessed a steady loss of university staff, leading to low scientific research output, and poor preparation of the next generation of African biotechnology scientists and bioinformaticians. Consequently, increased funding on research studies utilizing NGS technologies while also focusing on capacity building and infrastructural development for increased awareness is highly needed.

References

Abd A, Lee K, Park B et al (2018) Comparative study of the airborne microbial communities and their functional composition in fine particulate matter (PM$_{2.5}$) under non-extreme and extreme PM$_{2.5}$ conditions. Atmos Environ 194:82–92. https://doi.org/10.1016/j.atmosenv.2018.09.027

Abiola I, Abass A, Duodu S, Mosi L (2018) Characterization of culturable airborne bacteria and antibiotic susceptibility profiles of indoor and immediate-outdoor environments of a research institute in Ghana. AAS Open Res 1:17. https://doi.org/10.12688/aasopenres.12863.2

Adams RI, Bhangar S, Dannemiller KC et al (2016) Ten questions concerning the microbiomes of buildings. Build Environ 109:224–234. https://doi.org/10.1016/j.buildenv.2016.09.001

Adams RI, Miletto M, Taylor JW, Bruns TD (2013) Dispersal in microbes: fungi in indoor air are dominated by outdoor air and show dispersal limitation at short distances. ISME J 7:1262–1273. https://doi.org/10.1038/ismej.2013.28

Amato P, Besaury L, Joly M et al (2019) Metatranscriptomic exploration of microbial functioning in clouds. Sci Rep 9:1–12. https://doi.org/10.1038/s41598-019-41032-4

Andualem Z, Gizaw Z, Bogale L, Dagne H (2019) Indoor bacterial load and its correlation to physical indoor air quality parameters in public primary schools. Multidiscip Respir Med 14:1–7. https://doi.org/10.1186/s40248-018-0167-y

Armstrong JL, Dills RL, Yu J et al (2014) A sensitive LC-MS/MS method for measurement of organophosphorus pesticides and their oxygen analogs in air sampling matrices. J Environ Sci Health B 49:102–108. https://doi.org/10.1002/cncr.27633.Percutaneous

Ashmore MR, Dimitroulopoulou C (2009) Personal exposure of children to air pollution. Atmos Environ 43:128–141. https://doi.org/10.1016/J.ATMOSENV.2008.09.024

Badihi-Mossberg M, Buchner V, Rishpon J (2007) Electrochemical biosensors for pollutants in the environment. Electroanalysis 19(19–20):2015–2028. https://doi.org/10.1002/elan.200703946

Barberán A, Dunn RR, Reich BJ et al (2015a) The ecology of microscopic life in household dust. Proc R Soc B Biol Sci 282:20151139. https://doi.org/10.1098/rspb.2015.1139

Barberán A, Ladau J, Leff JW et al (2015b) Continental-scale distributions of dust-associated bacteria and fungi. Proc Natl Acad Sci 112:5756–5761. https://doi.org/10.1073/pnas.1420815112

Barnes J, Bender J, Lyons T, Borland A (1999) Natural and man-made selection for air pollution resistance. J Exp Bot 50:1423–1435. https://doi.org/10.1093/jxb/50.338.1423

Be NA, Thissen JB, Fofanov VY et al (2014) Metagenomic analysis of the airborne environment in urban spaces. Microb Ecol 69:346–355. https://doi.org/10.1007/s00248-014-0517-z

Bernstein JA, Alexis N, Bacchus H et al (2008) The health effects of nonindustrial indoor air pollution. J Allergy Clin Immunol 121:585–591. https://doi.org/10.1016/j.jaci.2007.10.045

Bernstein JA, Alexis N, Barnes C et al (2004) Health effects of air pollution. J Allergy Clin Immunol 114:1116–1123. https://doi.org/10.1016/j.jaci.2004.08.030

Beuadry C, Mouton J, Prozesky H, Beaudry C (2018) The next generation of scientists in Africa. African Minds 1–216. ISBN Paper 978-1-928331-93-3

Bowers RM, Clements N, Emerson JB et al (2013) Seasonal variability in bacterial and fungal diversity of the near-surface atmosphere. Environ Sci Technol 47:12097–12106. https://doi.org/10.1021/es402970s

Bowers RM, Sullivan AP, Costello EK et al (2011) Sources of bacteria in outdoor air across cities in the midwestern United States. Appl Environ Microbiol 77:6350–6356. https://doi.org/10.1128/AEM.05498-11

Bragoszewska E, Biedroń I, Kozielska B, Pastuszka JS (2018) Microbiological indoor air quality in an office building in Gliwice, Poland: analysis of the case study. Air Qual Atmos Health 11:729–740. https://doi.org/10.1007/s11869-018-0579-z

Bragoszewska E, Mainka A, Pastuszka JS et al (2018) Assessment of bacterial aerosol in a preschool, primary school and high school in Poland. Atmosphere (Basel) 9:1–15. https://doi.org/10.3390/atmos9030087

Brodie EL, DeSantis TZ, Parker JPM et al (2007) Urban aerosols harbor diverse and dynamic bacterial populations. Proc Natl Acad Sci 104:299–304. https://doi.org/10.1073/pnas.0608255104

Bünger J, Antlauf-Lammers M, Schulz TG et al (2000) Health complaints and immunological markers of exposure to bioaerosols among biowaste collectors and compost workers. Occup Environ Med 57:458–464. https://doi.org/10.1136/oem.57.7.458

Burge HA, Solomon WR, Muilenberg ML (1982) Evaluation of indoor plantings as allergen exposure sources. J Allergy Clin Immunol 70:101–108. https://doi.org/10.1016/0091-6749(82)90236-6

Cabo Verde S, Almeida SM, Matos J et al (2015) Microbiological assessment of indoor air quality at different hospital sites. Res Microbiol 166:557–563. https://doi.org/10.1016/j.resmic.2015.03.004

Cai W, Borlace S, Lengaigne M et al (2014) Increasing frequency of extreme El Niño events due to greenhouse warming. Nat Clim Chang 4:111–116. https://doi.org/10.1038/nclimate2100

Cao C, Jiang W, Wang B et al (2014) Inhalable microorganisms in Beijing's PM2.5 and PM10 pollutants during a severe smog event. Environ Sci Technol 48:1499–1507. https://doi.org/10.1021/es4048472

Chai F, Gao J, Chen Z et al (2014) Spatial and temporal variation of particulate matter and gaseous pollutants in 26 cities in China. J Environ Sci 26:75–82. https://doi.org/10.1016/S1001-0742(13)60383-6

Chase J, Fouquier J, Zare M et al (2016) Geography and location are the primary drivers of office microbiome composition. mSystems 1:1–18. https://doi.org/10.1128/mSystems.00022-16. Editor

Cohen AJ, Brauer M, Burnett R et al (2017) Estimates and 25-year trends of the global burden of disease attributable to ambient air pollution: an analysis of data from the global burden of diseases study 2015. Lancet 389:1907–1918. https://doi.org/10.1016/S0140-6736(17)30505-6

Cohen AJ, Ross Anderson H, Ostro B et al (2005) The global burden of disease due to outdoor air pollution. J Toxicol Environ Health A 68:1301–1307

Curtis L, Rea W, Smith-Willis P et al (2006) Adverse health effects of outdoor air pollutants. Environ Int 32:815–830. https://doi.org/10.1016/j.envint.2006.03.012

Cyprowski M, Sowiak M, Soroka PM et al (2008) Assessment of occupational exposure to fungal aerosols in wastewater treatment plants. Med Pr 59:365–371

Damialis A, Gioulekas D (2006) Airborne allergenic fungal spores and meteorological factors in Greece: forecasting possibilities. Grana 45:122–129. https://doi.org/10.1080/00173130600601005

Das S, Gupta-Bhattacharya S (2008) Enumerating outdoor aeromycota in suburban West Bengal, India, with reference to respiratory allergy and meteorological factors. Ann Agric Environ Med 15:105–112. https://doi.org/10.1080/0305707032000094965

Deloge-Abarkan M, Ha TL, Robine E et al (2007) Detection of airborne legionella while showering using liquid impingement and fluorescent in situ hybridization (FISH). J Environ Monit 9:91–97. https://doi.org/10.1039/b610737k

Denning DW, Pashley C, Hartl D et al (2014) Fungal allergy in asthma–state of the art and research needs. Clin Transl Allergy 4:14

Després VR, Nowoisky JF, Klose M et al (2007) Characterization of primary biogenic aerosol particles in urban, rural, and high-alpine air by DNA sequence and restriction fragment analysis of ribosomal RNA genes. Biogeosciences 4:1127–1141

Ding X, Fronczek CF, Yoon JY (2015) Biosensors for monitoring airborne pathogens. J Lab Autom 20:390–410. https://doi.org/10.1177/2211068215580935

Djikeng A, Ommeh S, Sangura S, Isaac N, Ngara M (2012) Genomics and potential downstream applications in the developing world. In: Nelson KE, Jones-Nelson B (eds) Genomics applications 335 for the developing world. NY, USA, pp 335–356. https://doi.org/10.1007/978-1-4614-2182-5_20

Dong L, Qi J, Shao C et al (2016) Concentration and size distribution of total airborne microbes in hazy and foggy weather. Sci Total Environ 541:1011–1018

Doyle ME, Mazzota AS (2000) Review of studies on the thermal resistance of salmonellae. J Food Prot 63:779–795

Du P, Du R, Ren W et al (2018) Variations of bacteria and fungi in PM 2. 5 in Beijing, China. Atmos Environ 172:55–64. https://doi.org/10.1016/j.atmosenv.2017.10.048

Dungan RS, Leytem AB (2009) Qualitative and quantitative methodologies for determination of airborne microorganisms at concentrated animal-feeding operations. World J Microbiol Biotechnol 25:1505–1518. https://doi.org/10.1007/s11274-009-0043-1

Emmerich SJ, Teichman KY, Persily AK (2017) Literature review on field study of ventilation and indoor air quality performance verification in high-performance commercial buildings in North America. Sci Technol Built Environ 23:1159–1166. https://doi.org/10.1080/23744731.2016.1274627

European Commission (2001) Ambient air pollution by polycyclic aromatic hydrocarbons (PAH). Position Paper

Fahlgren C, Bratbak G, Sandaa R-A et al (2011) Diversity of airborne bacteria in samples collected using different devices for aerosol collection. Aerobiologia (Bologna) 27:107–120. https://doi.org/10.1007/s10453-010-9181-z

Fairs A, Agbetile J, Hargadon B et al (2010) IgE sensitization to Aspergillus fumigatus is associated with reduced lung function in asthma. Am J Respir Crit Care Med 182:1362–1368

Filali Ben Sidel F, Bouziane H, del Mar Trigo M et al (2015) Airborne fungal spores of Alternaria, meteorological parameters and predicting variables. Int J Biometeorol 59:339–346. https://doi.org/10.1007/s00484-014-0845-1

Fischer G, Albrecht A, Jäckel U, Kämpfer P (2008) Analysis of airborne microorganisms, MVOC and odour in the surrounding of composting facilities and implications for future investigations. Int J Hyg Environ Health 211:132–142. https://doi.org/10.1016/j.ijheh.2007.05.003

Fracchia L, Pietronave S, Rinaldi M, Martinotti MG (2006) The assessment of airborne bacterial contamination in three composting plants revealed site-related biological hazard and seasonal variations. J Appl Microbiol 100:973–984. https://doi.org/10.1111/j.1365-2672.2006.02846.x

Gandolfi I, Bertolini V, Bestetti G et al (2015) Spatio-temporal variability of airborne bacterial communities and their correlation with particulate matter chemical composition across two urban areas. Appl Microbiol Biotechnol 99:4867–4877. https://doi.org/10.1007/s00253-014-6348-5

Getachew H, Derbie A, Mekonnen D (2018) Surfaces and air bacteriology of selected wards at a referral hospital, Northwest Ethiopia: a cross-sectional study. Int J Microbiol 2018:1–7. https://doi.org/10.1155/2018/6413179

Ghorani-Azam A, Riahi-Zanjani B, Balali-Mood M (2016) Effects of air pollution on human health and practical measures for prevention in Iran. J Res Med Sci 21(65). https://doi.org/10.4103/1735-1995.189646

Gizaw Z, Gebrehiwot M, Yenew C (2016) High bacterial load of indoor air in hospital wards: the case of University of Gondar Teaching Hospital, Northwest Ethiopia. Multidiscip Respir Med 11:1–7. https://doi.org/10.1186/s40248-016-0061-4

Gou H, Lu J, Li S et al (2016) Assessment of microbial communities in PM1 and PM10 of Urumqi during winter. Environ Pollut 214:202–210. https://doi.org/10.1016/J.ENVPOL.2016.03.073

Hall RJ, Leblanc-Maridor M, Wang J et al (2013) Metagenomic detection of viruses in aerosol samples from workers in animal slaughterhouses. PLoS One 8:e72226. https://doi.org/10.1371/journal.pone.0072226

Hanson B, Zhou Y, Bautista EJ et al (2016a) Microbiome in indoor dust and outdoor air samples: a pilot study. Environ Sci Process Impacts:713–724. https://doi.org/10.1039/c5em00639b

Hanson B, Zhou Y, Bautista EJ et al (2016b) Characterization of the bacterial and fungal microbiome in indoor dust and outdoor air samples: a pilot study. Environ Sci Process Impacts 18:713–724. https://doi.org/10.1039/c5em00639b

Herr CEW, zur Nieden A, Jankofsky M et al (2003) Effects of bioaerosol polluted outdoor air on airways of residents: a cross sectional study. Occup Environ Med 60:336–342. https://doi.org/10.1136/oem.60.5.336

Hewitt KM, Gerba CP, Maxwell SL, Kelley ST (2012) Office space bacterial abundance and diversity in three metropolitan areas. PLoS One 7:3–9. https://doi.org/10.1371/journal.pone.0037849

Ho H-M, Rao CY, Hsu H-H et al (2005) Characteristics and determinants of ambient fungal spores in Hualien, Taiwan. Atmos Environ 39:5839–5850. https://doi.org/10.1016/J.ATMOSENV.2005.06.034

Hoek G (2017) Methods for assessing long-term exposures to outdoor air pollutants. Curr Environ Health Rep 4:450–462. https://doi.org/10.1007/s40572-017-0169-5

Homa D, Haile E, Washe AP (2017) Spectrophotometric method for the determination of atmospheric Cr pollution as a factor to accelerated corrosion. J Anal Methods Chem 2017:1–9. https://doi.org/10.1155/2017/7154206

Hospodsky D, Qian J, Nazaroff WW et al (2012) Human occupancy as a source of indoor airborne bacteria. PLoS One. https://doi.org/10.1371/journal.pone.0034867

Hospodsky D, Yamamoto N, Peccia J (2010) Accuracy, precision, and method detection limits of quantitative PCR for airborne bacteria and fungi. Appl Environ Microbiol 76:7004–7012. https://doi.org/10.1128/AEM.01240-10

Hu J, Zhao F, Zhang XX et al (2018) Metagenomic profiling of ARGs in airborne particulate matters during a severe smog event. Sci Total Environ 615:1332–1340. https://doi.org/10.1016/j.scitotenv.2017.09.222

Hubad B, Lapanje A (2013) The efficient method for simultaneous monitoring of the culturable as well as nonculturable airborne microorganisms. PLoS One. https://doi.org/10.1371/journal. pone.0082186

Ijaz MK, Zargar B, Wright KE et al (2016) Generic aspects of the airborne spread of human pathogens indoors and emerging air decontamination technologies. Am J Infect Control 44:S109–S120. https://doi.org/10.1016/j.ajic.2016.06.008

Järup L (2003) Hazards of heavy metal contamination. Br Med Bull 68:167–182. https://doi.org/10.1093/bmb/ldg032

Julvez J, Sunyer J, Morales E et al (2009) Association of early-life exposure to household gas appliances and indoor nitrogen dioxide with cognition and attention behavior in preschoolers. Am J Epidemiol 169:1327–1336. https://doi.org/10.1093/aje/kwp067

Kaarakainen P, Rintala H, Meklin T et al (2011) Concentrations and diversity of microbes from four local bioaerosol emission sources in Finland. J Air Waste Manag Assoc 61(12):1382–1392. https://doi.org/10.1080/10473289.2011.628902

Kalwasińska A, Burkowska A, Wilk I (2012) Microbial air contamination in indoor environment of a University Library. Ann Agric Environ Med 19:25–29

Karwowska E (2005) Microbiological air contamination in farming environment. Pol J Environ Stud 14:445–449

Katoto PDMC, Byamungu L, Brand AS et al (2019) Ambient air pollution and health in sub-Saharan Africa: current evidence, perspectives and a call to action. Environ Res 173:174–188. https://doi.org/10.1016/j.envres.2019.03.029

Khan J, Ketzel M, Kakosimos K et al (2018) Road traffic air and noise pollution exposure assessment – a review of tools and techniques. Sci Total Environ 634:661–676. https://doi.org/10.1016/j.scitotenv.2018.03.374

Kim BJ, Seo JH, Jung YH et al (2013) Air pollution interacts with past episodes of bronchiolitis in the development of asthma. Allergy 68:517–523. https://doi.org/10.1111/all.12104

King MF, Noakes CJ, Sleigh PA (2015) Modeling environmental contamination in hospital single- and four-bed rooms. Indoor Air 25:694–707. https://doi.org/10.1111/ina.12186

Kiviranta H, Tuomainen A, Reiman M et al (1999) Exposure to airborne microorganisms and volatile organic compounds in different types of waste handling. Ann Agric Environ Med 6:39–44

Knutsen AP, Slavin RG (2011) Allergic bronchopulmonary aspergillosis in asthma and cystic fibrosis. Clin Dev Immunol 2011:843763. https://doi.org/10.1155/2011/843763

Ko G, Simmons OD 3rd, Likirdopulos CA et al (2008) Investigation of bioaerosols released from swine farms using conventional and alternative waste treatment and management technologies. Environ Sci Technol 42:8849–8857

Kumari P, Woo C, Yamamoto N, Choi H (2016) Variations in abundance, diversity and community composition of airborne fungi in swine houses across seasons. Sci Rep:1–11. https://doi.org/10.1038/srep37929

Kuske CR (2006) Current and emerging technologies for the study of bacteria in the outdoor air. Curr Opin Biotechnol 17:291–296. https://doi.org/10.1016/j.copbio.2006.04.001

Language B, Piketh SJ, Burger RP (2016) Correcting respirable photometric particulate measurements using a gravimetric sampling method. Clean Air Journal= Tydskrif vir Skoon Lug 26:10–14

Lee SH, Lee HJ, Kim SJ et al (2010) Identification of airborne bacterial and fungal community structures in an urban area by T-RFLP analysis and quantitative real-time PCR. Sci Total Environ 408:1349–1357. https://doi.org/10.1016/j.scitotenv.2009.10.061

Leung DYC (2015) Outdoor-indoor air pollution in urban environment: challenges and opportunity. Front Environ Sci 2:1–7. https://doi.org/10.3389/fenvs.2014.00069

Leung MHY, Lee PKH (2016) The roles of the outdoors and occupants in contributing to a potential pan-microbiome of the built environment: a review. Microbiome 4(21). https://doi.org/10.1186/s40168-016-0165-2

Levetin E, Shaughnessy R, Rogers CA, Scheir R (2001) Effectiveness of germicidal UV radiation for reducing fungal contamination within air-handling units. Appl Environ Microbiol 67:3712–3715. https://doi.org/10.1128/AEM.67.8.3712-3715.2001

Lignell U, Meklin T, Rintala H et al (2008) Evaluation of quantitative PCR and culture methods for detection of house dust fungi and streptomycetes in relation to moisture damage of the house. Lett Appl Microbiol 47:303–308. https://doi.org/10.1111/j.1472-765X.2008.02431.x

Lim V (2011) Occupational infections. Malays J Pathol 31:1–9. https://doi.org/10.1002/9781444329629.ch4

Liu H, Hu Z, Zhou M et al (2019a) The distribution variance of airborne microorganisms in urban and rural environments. Environ Pollut 247:898–906. https://doi.org/10.1016/J.ENVPOL.2019.01.090

Liu H, Zhang X, Zhang H et al (2018) Effect of air pollution on the total bacteria and pathogenic bacteria in different sizes of particulate matter. Environ Pollut 233:483–493. https://doi.org/10.1016/j.envpol.2017.10.070

Liu M, Liu J, Ren J et al (2019b) Bacterial community in commercial airliner cabins in China. Int J Environ Health Res:1–12. https://doi.org/10.1080/09603123.2019.1593329

Lymperopoulou DS, Adams RI, Lindow SE (2016) Contribution of vegetation to the microbial composition of nearby outdoor air. Appl Environ Microbiol 82:3822–3833. https://doi.org/10.1128/AEM.00610-16.

Mabahwi NAB, Leh OLH, Omar D (2014) Human health and wellbeing: human health effect of air pollution. Procedia - Soc Behav Sci 153:221–229. https://doi.org/10.1016/j.sbspro.2014.10.056

Malebo NJ, Shale K (2013) MALDI biotyper characterization of microorganisms colonizing heating ventilation air-conditioning systems at a South African hospital. Life Sci J 10:413–441

Mandal J, Brandl H (2011) Bioaerosols in indoor environment – a review with special reference to residential and occupational locations. Open Environ Biol Monit J 4(1):83–96

Marks GB, Ezz W, Aust N et al (2010) Respiratory health effects of exposure to low-NOxunflued gas heaters in the classroom: a double-blind, cluster-randomized, crossover study. Environ Health Perspect 118:1476–1482. https://doi.org/10.1289/ehp.1002186

Maron PA, Mougel C, Lejon DPH et al (2006) Temporal variability of airborne bacterial community structure in an urban area. Atmos Environ 40:8074–8080. https://doi.org/10.1016/j.atmosenv.2006.08.047

Matinyi S, Enoch M, Akia D et al (2018) Contamination of microbial pathogens and their antimicrobial pattern in operating theatres of peri-urban eastern Uganda: a cross-sectional study. BMC Infect Dis 18:460. https://doi.org/10.1186/s12879-018-3374-4

Miletto M, Lindow SE (2015) Relative and contextual contribution of different sources to the composition and abundance of indoor air bacteria in residences. Microbiome 3(61). https://doi.org/10.1186/s40168-015-0128-z

Montagna MT, De Giglio O, Cristina ML et al (2017) Evaluation of legionella air contamination in healthcare facilities by different sampling methods: an Italian multicenter study. Int J Environ Res Public Health. https://doi.org/10.3390/ijerph14070670

Napoli C, Marcotrigiano V, Montagna MT (2012) Air sampling procedures to evaluate microbial contamination: a comparison between active and passive methods in operating theatres. BMC Public Health 12(1). https://doi.org/10.1186/1471-2458-12-594

Nazaroff WW (2004) Indoor particle dynamics. Indoor Air 14:175–183. https://doi.org/10.1111/j.1600-0668.2004.00286.x

Newson R, Strachan D, Corden J, Millington W (2000) Fungal and other spore counts as predictors of admissions for asthma in the Trent region. Occup Environ Med 57:786–792. https://doi.org/10.1136/oem.57.11.786

Nigam VK, Shukla P (2015) Enzyme based biosensors for detection of environmental pollutants – a review. J Microbiol Biotechnol 25:1773–1781. https://doi.org/10.4014/jmb.1504.04010

Nkhebenyane J, Theron MM, Venter P, Lues JFR (2012) Antibiotic susceptibility of bacterial pathogens isolated from food preparation areas in hospice kitchens. Afr J Microbiol Res 6:2649–2653. https://doi.org/10.5897/ajmr11.1039

Oluwakemi Omolola A (2019) Indoor airborne microbial load of selected offices in a tertiary institution in South-Western Nigeria. J Health Environ Res 4:113. https://doi.org/10.11648/j.jher.20180403.15

Osman ME, Ibrahim HY, Yousef FA et al (2018) A study on microbiological contamination on air quality in hospitals in Egypt. Indoor Built Environ 27:953–968. https://doi.org/10.1177/1420 326X17698193

Ou F, McGoverin C, Swift S, Vanholsbeeck F (2017) Absolute bacterial cell enumeration using flow cytometry. J Appl Microbiol 123:464–477. https://doi.org/10.1111/jam.13508

Özden O, Dogeroglu T, Kara S (2008) Assessment of ambient air quality in Eskisehir, Turkey. Environ Int 34:678–687

Pasquarella C, Pitzurra O, Savino A (2000) The index of microbial air contamination. J Hosp Infect 46:241–256. https://doi.org/10.1053/jhin.2000.0820

Pearce DA, Bridge PD, Hughes KA et al (2009) Microorganisms in the atmosphere over Antarctica. FEMS Microbiol Ecol. https://doi.org/10.1111/j.1574-6941.2009.00706.x

Peden D, Diaz-Sanchez D, Tarlo SM et al (2004) Health effects of air pollution. J Allergy Clin Immunol 114:1116–1123. https://doi.org/10.1016/j.jaci.2004.08.030

Peter SG (2014) Comparative analysis of airborne microbial concentrations in the indoor environment of two selected clinical laboratories. IOSR J Pharm Biol Sci 8:13–19. https://doi.org/10.9790/3008-0841319

Pöschl U (2005) Atmospheric aerosols: composition, transformation, climate and health effects. Angew Chemie Int Ed 44:7520–7540. https://doi.org/10.1002/anie.200501122

Prazmo Z, Krysinska-Traczyk E, Skorska C et al (2003) Exposure to bioaerosols in a municipal sewage treatment plant. Ann Agric Environ Med 10:241–248

Prifti E, Zucker JD (2015) The new science of Metagenomics and the challenges of its use in both developed and developing countries. In: Morand S, Dujardin JP, Lefait-Robin R, et al. (eds) Socio-ecological dimensions of infectious diseases in Southeast Asia. Singapore, Springer, pp 191–216. https://doi.org/10.1007/978-981-287-527-3_12

Prigione V, Lingua G, Filipello Marchisio V (2004) Development and use of flow cytometry for detection of airborne fungi. Appl Environ Microbiol 70:1360–1365. https://doi.org/10.1128/AEM.70.3.1360-1365.2004

Prussin AJ, Marr LC (2015) Sources of airborne microorganisms in the built environment. Microbiome 3(78). https://doi.org/10.1186/s40168-015-0144-z

Rajagopalan S, Brook RD (2012) The indoor-outdoor air-pollution continuum and the burden of cardiovascular disease: an opportunity for improving global health. Glob Heart 7:207–213. https://doi.org/10.1016/j.gheart.2012.06.009.

Reanprayoon P, Yoonaiwong W (2012) Airborne concentrations of bacteria and fungi in Thailand border market. Aerobiologia (Bologna) 28:49–60. https://doi.org/10.1007/s10453-011-9210-6

Recer GM, Browne ML, Horn EG et al (2001) Ambient air levels of Aspergillus fumigatus and thermophilic actinomycetes in a residential neighborhood near a yard-waste composting facility. Aerobiologia (Bologna) 17:99–108. https://doi.org/10.1023/A:1010816114787

Rejc T, Kukec A, Bizjak M, GodičTorkar K (2019) Microbiological and chemical quality of indoor air in kindergartens in Slovenia. Int J Environ Health Res:1–14. https://doi.org/10.1080/0960 3123.2019.1572870

Richardson M, Gottel N, Gilbert JA et al (2019) Concurrent measurement of microbiome and allergens in the air of bedrooms of allergy disease patients in the Chicago area. Microbiome 7:1–10

Robinson VK, Wemedo SA (2019) Molecular characterization of indoor air microorganisms of a model primary health Care in Port Harcourt, Rivers State, Nigeria. Asian J Bitechnol Genet Eng 2:1–9

Salem AA, Soliman AA, El-Haty IA (2009) Determination of nitrogen dioxide, sulfur dioxide, ozone, and ammonia in ambient air using the passive sampling method associated with ion chromatographic and potentiometric analyses. Air Qual Atmos Health 2:133–145. https://doi.org/10.1007/s11869-009-0040-4

Sanchez-Monedero MA, Aguilar MI, Fenoll R, Roig A (2008) Effect of the aeration system on the levels of airborne microorganisms generated at wastewater treatment plants. Water Res 42:3739–3744. https://doi.org/10.1016/j.watres.2008.06.028

Saramanda G, Byragi Reddy T, Kaparapu J (2016) Microbiological indoor and outdoor air quality of selected places in Visakhapatnam City, India. Int J Curr Res 08(04):29059–29062

Schecter A, Birnbaum L, Ryan JJ, Constable JD (2006) Dioxins: an overview. Environ Res 101:419–428. https://doi.org/10.1016/j.envres.2005.12.003

Schlatter DC, Schillinger WF, Bary AI et al (2018) Dust-associated microbiomes from dryland wheat fields differ with tillage practice and biosolids application. Atmos Environ 185:29–40. https://doi.org/10.1016/j.atmosenv.2018.04.030

Schmidt MG, Attaway HH, Terzieva S et al (2012) Characterization and control of the microbial community affiliated with copper or aluminum heat exchangers of HVAC systems. Curr Microbiol 65:141–149. https://doi.org/10.1007/s00284-012-0137-0

Schwela D (2012) Review of urban air quality in Sub-Saharan Africa region – air quality profile of SSA countries (English). World Bank, Washington, DC. http://documents.worldbank.org/curated/en/936031468000276054/Review-of-urban-air-quality-in-Sub-Saharan-Africa-region-air-quality-profile-of-SSA-countries

She Q, Peng X, Xu Q et al (2017) Air quality and its response to satellite-derived urban form in the Yangtze River Delta, China. Ecol Indic 75:297–306. https://doi.org/10.1016/J.ECOLIND.2016.12.045

Shin SK, Kim J, Ha SM et al (2015) Metagenomic insights into the bioaerosols in the indoor and outdoor environments of childcare facilities. PLoS One 10:1–17. https://doi.org/10.1371/journal.pone.0126960

Simwela A, Xu B, Mekondjo SS, Morie S (2018) Air quality concerns in Africa: a literature review. Int J Sci Res Publ. https://doi.org/10.29322/ijsrp.8.5.2018.p7776

Śmiełowska M, Marć M, Zabiegała B (2017) Indoor air quality in public utility environments – a review. Environ Sci Pollut Res 24:11166–11176. https://doi.org/10.1007/s11356-017-8567-7

Sommer U, Moustaka-gouni M (2017) Variability of airborne bacteria in an urban Mediterranean area. Atmos Environ 157:101–110. https://doi.org/10.1016/j.atmosenv.2017.03.018

Srivastava A, Mazumdar D (2011) Monitoring and reporting VOCs in ambient air. INTECH Open Access Publisher. https://doi.org/10.5772/16774

Stetzenbach LD, Buttner MP, Cruz P (2004) Detection and enumeration of airborne biocontaminants. Curr Opin Biotechnol 15:170–174. https://doi.org/10.1016/j.copbio.2004.04.009

Stryjakowska-Sekulska M, Piotraszewska-Pajak A, Szyszka A et al (2007) Microbiological qualtiy of indoor air in university rooms. Polish J Environ Stud 16:623–632

Sudharsanam S, Srikanth P, Sheela M, Steinberg R (2008) Study of the indoor air quality in hospitals in South Chennai, India – microbial profile. Indoor Built Environ 17:435–441. https://doi.org/10.1177/1420326X08095568

Taha MPM, Drew GH, Longhurst PJ et al (2006) Bioaerosol releases from compost facilities: evaluating passive and active source terms at a green waste facility for improved risk assessments. Atmos Environ 40:1159–1169. https://doi.org/10.1016/j.atmosenv.2005.11.010

Tang JW, Li Y, Eames I et al (2006) Factors involved in the aerosol transmission of infection and control of ventilation in healthcare premises. J Hosp Infect 64:100–114. https://doi.org/10.1016/j.jhin.2006.05.022

Tarwater PM, Green CF, Gibbs SG (2010) Air biocontamination in a variety of agricultural industry environments in Egypt: a pilot study. Aerobiologia:223–232. https://doi.org/10.1007/s10453-010-9158-y

Täubel M, Sulyok M, Vishwanath V et al (2011) Co-occurrence of toxic bacterial and fungal secondary metabolites in moisture-damaged indoor environments. Indoor Air 21:368–375. https://doi.org/10.1111/j.1600-0668.2011.00721.x

Terzieva S, Donnelly J, Ulevicius V et al (1996) Comparison of methods for detection and enumeration of airborne microorganisms collected by liquid impingement. Appl Environ Microbiol 62:2264–2272

Tham KW (2016) Indoor air quality and its effects on humans – a review of challenges and developments in the last 30 years. Energ Buildings 130:637–650. https://doi.org/10.1016/j.enbuild.2016.08.071

Thompson EM (2008) High-performance liquid chromatography/mass spectrometry (LC/MS). AMC technical briefs v3 R Soc Chem ISSN 1757–5958

Tong X, Leung MHY, Wilkins D, Lee PKH (2017) City-scale distribution and dispersal routes of mycobiome in residences. Microbiome 5:1–13. https://doi.org/10.1186/s40168-017-0346-7

Tringe SG, Zhang T, Liu X et al (2008) The airbone metagenome in an indoor urban environment. PLoS One 3:e1862. https://doi.org/10.1371/journal.pone.0001862

Tsai W (2016) Toxic volatile organic compounds (VOCs) in the atmospheric environment: regulatory aspects and monitoring in Japan and Korea. Environments. https://doi.org/10.3390/environments3030023

UNECA (2017) Economic report on Africa 2017: an overview of urbanization and structural transformation in Africa. Addis Ababa

Vainio H (2015) Epidemics of asbestos-related diseases – something old, something new. Scand J Work Environ Health 41:1–4. https://doi.org/10.5271/sjweh.3471

Valeriani F, Cianfanelli C, Gianfranceschi G et al (2017) Monitoring biodiversity in libraries: a pilot study and perspectives for indoor air quality. J Prev Med Hyg 58:E238–E251

Vincken W, Roels P (1984) Hypersensitivity pneumonitis due to Aspergillus fumigatus in compost. Thorax 39:74

Waters CN, Zalasiewicz J, Summerhayes C et al (2016) The Anthropocene is functionally and stratigraphically distinct from the Holocene. Science. 351:aad2622. https://doi.org/10.1126/science.aad2622

Wei M, Xu C, Xu X et al (2019) Characteristics of atmospheric bacterial and fungal communities in PM 2.5 following biomass burning disturbance in a rural area of North China Plain. Sci Total Environ 651:2727–2739. https://doi.org/10.1016/j.scitotenv.2018.09.399

Wei X, Lyu S, Yu Y et al (2017) Phylloremediation of air pollutants: exploiting the potential of plant leaves and leaf-associated microbes. Front Plant Sci 8:1–23. https://doi.org/10.3389/fpls.2017.01318

WHO (2018) Air pollution and child health. WHO, Geneva (Switzerland)

WHO (2019) Factsheet: Ambient (outdoor) air quality and health.

Whon TW, Kim M-S, Roh SW et al (2012) Metagenomic characterization of airborne viral DNA diversity in the near-surface atmosphere. J Virol 86:8221–8231. https://doi.org/10.1128/JVI.00293-12

Wichmann J, Lind T, Nilsson MA-M, Bellander T (2010) PM2.5, soot and NO2 indoor–outdoor relationships at homes, pre-schools and schools in Stockholm, Sweden. Atmos Environ 44:4536–4544. https://doi.org/10.1016/J.ATMOSENV.2010.08.023

Wu Y, Chen A, Luhung I et al (2016) Bioaerosol deposition on an air-conditioning cooling coil. Atmos Environ 144:257–265. https://doi.org/10.1016/j.atmosenv.2016.09.004

Yamamoto N, Hospodsky D, Dannemiller KC et al (2015) Indoor emissions as a primary source of airborne allergenic fungal particles in classrooms. Environ Sci Technol 49:5098–5106. https://doi.org/10.1021/es506165z

Yassin MF, Almouqatea S (2010) Assessment of airborne bacteria and fungi in an indoor and outdoor environment. Int J Environ Sci Technol 7:535–544. https://doi.org/10.1007/BF03326162

Yooseph S, Andrews-Pfannkoch C, Tenney A et al (2013) A metagenomic framework for the study of airborne microbial communities. PLoS One. https://doi.org/10.1371/journal.pone.0081862

Yousefi V, Rama DBK (1992) Monitoring of air for microbial and metal contamination at selected sites in the vicinity. Sci Total Environ 116:159–167

Zeng QY, Westermark SO, Rasmuson-Lestander Å, Wang XR (2006) Detection and quantification of Cladosporium in aerosols by real-time PCR. J Environ Monit 8:153–160. https://doi.org/10.1039/b509515h

Zhai Y, Li X, Wang T et al (2018) A review on airborne microorganisms in particulate matters: composition, characteristics and influence factors. Environ Int 113:74–90

Zhang Q, Zhu Y (2012) Characterizing ultrafine particles and other air pollutants at five schools in South Texas. Indoor Air 22:33–42. https://doi.org/10.1111/j.1600-0668.2011.00738.x

Zhen Q, Deng Y, Wang Y et al (2017) Meteorological factors had more impact on airborne bacterial communities than air pollutants. Sci Total Environ 601–602:703–712. https://doi.org/10.1016/j.scitotenv.2017.05.049

Zhou Y, Cheng S, Chen D et al (2015) Temporal and spatial characteristics of ambient air quality in Beijing, China. Aerosol Air Qual Res 1868–1880. https://doi.org/10.4209/aaqr.2014.11.0306

Zhu H, Phelan PE, Duan T et al (2003) Experimental study of indoor and outdoor airborne bacterial concentrations in Tempe, Arizona, USA. Aerobiologia 19:201–211

Chapter 6
Bacterial Contamination on Household Latrine Surfaces: A Case Study in Rural and Peri-Urban Communities in South Africa

Natasha Potgieter, Ugonna Aja-Okorie, Rendani L. Mbedzi, and Afsatou N. Traore-Hoffman

6.1 Introduction

Globally, 4.5 million people use a sanitation system which does not provide adequate protection (WHO 2019). The Millennium Development Goals were replaced at the end of 2015 by 17 Sustainable Development Goals (SDGs) with Goal 6 highlighting aspects on sanitation and setting an agenda for the water and sanitation (WASH) professionals to work towards 2030 (United Nations 2015a, b).

Safe sanitation is a fundamental human right which ensures dignity, prevents infection and improves social well-being (WHO, 2019). The World Health Organisation (WHO 2019) defines safe sanitation as access to and use of facilities and services for the safe disposal of human urine and faeces. The lack of safe sanitation contributes to diarrhoea (a leading cause of disease and death among children under 5 years in low- and middle- income countries) (UNWater 2008a, b; Richard et al. 2013; Danaei et al. 2016; UNICEF, WHO, World Bank 2018), neglected tropical diseases (Ziegelbauer et al. 2012) and environmental enteric dysfunction (Humphrey 2009; Keusch et al. 2013; Crane et al. 2015).

Rural areas in many low- and middle-income countries (LMICs) do not have safe sanitation while rapid urbanization is putting a strain on cities that are struggling to cope with the scale of sanitation needs. Globally, the cost and maintenance of sanitation systems are challenging and ensuring public health requires the continued adaption of systems (United Nations 2015a, b). The safe disposal of human excreta implies that not only do people have to excrete hygienically but that their excreta must be appropriately contained and treated to avoid affecting their (or anybody else's) health (Mara et al. 2010). Improved hygiene practices at households can help to decrease the

N. Potgieter (✉) · U. Aja-Okorie · R. L. Mbedzi · A. N. Traore-Hoffman
Department of Microbiology, School of Mathematical and Natural Sciences, University of Venda, Thohoyandou, Limpopo Province, South Africa
e-mail: natasha.potgieter@univen.ac.za

© Springer Nature Switzerland AG 2020
A. L. K. Abia, G. R. Lanza (eds.), *Current Microbiological Research in Africa*,
https://doi.org/10.1007/978-3-030-35296-7_6

risks linked with infective organisms such as pathogenic *Escherichia coli* that are spread via the faecal-oral transmission route other than via contaminated water (Prüss-Üstün et al. 2016). Pathogenic *E. coli* strains have been identified as the leading human diarrhoeagenic pathogens worldwide, causing infection, particularly in young children in developing countries (Porat et al. 1998).

Generally, the toilet in the household is in constant use throughout the day and provides an ideal environment for the spread of skin, gut and respiratory organisms through hands and surfaces between family members (International Scientific Forum on Home Hygiene 2002). Very few studies are seen on the role of toilet seats in rural and peri-urban households (with inadequate or no access to safe sanitation) in the transmission of diseases within these households and the community. In Cambodia, Sinclar and Gerba (2010) found that a flush toilet is a barrier to microbial contamination in the transmission pathway in rural households. Jeon et al. (2013) found that differences in the survivability of bacteria depend on the moisture or temperature of the surfaces and the frequency of transmission. They also showed that the phylum and genus compositions found on different surfaces within households varied and were unique. This study further showed that human skin microbiome could be a significant source of bacteria transmission by touch or exposure within a household (Jeon et al. 2013). In a recent study in Kathmandu, Nepal, by McGinnes et al. (2019), pay-per-use community toilets were assessed and compared to pit latrines and cistern flush toilets in private households. This study showed no difference in the contamination levels between privately owned toilets and community toilets.

South Africa presently has less than 70% overall sanitation access rate, which does not truly reflect on the situation in rural and peri-urban communities (Prüss-Üstün et al. 2016). Despite effective policies and programmes, the peri-rural and rural areas of South Africa are still facing health-related problems associated with poor hygiene and sanitation. Insufficient water resources in South Africa add to the problem of providing all people with waterborne sewerage. Many of the households in rural and peri-urban communities are using either VIP toilets, self-constructed pit latrines or practice open defaecation due to the absence of sanitation facilities. Although the provision of decent toilets is essential, hygiene aspects must also be addressed to decrease health issues in these communities, especially because of a high number of HIV-positive individuals residing in these communities. There is, at present, very little data available on the impact inadequate sanitation facilities have on the spread of diseases in rural households of South Africa. This chapter reports a case study that used *E. coli* to determine the potential health risks associated with poor sanitation in rural and peri-urban households. Although specific to selected communities in South Africa, the approach and results should apply to other communities in other parts of Africa and other tropical countries in general.

6.2 Materials and Methods

6.2.1 Study Sites and Sample Collection

The study was carried out in peri-urban communities in the Gauteng and Mpumalanga Provinces and in rural communities in the Limpopo Province of South Africa. A consent form was drafted and distributed to the households. The caregiver in each household signed the form after the study activities and purpose were explained to them.

In the rural communities, a total of 130 latrines were assessed from two villages. In the peri-urban areas, a total of 72 households were selected to participate in this study. The seats of each latrine were swabbed separately using sterile swabs which were separately placed into sterile 100 mL phosphate-buffered saline (PBS) solution (pH 7.4) in 4 °C cooler boxes and transported to the laboratory for further analysis.

6.2.2 Microbiological Assessment of Samples

Using *E. coli* as a model pathogen, the Colilert® Quanti-tray/2000 system was used for the enumeration of viable *E. coli* cells from samples. The PBS/cotton samples were vortexed for 1 min each and treated according to the Quanti-tray method described by the manufacturer (IDDEX). All sample were examined under a long wave (360 nm) ultraviolet light, and all fluorescent wells were counted as *E. coli* positive (Omar et al. 2010).

6.2.3 DNA Extraction

A total of 2 mL of the media was removed from up to ten positive *E. coli* wells of the Colilert-Trays/2000 with sterile 1-mL Neomedic disposable syringes with a mounted needle (Kendon Medical Suppliers, South Africa) and aliquoted into 2 mL sterile Eppendorf tubes. The tubes were centrifuged for 5 min at $13,000 \times g$ to pellet the cells and the supernatant discarded. DNA was extracted from the collected bacterial cells using an adapted version of the guanidium thiocyanate/silica method reported by Boom et al. (1990) using homemade spin columns prepared as reported by Borodina et al. (2003). The changes to the DNA extraction method included the addition of 250 µL of 100% ethanol to the lysis buffer to enhance the binding of DNA to the celite. The celite containing the bound DNA was loaded into spin columns before doing the washing steps. DNA was eluted from the celite with 100 µL Qiagen elution buffer (Southern Cross Biotechnology, South Africa). The extracted DNA was used as a template in all PCR assays (Omar et al. 2010).

6.2.4   Molecular Assessment of Samples

The procedure and primers used were published by Omar et al. (2010). Briefly, all
m-PCR assays were performed in a Biorad Mycycler™ Thermal cycler in a total
volume of 20 µL. A multiplex PCR kit (Qiagen) was used for the m-PCR protocol.
Each reaction consisted of 1X Qiagen® PCR multiplex mix (containing HotstartTaq®
DNA polymerase, multiplex PCR buffer and dNTP mix); 2 µL of the primer mixture (0.1 µM of *mdh* and *lt* primers (Forward (F) and reverse (R), 0.2 µM of *ial* and
eagg primers (F and R), 0.3 µM of *eaeA* and *stx2* primers (F and R), 0,5 µm of *stx1*
and *st* primers (F and R), 4 µL of sample DNA and 4 µL PCR grade water. The reaction was subjected to an initial activation step at 95 °C for 15 min, followed by
35 cycles consisting of denaturing at 94 °C for 45 s, annealing at 55 °C for 45 s,
extension at 68 °C for 2 min and final elongation at 72 °C for 5 min (Omar et al.
2010). DNA was analysed using a 2.5% (w/v) agarose gel in TAE buffer (40 mmol⁻¹
Tris-acetate; 2 mmol⁻¹ EDTA, pH 8.3) with 0.5 µg/mL ethidium bromide.
Electrophoresis was conducted for 1–2 h in an electric field strength of 8 V/cm gel
and the DNA visualized with UV light (Gene Genius Bio Imaging system, Vacutec®).
DNA fragments were estimated by electrophoresis using either a 1 kB or 100 bp
molecular marker (Fermentas®) (Fig. 6.1).

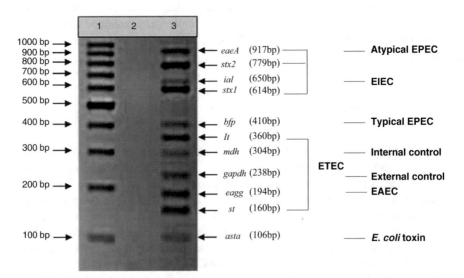

Fig. 6.1 Agarose gel showing the mPCR detecting the housekeeping gene (internal control) and
virulent genes of *Escherichia coli*. Lane 1 indicates the 100 bp Fermentas O'Generuler DNA ladder run and lane 3 indicates the mPCR gene distribution (Omar et al. 2010)

6.3   Results and Discussion

In the rural study area, only 130 (24%) toilet seat samples were obtained from a total of 540 households in two villages. Village 1 had 480 households with 111 (23%) toilets. Village 2 had 60 households and only 19 (32%) toilets. In the peri-urban areas, a total of 72 households were selected to participate in this study, of which only 59 (82%) households had a latrine because 23 (39.0%) households had to share toilets and 12 (20.4%) households did not have a latrine.

The South African water quality guideline for total coliforms (TC) is 10 TC CFU/100 mL (DWAF 1998), and this limit was used as a guideline in this study. In both the rural and peri-urban latrines tested, the minimum TC most probable number (MPN)/100 mL ranged from <1 MPN/100 mL to a maximum count of 2419.6 MPN/100 mL. In the peri-urban study area, a total of 57 (96.6%) toilet seats out of 59 contained TC, and the average TC count was 985.6 MPN/100 mL. The results also showed that 2 (3.4%) samples had <1 MPN/100 mL TC, 4 (6.8%) samples had between <1 and 19 MPN/100 mL TC, while 53 (92.9%) of toilet seat samples had TC counts higher than 10 CFU/100 mL. In the rural study area, a total of 126 (95%) toilet seats out of 130 contained TC. From these samples, only 4 (3.1%) samples had <1 MPN/100 mL TC, while 15 (11.5%) of toilet seat samples had TC counts higher than 2420 MPN/100 mL.

The South African water quality guideline for *Escherichia coli* (EC) is <1 EC MPN/100 mL (DWAF 1998) and used as a guideline in this study to implicate health risk. In both the rural and peri-urban latrines tested, the minimum EC count/100 mL ranged from <1 MPN/100 mL to a maximum count of 2419.6 MPN/100 mL. In the peri-urban study area, a total of 45 (76.3%) toilet seats out of 59 tested positive for EC counts. In the rural study area, a total of 90 (68%) toilet seats out of 130 tested positive for EC counts. Of these samples, 14 (16%) had EC counts higher than 300 MPN/100 mL.

The majority of TC and EC counts in this study were higher than the South African guideline counts, which confirmed a potential transmission risk to the people in the study. It has also been shown that the total coliform group includes bacteria of faecal origin and indicates the possible presence of bacterial pathogens such as *Salmonella* spp., *Shigella* spp., *Vibrio cholerae*, *Campylobacter jejuni*, *C. coli, Yersinia enterocolitica* and pathogenic *E. coli*, especially when detected in conjunction with other faecal coliforms (DWAF 1998). Therefore, one can make a case that these organisms can cause diseases such as gastroenteritis, salmonellosis, dysentery, cholera and typhoid fever (DWAF 1998). The CDC (2008) has stated that the spread of these bacteria cannot be stopped unless some form of disinfection process is implemented. Furthermore, the results from this study showed that the presence of *E. coli* on toilet seats indicated the risk that these households were exposed to diarrhoeal diseases (DWAF 1998).

The results of this study corroborate the findings of Potgieter et al. (2011) in a study done in Zimbabwe which showed that 88% of the urban toilets, 83% of the peri-urban toilets and 73% of the rural toilet seat swabs were positive for total coliforms and *E. coli*. The study of Potgieter et al. (2011) also concluded that unsanitary facilities and lack of adequate hygiene practices could be the reason for these results, which might also be the case in the current study population (but was not part of the objectives to investigate).

Molecular assessment of the EC-positive samples indicated that different diarrhoeagenic EC strains were present on the seats and could be responsible for the transmission of diarrhoeal diseases in the households and communities due to unhygienic practices (Table 6.1).

In the peri-urban study area, all the 45 EC-positive samples were confirmed to be *Escherichia coli* using the Commensal EC gene, while 26 (57.8%) of the samples tested positive for EHEC, 39 (86.7%) for ETEC, 10 (22.2%) for EIEC, 31 (68.9%) for EAEC and 44 (97.7%) for EPEC. In the rural study area, all 90 EC-positive samples were confirmed to be *Escherichia coli* using the Commensal EC genes, while only 45 samples had virulence genes. The pathotypes were distributed as follows: 1 (2%) EHEC, 14 (14%) ETEC, 5 (6%), 14 (16%) and 11 (12%) EPEC.

EPEC strains cause either bloody or watery diarrhoea and are linked to infant diarrhoea (Matar et al. 2002). A study conducted by Potgieter et al. (2011) in Zimbabwe found *Escherichia coli* strains from toilet seat samples where atypical EPEC was the most prevalent pathogenic *E. coli* identified in urban areas at 25%, followed by typical EPEC at 24%. EAEC was also fairly prevalent at 18%, with ETEC following at 10%. Almost 83% of *E. coli* strains identified in urban areas were pathogenic, compared to 40% in peri-urban areas and 69% in rural areas. The study concluded that the higher presence of pathogenic *E. coli* strains in urban areas could contribute to higher diarrhoea prevalence (Potgieter et al. 2011). The prevalence of EPEC indicates the risk of infantile diarrhoea among children, particularly in developing countries (Levine 1987; Donnenberg and Kaper 1992). ETEC causes diarrhoea in infants and travellers, especially in regions with poor sanitation (Huerta et al. 2000). EHEC is responsible for the development of haemorrhagic colitis (Riley et al. 1983) and haemolytic uraemic syndrome (HUS) in humans (Karmali 1989). At least 10% of individuals who contract enterohaemorrhagic diarrhoea develop haemolytic uraemic syndrome, which leads to kidney failure and death (Lightfoot 2003). EIEC has been recognized as a causative agent of diarrhoea in human since 1971 (Dupont et al. 1971). EIEC has been named thus because of the organism's ability to invade the epithelium of host colonic mucosa

Table 6.1 Pathogenic *E. coli* strains isolated from the toilet seats in rural and peri-urban households

	Entero-haemorrhagic *E. coli* (EHEC)	Entero-toxigenic *E. coli* (ETEC)	Entero-invasive *E. coli* (EIEC)	Entero-aggregative *E. coli* (EAEC)	Entero-pathogenic *E. coli* (EPEC)
Peri-urban toilet seat swab samples	26 (57.8%)	39 (86.7%)	10 (22.2%)	31 (68.9%)	44 (97.7%)
Rural toilet seat swab samples	1 (2%)	14 (14%)	5 (6%)	14 (16%)	11 (12%)

during pathogenesis (Dupont et al. 1971). EAEC causes persistent diarrhoea in developing countries (Nataro and Kaper 1998). In South Africa, ETEC and EPEC are the causative agents in 8–42% of diarrhoea incidences (Arvida et al. 2004). Therefore, the prevalence of these strains on toilet seats indicates a potential health risk.

While the current study detected the presence of diarrhoea-causing pathotypes of *E. coli* on toilet seats in the studied area, it should be noted that the guidelines used to infer the risk were those for drinking water quality in South Africa. Thus, poor sanitation habits would therefore lead to the transfer of the pathogens to hands and subsequently to the mouth, either directly or through the contamination of water and food.

6.4   Conclusions

The case studies in rural and peri-urban households in South Africa were used to indicate the prevalence of pathogenic *E. coli* strains on toilet seats which could potentially be harmful to older adults, young children and other vulnerable individuals in South Africa where there is a high incidence of HIV (UNAIDS 2018). In order to do this toilet seat samples were collected using sterile swabs, and total coliforms and *E. coli* were isolated. Different combinations of *E. coli* strains were also detected, and this indicated a high potential health risk to the members of these study households. The prevalence of pathogenic *E. coli* found in samples collected from toilet seats indicates an important transmission route due to poor hygiene practices. Most people in these rural and peri-urban communities seem to lack information on hygiene practices, and it poses a health risk to young children, older adults and immune-compromised individuals such as HIV/AIDS patients. It is therefore important to equipped communities with the knowledge, combined with proper training on how to build their household toilet. Additional personal hygiene education (washing hands with soap after visiting the toilet or before preparing food) and household hygiene education, which includes keeping the toilet clean, safe disposal of refuse and solid wastes, must be part of the educational messages from the Department of Health.

References

Arvida S, du Plessis M, Keddy K (2004) Review article: Enterovirulent E. coli. South Afr J Epidemiol Infect 19:23–33

Boom R, Sol CJA, Salimans MMM et al (1990) Rapid and simple method for purification of nucleic acids. J Clin Microbiol 28:495–503

Borodina TA, Lehrach H, Soldatov AV (2003) DNA purification on homemade silica spin-columns. Anal Biochem 321:135–137

CDC (2008) Guideline for Disinfection and Sterilization in Healthcare Facilities, 2008. http://www.cdc.gov/hicpac/pdf/guidelines/Disinfection_Nov_2008.pdf (Accessed June 2010)

Crane RJ, Jones KD, Berkley JA (2015) Environmental enteric dysfunction: an overview. Food Nutr Bull 36(Suppl 1):S76–S87

Danaei G, Andrews KG, Sudfeld CR et al (2016) Risk factors for childhood stunting in 137 developing countries: a comparative risk assessment analysis at global, regional, and country levels. PLoS Med 13(11):e1002164

Donnenberg MS, Kaper JB (1992) Enteropathogenic Escherichia coli. Infect Immun 60:3953–3961

DuPont HL, Formal SB, Hornick RB et al (1971) Pathogenesis of Escherichia coli in diarrhea. New Engl J Med 285:1–9

DWAF (1998) Quality of domestic water supplies: Assessment Guide second edition. Water Research Commission: Pretoria, South Africa ISBN 186845 4169. P1–104

Huerta M, Grotto I, Gdalevich M et al (2000) A waterborne outbreak of gastroenteritis in the Golan Heights due to Enterotoxigenic Escherichia coli. Infect 28:267–271

Humphrey JH (2009) Child undernutrition, tropical enteropathy, toilets, and handwashing. Lancet 374:1032–1035

International Scientific Forum on Home Hygiene (2002) The infection potential in the domestic setting and the role of hygiene practice in reducing Infection. http://www.ifhhomehygiene.org/Intergra-tedC

Jeon YS, Chun J, Kim BS (2013) Identification of household bacterial community and analysis of species shared with human microbiome. Curr Microbiol 67:557–563

Karmali MA (1989) Infection by verocytotoxin-producing Escherichia coli. Clin Microbiol Rev 2:15–38

Keusch GT, Rosenberg IH, Denno DM et al (2013) Implications of acquired environmental enteric dysfunction for growth and stunting in infants and children living in low- and middle-income countries. Food Nutr Bull 34(3):357–364

Levine MM (1987) Escherichia coli that cause diarrhea: enterotoxigenic, enteropathogenic, enteroinvasive, enterohemorrhagic, and enteroadherent. J Infect Dis 155:377–389

Lightfoot NF (2003) Bacteria of potential health concern. In: Batram J, Cotruvo J, Exner M et al (eds) Heterotrophic plate counts and drinking-water safety. IWA Publishing, London. Chapter 5, p P69

Mara D, Lane J, Scott B, Trouba D (2010) Sanitation and health. PLoS Med 7(11):1–7

Matar GM, Abdo D, Khneisser I et al (2002) The multiplex PCR based detection and genotyping of diarrhoegenic E. coli in diarrhoeal stools. Ann Trop Med Parasitol 96:317–324

McGinnes S, Marini D, Amatya P, Murphy HM (2019) Bacterial contamination on latrine surfaces in community and household latrines in Kathmandu, Nepal. Int J Environ Res Public Health 16:257

Nataro JP, Kaper JB (1998) Diarrhoeagenic Escherichia coli. Clin Microbiol Rev 11:142–201

Omar KB, Potgieter N, Barnard TG (2010) Rapid screening method for detection of pathogenic E. coli. Water Sci Technol Water Supply 10(1):7–12

Porat N, Levy A, Fraser D et al (1998) Prevalence of intestinal infections caused by diarrheagenic Escherichia coli in Bedouin infants and young children in southern Israel. Pediatr Infect Dis J 17:482–488

Potgieter N, Mpofu TB, Barnard TG (2011) The impact water, sanitation and hygiene infrastructures have on people living with HIV and AIDS in Zimbabwe. In: Zajac V (ed) Microbes, viruses and parasites in AIDS process. ISBN: 978-953-307-601-0. InTech. http://www.intechopen.com/articles/show/title/the-impact-water-sanitation-and-hygiene-infrastructure-have-on-people-living-with-hiv-and-aids-in-z (Accessed February 2012)

Prüss-Üstün A, Bros R, Gore F, Bartram J (2016) Safe water, better health: costs, benefits and sustainability of interventions to protect and promote health. World Health Organisation, Geneva. http//www.whqlibdoc.who.int/publications/2008/9789241596435_eng.pdf (Accessed July 2018)

Richard SA, Black RE, Gilman RH, Childhood Malnutrition and Infection Network et al (2013) Diarrhea in early childhood: short-term association with weight and long-term association with length. Am J Epidemiol 178(7):1129–1138

Riley LW, Remis RS, Helgerson SD et al (1983) Hemorrhagic colitis associated with a rare Escherichia coli serotype. New Engl J Med 308:681–685

Sinclar RG, Gerba CP (2010) Microbial contamination in kitchens and bathrooms of rural Cambodian village households. Lett Appl Microbiol 52:144–149

UNAIDS (2018) AIDS info (Accessed October 2018)

UNICEF, WHO, World Bank (2018) Joint child malnutrition estimates – levels and trends (2018 edition). Global Database on Child Growth and Malnutrition

United Nations (2015a) General Assembly Resolution 70/169: The human rights to safe drinking water and sanitation. United Nations, New York, USA

United Nations (2015b) General Assembly Resolution 70/1. Transforming our world: the 2030 Agenda for Sustainable Development. United Nations, New York, USA

UNWater (2008a) FactSheet 2: Sanitation is vital for good health. http://www.sanitationdrive2015.org/factsheets/sanitatin_is_vital_good_health.pdf (Accessed January 2012)

UNWater (2008b) Gender, water and sanitation: a policy brief. http//www.unwater.org/downloads/unwpolbrief230606.pdf (Accessed January 2012)

WHO (2019) WHO Global Water, Sanitation and Hygiene Annual Report 2018. Geneva: World Health Organization; 2019 (WHO/CED/PHE/WSH/19.147). Licence: CC BY-NC-SA 3.0 IGO

Ziegelbauer K, Speich B, Mäusezahl D et al (2012) Effect of sanitation on soil-transmitted helminth infection: systematic review and meta-analysis. PLoS Med 9(1):e1001162

Chapter 7
The Impact and Control of Emerging and Re-Emerging Viral Diseases in the Environment: An African Perspective

Juliet Adamma Shenge and Adewale Victor Opayele

7.1 Introduction

The emergence of new infectious diseases is characterized by a sudden invasion or spread of diseases typically known to be rare or uncommon (Morens et al. 2008). Emerging viruses are newly recognized or newly evolved viruses or those that occurred previously in a population but show increased incidence or expansion into new geographical, host or vector range (WHO 2004). Re-emerging viruses, on the other hand, include those viruses that had previously decreased in incidence but are currently experiencing an upsurge (Morse 1995). These agents present significant challenges to researchers regarding their epidemiology, vaccine design and eradication (Ahmed et al. 2014).

Approximately 80% of viruses that infect humans are zoonotic (Taylor et al. 2001). Most of these viruses perpetuate naturally in animal reservoirs that include numerous mammalian, avian and invertebrate vectors or intermediate hosts (Woolhouse and Gowtage-Sequeria 2005; Kilpatrick and Randolph 2012). Farm and wild animals are the primary sources of novel zoonotic viruses causing disease in human. This gives rise to about 60% of known human pathogens and up to 75% of emerging human infections (Cleaveland et al. 2001). Outbreaks of several viral haemorrhagic diseases have been reported in Africa, majorly caused by viruses in the families *Filoviridae, Arenaviridae, Flaviviridae* and *Bunyaviridae*. Animal reservoirs have been involved in the emergence of these viruses in humans. Hence, the transmission of these infectious agents from animals to humans, or vice versa, has resulted in the sustained periodic dissemination in the human population.

J. A. Shenge (✉) · A. V. Opayele
Department of Virology, College of Medicine, University of Ibadan, Ibadan, Nigeria

© Springer Nature Switzerland AG 2020
A. L. K. Abia, G. R. Lanza (eds.), *Current Microbiological Research in Africa*,
https://doi.org/10.1007/978-3-030-35296-7_7

7.2   Environmental Sources and Patterns of Emergence of New Viral Agents

The emergence patterns of new viral agents and re-emergence of already known ones have been continuously changing, mostly due to several factors that could be climatic, agricultural, anthropological or a combination of two or more of these factors (Andrew and Gregory 2008). Viruses cannot replicate outside the living host but can survive in the environment for a long period. For instance, enteroviruses are very stable in harsh environments. Several waterborne viruses have been isolated from contaminated sources such as sewage. Such enteric viruses include norovirus, hepatitis A and E viruses, human enterovirus A-D, and human adenovirus A-G. These viruses have been implicated in gastrointestinal problems and responsible for infections such as hepatitis, conjunctivitis, poliomyelitis and several others (La Rosa et al. 2012). Other known sources of emerging viral diseases are wild and domestic animals responsible for most zoonotic viral infections (Woolhouse et al. 2005). Again, human involvement in numerous activities to harness natural resources, coupled with other essential factors such as increased residencies, close contacts with animal reservoirs in the wild, poor land use leading to degradation and alteration of natural habitations, have been implicated in the emergence of new infectious diseases. A host of disease-causing microorganisms including viruses in the environment in Africa are not often detected, probably due to inadequate evaluation of the environment. Residential areas sited close to dumpsites and stagnant water sources face hazards of air pollution which poses serious threats to human health. This is commonplace in most urban cities in the continent. Humans interact with such environments through activities such as scavenging at dumpsites, fishing and washing in contaminated water making these sites primary sources of infections, which could subsequently spread through the community (van Doorn 2014). Exchanges through trade, coupled with human and animal travels, have also contributed to the spread of virus-carrying vectors from one region to another (Tatem et al. 2006; Hufnagel et al. 2004). Human-animal-environment connections have resulted in cyclical events that promote infectious disease epidemiology. Zoonotic infectious diseases are most implicated in this pattern, depicted by disease transmission from animals to humans or vice versa and result in disease outbreaks among the human population. It is believed that new viral agents emerge when known viruses from animals switch hosts and become established in humans. This occurrence can lead to severe epidemics when the human-human transmission is achieved, especially with highly pathogenic viruses (Olsen et al. 2006). The viral agents can be airborne and so transmitted through droplets, as in the case of influenza viruses. They can also be transmitted through direct animal bites and scratches such as the rabies virus in dogs and cats; other viral agents such as HIV-1 in primates and SARS-CoV in bats undergo host switching over time to cause epidemics in humans (Parish et al. 2008). Thus, restricted human contact with primates and other virus reservoirs in the wild is notably a significant obstacle to the establishment of viral infections and the occurrence of outbreaks in the human population.

Infectious disease agents may resurface with fresh features which may include altered or mutated genes resulting in resistance to several antimicrobial agents as in the case of HIV/AIDS, viral hepatitis and influenza. Evolutionary studies have shown that infectious agents that occasionally cause endemics originated and stabilized with time in a given environment (Center for Disease Control and Prevention 2018).

7.3 Drivers of Emerging Viral Diseases

Increases in the movement of people, climate change, and viral evolution involving cross-species at the human-animal interface lead to the emergence of new viruses (Parish et al. 2008; Ippolito and Rezza 2017). Environmental factors such as inadequate water supply, sanitation, food and climatic changes, as well as lapses associated with the collection, treatment and disposal of solid waste in developing nations often result to pollution of soil, water and air. This creates breeding sites for biological vectors of viral diseases such as insects and rodents (Obi and Shenge 2018). A likely trend is the dispersal of insect vectors such as *Aedes* mosquitoes implicated in the spread of many arboviruses which have spread to new locations by adopting new breeding sites in tyres, overcrowded settings and stagnant water pools, mainly because of population growth and expansion. Most are as a result of massive settlements and residential building in unapproved sites such as marshy areas and near dumpsites lacking basic sanitation and proper waste disposal facilities (Marston et al. 2014). In such areas, solid and open waste such as tyres, cans, plastic bottles, and nylon bags that collect water become breeding sites for vectors of viral agents.

Climatic factors like temperature, rainfall and relative humidity influence the incidence, duration and intensity of the emergence of epidemic forms of viral diseases. For example, it has been observed that the incidence of Rift Valley fever in East Africa at a time increased as a result of long periods of heavy rainfall, which created increased breeding sites for the vectors of the disease (Davies et al. 1985). Microbial adaptation in any environment is driven by factors which range from host to agent and ecological influences. Hosts factors include gender, age, and immune status, among others, which may determine the outcome of infection in an exposed host. Additionally, human activities and individual behaviours may influence the transmission of viral diseases. Evolution of viruses resulting in cross-species transmission has also been tagged as a driver of emerging viruses. This is mainly seen among rapidly evolving viruses such as RNA viruses, which are prone to error during replication (Drake and Holland 1999), due to lack of proofreading machinery by the RNA-dependent RNA polymerases (Smith et al. 1997). This results in nucleotide substitutions (mutations) due to rapid replication and synthesis of enormous virus quantity within a short time (Moya et al. 2004). The distribution of viral variants in the human population can be due to epidemics through different means of transmission (Parish et al. 2014). Such diversities have been observed in some RNA viruses such as the HIV and hepatitis C virus (Argentini et al. 2009; Shenge et al. 2018).

Another critical driver of the emergence of viral diseases is host switching, where a virus that initially infected an animal host shifts to a new host and subsequently gets established within the human population after it is successfully transmitted (Lindstrom et al. 2004). Host switching enables these viral agents to adapt as a result of recombination and assortment in their genomes, which may lead to the emergence of entirely new strains of the virus or strains with reassorted gene segments. This trend was observed in the 1957 H2N2, and 1968 H3N2 Influenza, human pandemic viruses containing gene segments of haemagglutinin (HA) and polymerase (PB1) obtained from the avian influenza virus (Webby and Webster 2001).

7.4 Epidemiology, Transmission and Impact of some Emerging Viral Diseases in Africa

7.4.1 Rift Valley Fever

Rift Valley fever (RVF) is a devastating mosquito-borne zoonotic disease that causes illnesses in humans and animals (Hartman 2017). The virus is a Phlebovirus in the family *Bunyaviridae*. It has a tri-segmented, single-stranded RNA genome in which two segments, L (large) and M (medium), have a negative polarity, while the third, S (small), is ambisense (Ly and Ikegami 2016). The virus was first isolated from sheep in Kenya in 1931 and has since spread to other countries (Pepin et al. 2010). Large outbreaks of the disease have occurred in many African countries, Madagascar and the Arabian Peninsula, and have usually been indicated by a sweep of abortion in livestock (Mariner 2018). The RVF virus is transmitted through bites of various species of *Aedes* and *Culex* mosquitoes which serve as reservoirs and amplifying vectors, respectively. The virus can also be transmitted through contact with infective animal tissues such as blood, aborted foetus and other body fluids (Nyakarahuka et al. 2018). Humans have also been infected through drinking of raw, unpasteurized milk from infected animals (Ng'ang'a et al. 2016).

The RVF virus is maintained during the dry season in desiccation-resistant vertically infected eggs of several *Aedes* species (Lumley et al. 2017). RVF epizootics and epidemics usually occur during periods of above normal and prolonged rainfall. The rainfall events result in flooding of dambo-type grassland depressions at greater depths and for extended periods compared to interepizootic periods. Consequently, previously dormant infected *Aedes* eggs are induced to hatch, causing the emergence and survival of at least one large generation of virus-infected *Aedes* mosquitoes (Williams et al. 2016). These mosquitoes then transmit the virus to ungulate livestock, especially sheep, cattle and goats which serve as amplifying vertebrate hosts that may begin the epizootic cycle. After infection from primary *Aedes* vectors, livestock species, especially sheep, suffer significant morbidity, mortality and up to 100% abortion. These amplifying vertebrates develop high viraemia, enough to infect secondary mosquito vector species (Lutomiah et al. 2014). Secondary mosquito species including *Culex* and *Mansonia* species then serve as the primary horizontal vectors of the RVF virus between viraemic domestic animals and humans. The pace of the epizootic may be further compounded because virus-infected moribund ungulates display little mosquito avoidance behaviour and are preferentially fed upon by secondary vectors (Linthicum et al. 2016).

Notable epidemics of the virus include the 1977 outbreak in Egypt, where about 200,000 human cases and 598 deaths occurred (Meegan et al. 1979). In 2006, an outbreak with over 1000 human cases and 300 deaths occurred in Somalia, Kenya and Tanzania (Dar et al. 2013). Other series of large-scale RVF outbreaks include Sudan (2007), Madagascar (2008), southern Africa (2008–2011), Mauritania (2010, 2012) and Saudi Arabia and Yemen (2000) (Clark et al. 2018). Subclinical circulation of the virus in humans and animals has also been documented in many African countries, including Nigeria (Opayele et al. 2018). This virus severely affects the health and economy of thousands of humans and livestock in the affected countries (Linthicum et al. 2016).

7.4.2 Yellow Fever

Yellow fever, caused by the yellow fever virus (YFV), is another mosquito-borne viral haemorrhagic disease with high mortality. It mostly occurs in Africa and South America. Historically, it caused large outbreaks in Europe and North America when it was introduced during the slave trade (Holbrook 2017).

The yellow fever virus possesses a positive-sense, single-stranded RNA genome. It is the prototype of the genus *Flavivirus*, family *Flaviviridae*, which comprises approximately 70 viruses, most of which are arthropod borne. Other major human and veterinary pathogens in the genus include dengue, Japanese encephalitis and West Nile viruses (McLinden et al. 2017).

Yellow fever is a zoonotic disease, maintained in nature by non-human primates and diurnally active mosquitoes that breed in tree holes in the forest canopy (*Haemagogus* spp. in the Americas, and *Aedes* spp. in Africa). People are exposed to infected mosquitoes when they encroach on this cycle during occupational or recreational activities (Hanley et al. 2013). In the warm, humid savanna regions of Africa, tree-hole-breeding *Aedes* mosquitoes reach very high densities and are implicated in the endemic and epidemic transmission of viruses from monkeys to humans. *Aedes aegypti*, a domestic mosquito species that breeds in human-made containers, transmits the yellow fever virus among humans in cities (Possas et al. 2018). The virus is maintained over the dry season by vertical transmission in mosquitoes. Transovarially infected eggs survive in dry tree-holes and hatch infectious mosquitoes when the wet season begins (Monath 2001).

Approximately 200,000 cases of yellow fever occur annually worldwide, 90% of which occurs in Africa. A striking resurgence of the disease has occurred since the

1980s in both sub-Saharan Africa and South America (Monath and Vasconcelos 2015). Studies on yellow fever outbreaks across Africa revealed that epidemics of the disease were more frequent and more extensive in west and east African countries than countries in central Africa. This is due to the influence of climate and the environment on both mosquito vectors and the virus (Hamlet et al. 2018). Also, the periodicity of an upsurge in yellow fever activities in west African countries is between 5 and 20 years while much longer intervals of up to 45 years have been observed in parts of east Africa. Some countries have also sustained epidemics across multiple years, for example, in Ghana (1977–1983), Guinea (2000–2005), and Nigeria (1986–1994) (Monath and Vasconcelos 2015). In December 2015, another urban outbreak of yellow fever was declared in Angola and soon after in the Democratic Republic of Congo (DRC). These outbreaks ended in 2017 with 7334 suspected cases and 393 deaths (Kraemer et al. 2017). In 2016, another outbreak occurred in Uganda, and since then, several sporadic yellow fever cases have been reported in Chad, Ghana, the Republic of Congo, Guinea and Nigeria (Domingo et al. 2018).

Mosquito vectors capable of transmitting yellow fever exist in regions where the disease does not presently occur, such as Asia (Barnett 2007). Vector-control strategies that were once successful for the elimination of yellow fever from many affected regions have failed, leading to the re-emergence of the disease. Consequently, the administration of vaccines is still the most effective method of prevention of the disease, coupled with the prevention of mosquito bites. Effective vaccines against yellow fever have been discovered since 1937 (Theiler and Smith 1937) and have been responsible for the significant reduction in occurrences of the disease worldwide (Garske et al. 2014). Till date, the yellow fever 17D vaccine is very effective despite its use for over seven decades probably due to the genetic stability of the yellow fever virus structural proteins which the host immune cells target (Baba and Ikusemoran 2017). However, despite this success, effective administration of the vaccine is still a challenge and has been contributing to the resurgence of the disease in many African countries. There have been occasional issues of supply and demand, as experienced in Angola and Democratic Republic of Congo in 2016 (Barrett 2017). Inadequate vaccine coverage is also a problem in many countries where the disease is endemic (Shearer et al. 2017).

7.4.3   Lassa Fever

Lassa fever (LF) is a rodent-borne disease associated with an acute and potentially fatal haemorrhagic illness. It is caused by the Lassa virus (LASV), a member of the family *Arenaviridae*. The viral genome consists of two single-stranded RNA segments, a small (S) and large (L). Both genomic segments have an ambisense gene organization, encoding two genes in the opposite orientation (Günther and Lenz 2004). The virus was first isolated in 1969 from a nurse who presumably acquired the infection from one of her patients in Lassa town, northeastern Nigeria (Frame et al. 1970). The Lassa virus was initially thought to be endemic only in the West African countries of Sierra Leone, Guinea, Liberia and Nigeria (Monath et al. 1974; Lukashevich et al. 1993; Frame et al. 1984; Tomori et al. 1988; O'Hearn et al. 2016). However, its presence has also been reported in other countries including Ivory Coast, Mali, Ghana, Senegal, Burkina Faso, Gambia and the Central African Republic (Frame 1975; Safronetz et al. 2010).

Lassa virus is transmitted through contact with body fluids or excreta, or inhalation of aerosols produced by infected rodents. Hunting and consumption of peridomestic rodents as a source of food have also been reported as another route of virus transmission to humans (Ter Meulen et al. 1996). The infection also spreads between humans through contact with body fluids of infected persons or contaminated medical equipment (Fisher-Hoch et al. 1995). The Lassa virus infects approximately 500,000 people and causes about 5000 deaths worldwide (Mateer et al. 2018). In endemic regions, seropositivity for Lassa virus-specific antibodies could be as high as 55% in populations residing around forested areas harbouring large populations of reservoir rodents (Lukashevich et al. 1993). Most times, early diagnosis of the infection is missed as most early symptoms are mild and resemble those of other diseases like typhoid fever and malaria in endemic areas (Safronetz et al. 2010).

The multimammate mouse, *Mastomys natalensis*, is the primary host for the LASV (Lecompte et al. 2006). The virus has little or no adverse effect on its rodent reservoir under natural circumstances (Mariën et al. 2017). *Mastomys natalensis* is widely distributed throughout West, Central and East Africa (Lopez and Mathers 2006), where they aggregate in houses in large numbers during the dry and harmattan season due to harsh environmental conditions and dispersing into gardens and farmlands when the rains return (Fichet-Calvet et al. 2007). Infected rodents remain carriers of the virus throughout life; the virus is shed in urine, faeces, saliva, respiratory secretion and exposed blood vessels through micro or macro trauma (Keenlyside et al. 1983). In endemic countries, other common rodent species such as *Rattus rattus* and *Mus minutoides* have also been implicated in Lassa virus transmission (Wulff et al. 1975).

Lassa fever infects humans of all age groups and sexes. The disease is associated with a broad spectrum of clinical manifestations. The incubation period ranges from 7 to 21 days (McCormick et al. 1987). The clinical presentation is usually mild or asymptomatic in about 80% of infections (Richmond and Baglole 2003). The onset of the symptomatic disease is usually gradual, starting as a flu-like illness characterized by mild fever, weakness and general malaise. This may be accompanied by a headache, sore throat, muscle pain, chest pain, nausea, vomiting, diarrhoea, cough and abdominal pain (Bausch et al. 2001). In mild cases, the fever subsides, and the patient usually recovers. Other cases progress towards a more severe illness. Symptoms include haemorrhage, respiratory distress, facial oedema and fluid in the pulmonary cavity. Shock, seizures, tremor, disorientation and coma have also been reported during this stage of the disease, indicating a poor prognosis for the disease outcome (Mertens et al. 1973). Approximately 15–20% of hospitalized Lassa fever patients die from the illness, generally within 2 weeks after the onset of symptoms due to multi-organ complication and failure involving the liver, spleen or kidneys. Pregnant women are more likely to have severe illness due to infection with LASV than women who are not pregnant, with maternal case fatality rates as high as 80% and nearly 100% mortality in foetuses. Infection in infants can result in "swollen baby syndrome" with oedema, abdominal distension, bleeding and often death (McCormick et al. 1987). Neurological problems, including hearing loss and encephalopathy, have been shown to occur in patients who survive the disease (Mateer et al. 2018).

7.4.4   Ebola Haemorrhagic Fever (EHF)

The Ebola virus belongs to the family *Filoviridae* and the order Mononegavirales. The virus causes haemorrhagic fever, an important emerging viral infection in central Africa. The Ebola virus was first identified in 1976 following two outbreaks of haemorrhagic fever in northern Zaire (the present Democratic Republic of Congo) and southern Sudan. The most highly virulent strain (subtype) is Ebola-Zaire, which has a mortality rate of 88%. Between 2014 and 2016, several outbreaks of the Ebola virus disease were reported in parts of west Africa, where more than 28,000 cases were confirmed with over 11,000 deaths in Liberia, Guinea, Sierra Leone, Democratic Republic of Congo (DRC) and Nigeria (Baize et al. 2014). As of August 31, 2018, about 120 cases and 78 deaths had been confirmed from DRC alone (WHO 2018).

Contact with infected live animals such as bats, their excretions or carcass of infected animals causes primary infection, which subsequently results in sustained person-to-person transmission as have been observed in Ebola and Marburg viruses over the years (Groseth et al. 2007).

7.4.5   Other Emerging Viral Diseases

Viral disease outbreaks occur periodically with the potential to increase in incidence. Their unique nature allows the viruses to change their form and evade the hosts' immune defence to produce persistent or latent infections, or worse still, an outbreak in any population. This attribute has enabled the emergence of novel or pandemic variants among several viral agents, including influenza virus (Meseko et al. 2013, 2014). Typically, new infections emerge as a result of changes in existing strains that can lead to resistance or due to some environmental factors that support the growth of emerging infectious agents, in a particular geographical zone. A typical example is the emergence of several subtypes of the influenza virus over time, as a result of re-assortment and mutation in animal reservoirs, especially birds and swine, with major variants such as H1N1, H1N2 and H3N2, now being detected at the human-animal interface (Alexander and Brown 2000; Yu et al. 2009). Other newly identified viral agents include severe acute respiratory syndrome (SARS) virus and species or strains that have a significant burden on public health and global economies.

7.5 Methods of Detection and Identification of Emerging Viruses

The early detection of emerging viral agents is vital in the management and control of diseases. Several measures that facilitate clinical and epidemiological investigation of emerging viral diseases in any environment have been identified (Chan et al. 2017). Such measures involve rapid detection of viral agents using the most sensitive and specific laboratory assays, including viral antigen and antibody testing, isolation of viral agents using cell and tissue culture techniques, molecular characterization, including nucleic acid synthesis and amplification with polymerase chain reaction (PCR), genotyping and genomic sequencing. The use of quantitative molecular tools such as real-time reverse transcription PCR for enumerating RNA viruses, especially during outbreaks, has helped in the identification of new viral agents (Marston et al. 2014). These advanced techniques enable rapid diagnosis and identification of cases and contacts within a short period.

The successful detection of an emerging virus may depend on the type of procedure used, and this, in turn, can be determined by economic, human resources and laboratory robustness at the time. Serological techniques, including antigen and antibody testing, involve the use of labelled enzymes and their substrates as indicators in enzyme-linked immunosorbent assays (ELISA) (Whitehouse 2004). These methods are inexpensive and used mostly in clinical settings in Africa to detect viruses using serum or plasma obtained from blood. However, because the methods are time-consuming and not sensitive enough to detect occult and early infections, rapid detection of a viral agent may be eluded when using such methods, especially among individuals with nonspecific symptoms of the disease (Whitehouse 2004).

Molecular tools such as PCR and genome sequencing remain the gold standard for identifying emerging and re-emerging viruses (Rosenstierne et al. 2014). PCR is a very sensitive technique, which involves the amplification of the viral genetic material (DNA or RNA), even at very low concentrations, in the specimen for easy identification, regardless of the stage of infection. Methods such as real-time PCR complement the techniques involving antigen and antibody testing (Wang et al. 2011). It is best suited for early diagnosis in clinical samples and during viral diseases outbreaks such as Marburg or Ebola (Weidmann et al. 2007). Molecular techniques are, however, costly, requiring substantial funding to acquire equipment and reagents. Another limitation is the need for trained laboratory personnel. These limitations hinder the use of molecular techniques for the diagnosis of viral infections, including emerging and re-emerging ones, in most African countries given that the methods cannot be used for routine clinical diagnosis.

The detection of emerging viruses may also be achieved through the screening of vertebrate or arthropod hosts of tropical viral diseases. Samples from these animals (bats, rodents, ticks, non-human primates, birds and vectors such as mosquitoes and fleas) can be routinely screened for an array of viral agents using PCR and microarray analysis (Rosenstierne et al. 2014).

7.6 Prevention, Control and Management

Novel epidemics and increased incidence of emergent and re-emergent diseases highlight the need for the development of effective control strategies. A major challenge in controlling these diseases results from the fact that their transmission intensity is driven primarily by wildlife reservoirs. Therefore, one-sided disease prevention enacted either by the human or animal health sector is often inefficient (Lembo 2012). The following approaches have been recognized as suitable and effective in the control of zoonotic viral diseases.

7.6.1 Surveillance

There is a need for constant surveillance of emerging viral diseases in all countries around the world. Constant surveillance is needed to monitor trends in epidemiological patterns of disease occurrence. To achieve this, humans and animals need to be tested regularly to detect the presence of disease agents before the occurrence of epidemics. Considering the current ease of transportation especially through the air, it is possible for commercial aircraft to transport infected humans and vectors from one part of the world to another. For example, the Rift Valley fever introductions into Egypt in 1977 and the Arabian Peninsula in 2000 followed epizootics in Sudan in 1976 and the Horn of Africa in 1998, respectively. Also, the potential for such globalization of other arboviruses has been demonstrated in the case of West Nile and Chikungunya viruses over the past 15 years (Nash et al. 2001; Cassadou et al. 2014).

Effective surveillance in Africa is, however, a challenge due to the remoteness of some communities, especially in rural settings without access to good road infrastructure. For example, a study conducted in Zambia revealed that the lack of transportation facilities was among the leading contributing factors challenging the implementation of an integrated disease surveillance and response strategy in the country (Mandyata et al. 2017). Similarly, a review of the challenges faced by 18 African countries in implementing adequate surveillance schemes identified the lack of transportation as a significant compounding factor (Phalkey et al. 2013). This challenge has further been demonstrated by a recent study that reported that improved transportation networks led to improved case reporting, hence, disease surveillance in Uganda in 2016 (Nakiire et al. 2019).

7.6.2   Control in Animals

In order to prevent the transmission of viral agents to new locations through the movement of animals, quarantine is strongly advocated for some infectious diseases, especially when animals are moved over long distances, as with livestock import-export, pet trade and tourism. This has been the case with the prevention of spread of the Ebola virus (Gumusova et al. 2015). Immunization of exposed animals is also essential in the control of vaccine-preventable animal diseases. Vaccination campaigns have been shown to greatly assist in the control of numerous different zoonotic diseases such as rabies and the Rift Valley fever. However, vaccination can only be done for owned animals. With a high number of stray animals, like dogs, in Africa, both approaches (quarantining and immunization) may not prove effective for control of infections like rabies, for example (Leung and Davis 2017; Salomão et al. 2017).

7.6.3   Control of Vectors and Reservoirs

The control of vectors and reservoirs includes measures to prevent zoonotic pathogens from being transmitted to non-infected animals, humans and disease-free areas, through arthropod vectors, contaminated fomite and animal reservoirs. This may involve improving hygiene and control of the environment. For instance, the control of yellow fever during the construction of the Panama Canal in the early twentieth century is linked to the destruction of *Aedes* breeding sites which included draining of pools of standing water and grass cutting (Center for Disease Control and Prevention 2015). Arthropod vector control is another effective strategy in reducing sources of infection, as in the case of Rift Valley fever, yellow fever and other mosquito-borne zoonoses. Rodent control and avoidance of bush burning, setting traps in and around homes to reduce rat population and avoiding contact with rats are other effective methods of reducing contact with animal reservoirs of dangerous pathogens like Lassa fever virus. To achieve such effective control, adequate sanitation would be an absolute necessity. However, most African countries are far from achieving the set global sanitation goals of 2030 as detailed in the Sustainable Development Goals (Nhamo et al. 2019). Thus, without adequate sanitation, the control of vectors within African countries would be challenged, hence the control of emerging and re-emerging viral pathogens.

7.6.4   Prevention in Man

Humans are important in the control of zoonosis. Much of the success of disease control plans depends on health education. Public awareness of health risks connected with certain infections can greatly assist in reducing the spread of such diseases (Hasanov et al. 2018). Occupational health education should be explicitly directed at categories of workers that are exposed to certain diseases in the course of their duties. These include farmers, veterinarians, personnel in slaughterhouses and biological laboratories (Cripps 2000). Vaccination is also very useful in protecting at-risk individuals against diseases for which vaccines have been developed.

Proper food hygiene, including safe dietary habits, is of value in dealing with some diseases like Rift valley fever, which can be transmitted by the consumption of unpasteurized milk and dairy products (Ng'ang'a et al. 2016). The cooking of meat and meat products before consumption and high standards of hygiene in kitchens and catering facilities are also needed in the control of emerging diseases like Lassa fever. Improved sanitation from better sewage and drinking water treatment will also help in control and prevention strategies (Mara et al. 2010).

Often, emerging infectious diseases have the potential to spread rapidly in hospital settings. It is, therefore, important for healthcare workers to adhere strictly to infection control measures. Isolation of infected patients, barrier nursing of infected patients and the use of personal protective equipment (PPE) when caring for patients have been shown to reduce nosocomial outbreaks of Lassa fever and other haemorrhagic diseases (Helmick et al. 1986).

7.7   Conclusion

Africa is among the hot zones for zoonotic and emerging viral diseases, which may continue to result in the emergence of new viruses. Constant evaluation of water, soil, air and other sources for viral pollution, as well as proper disposal and treatment of wastewater, is suggested as a way forward to tackling the emergence of viral infections in Africa's environment.

Human activities, including agriculture and exploitation of forest resources, will continue to bring humans closer to sylvatic cycles of zoonotic viruses. These activities could also alter the natural population and geographic distributions of insect and animal, which serves as vectors or reservoirs for these viruses, thereby leading to the emergence of new viruses and re-emergence of known ones. Rapid globalization also increases the risk that infections that emerge in one part of the world can spread internationally in human and animal populations. Currently, vaccines for the prevention of many of these viruses are unavailable, and the public health delivery system is weak in many African countries where most of the emerging and remerging viruses are known to occur. There is, therefore, the need for consistent surveillance and monitoring of these diseases of epidemic and pandemic potential.

Surveillance and monitoring information will allow for timely alerts and effective response activities that will prevent major outbreaks in Africa and beyond. There is also the need to invest more resources in the development of vaccines for more of these emerging viruses. Increased investment in public health infrastructures, transportation and sanitation facilities are also required in the control of these viruses.

References

Ahmed SM, Hall AJ, Robinson AE, Verhoef L, Premkumar P, Parashar UD, Koopmans M, Lopman BA (2014) Global prevalence of norovirus in cases of gastroenteritis: a systematic review and meta-analysis. Lancet Infect Dis 14:725–730

Alexander D, Brown IH (2000) Recent Zoonoses caused by influenza a viruses. OIE Sci Tech Rev 19:197–225

Andrew P, Gregory EG (2008) Emerging viral diseases. Md Med 9(1):11–16

Argentini C, Genovese D, Dettori S, Rapicetta M (2009) HCV genetic variability: from quasispecies evolution to genotype classification. Future Microbiol 4:359–373

Baba MM, Ikusemoran M (2017) Is the absence or intermittent YF vaccination the major contributor to its persistent outbreaks in eastern Africa? Biochem Biophys Res Commun 492:548–557. https://doi.org/10.1016/j.bbrc.2017.01.079

Baize S, Pannetier D, Oestereich L, Reiger T, Koivogui L, Magassouba N, Soropogui B, Sow MS, Keita S, De Clerck H, Tiffany A, Dominguez G et al. (2014) Emergence of Zaire Ebola Virus Disease in Guinea. N Engl J Med 371:1418–1425

Barnett ED (2007) Yellow fever: epidemiology and prevention. Clin Infect Dis 44:850–856. https://doi.org/10.1086/511869

Barrett ADT (2017) Yellow fever live attenuated vaccine: a very successful live attenuated vaccine but still we have problems controlling the disease. Vaccine 35:5951–5955. https://doi.org/10.1016/j.vaccine.2017.03.032

Bausch DG, Demby AH, Coulibaly M et al (2001) Lassa fever in Guinea: I. epidemiology of human disease and clinical observations. Vector Borne Zoonotic Dis 1:269–281. https://doi.org/10.1089/15303660160025903

Cassadou S, Boucau S, Petit-Sinturel M et al (2014) Emergence of chikungunya fever on the French side of Saint Martin island, October to December 2013. Euro Surveill 19. https://doi.org/10.2807/1560-7917.ES2014.19.13.20752

Center for Disease Control and Prevention (2015) CDC – Malaria – About Malaria – History – The Panama Canal. https://www.cdc.gov/malaria/about/history/panama_canal.html. Accessed 9 Oct 2018

Center for Disease Control and Prevention (2018) Emerging Infectious Diseases. Emerging Infectious Disease (EID) Journal ISSN: 1080–6059.

Chan JF, Sridhar S, Yip CC, Lau SK, Woo PC (2017) The role of laboratory diagnostics in emerging viral infections: the example of the Middle East respiratory syndrome epidemic. J Microbiol 55(3):172–182

Clark MHA, Warimwe GM, Di Nardo A et al (2018) Systematic literature review of Rift Valley fever virus seroprevalence in livestock, wildlife and humans in Africa from 1968 to 2016. PLoS Negl Trop Dis 12:e0006627. https://doi.org/10.1371/journal.pntd.0006627

Cleaveland S, Laurenson MK, Taylor LH (2001) Diseases of humans and their domestic mammals: pathogen characteristics, host range and the risk of emergence. Philos Trans R Soc B Biol Sci 356(1411):991–999

Cripps PJ (2000) Veterinary education, zoonoses and public health: a personal perspective. Acta Trop 76:77–80. https://doi.org/10.1016/S0001-706X(00)00094-2

Dar O, McIntyre S, Hogarth S, Heymann D (2013) Rift Valley fever and a new paradigm of research and development for zoonotic disease control. Emerg Infect Dis 19. https://doi.org/10.3201/eid1902.120941

Davies FG, Linthicum KJ, James AD (1985) Rainfall and epizootic Rift Valley fever. Bull World Health Organ 63(5):941–943

Domingo C, Charrel RN, Schmidt-Chanasit J et al (2018) Yellow fever in the diagnostics laboratory review-article. Emerg Microbes Infect 7:129

Drake JW, Holland JJ (1999) Mutation rates among RNA viruses. PNAS 96(24):13910–13913

Fichet-Calvet E, Lecompte E, Koivogui L et al (2007) Fluctuation of abundance and Lassa virus prevalence in *Mastomys natalensis* in Guinea, West Africa. Vector Borne Zoonotic Dis 7:119–128. https://doi.org/10.1089/vbz.2006.052

Fisher-Hoch SP, Tomori O, Nasidi A et al (1995) Review of cases of nosocomial Lassa fever in Nigeria: the high price of poor medical practice. BMJ 311:857–859. https://doi.org/10.1136/bmj.311.7009.857

Frame JD (1975) Surveillance of Lassa fever in missionaries stationed in West Africa. Bull World Health Organ 52:593–598

Frame JD, Baldwin JM, Gocke DJ, Troup JM (1970) Lassa fever, a new virus disease of man from West Africa. I. Clinical description and pathological findings. Am J Trop Med Hyg 19:670–676. https://doi.org/10.1371/journal.pntd.0000548

Frame JD, Yalley-Ogunro JE, Hanson AP (1984) Endemic Lassa fever in Liberia. V. Distribution of Lassa virus activity in Liberia: hospital staff surveys. Trans R Soc Trop Med Hyg 78:761–763. https://doi.org/10.1016/0035-9203(84)90012-9

Garske T, Van Kerkhove MD, Yactayo S et al (2014) Yellow fever in Africa: estimating the burden of disease and impact of mass vaccination from outbreak and serological data. PLoS Med. https://doi.org/10.1371/journal.pmed.1001638

Groseth A, Feldmann H, Strong JE (2007) The ecology of Ebola virus. Trends Microbiol 15(9):408–416

Gumusova S, Sunbul M, Leblebicioglu H (2015) Ebola virus disease and the veterinary perspective. Ann Clin Microbiol Antimicrob 14(30). https://doi.org/10.1186/s12941-015-0089-x

Günther S, Lenz O (2004) Lassa virus. Crit Rev Clin Lab Sci 41:339–390. https://doi.org/10.1080/10408360490497456

Hamlet A, Jean K, Perea W et al (2018) The seasonal influence of climate and environment on yellow fever transmission across Africa. PLoS Negl Trop Dis 12:e0006284. https://doi.org/10.1371/journal.pntd.0006284

Hanley KA, Monath TP, Weaver SC et al (2013) Fever versus fever: the role of host and vector susceptibility and interspecific competition in shaping the current and future distributions of the sylvatic cycles of dengue virus and yellow fever virus. Infect Genet Evol 19:292–311. https://doi.org/10.1016/j.meegid.2013.03.008

Hartman A (2017) Rift Valley fever. Clin Lab Med 37:285–301. https://doi.org/10.1016/j.cll.2017.01.004

Hasanov E, Zeynalova S, Geleishvili M et al (2018) Assessing the impact of public education on a preventable zoonotic disease: rabies. Epidemiol Infect 146:227–235. https://doi.org/10.1017/S0950268817002850

Helmick CG, Webb PA, Scribner CL et al (1986) No evidence for increased risk of Lassa fever infection in hospital staff. Lancet 2:1202–1205

Holbrook MR (2017) Historical perspectives on Flavivirus research. Viruses 9(5):E97. https://doi.org/10.3390/v9050097

Hufnagel L, Brockmann D, Geisel T (2004) Forecast and control of epidemics in a globalized world. Proc Natl Acad Sci U S A 101:15124–15129

Ippolito G, Rezza G (2017) Preface - emerging viruses: from early detection to intervention. Adv Exp Med Biol 972:1–5

Keenlyside RA, McCormick JB, Webb PA et al (1983) Case-control study of Mastomys natalensis and humans in Lassa virus-infected households in Sierra Leone. Am J Trop Med Hyg 32:829–837. https://doi.org/10.4269/ajtmh.1983.32.829

Kilpatrick AM, Randolph SE (2012) Drivers, dynamics, and control of emerging vector-borne zoonotic diseases. Lancet 380(9857):1946–1955

Kraemer MUG, Faria NR, Reiner RC et al (2017) Spread of yellow fever virus outbreak in Angola and the Democratic Republic of the Congo 2015-16: a modelling study. Lancet Infect Dis 17:330–338. https://doi.org/10.1016/S1473-3099(16)30513-8

La Rosa G, Fratini M, della Libera S, Iaconelli M, Muscillo M (2012) Emerging and potentially emerging viruses in water environments. Ann Ist Super Sanita 48(4):397–406

Lecompte E, Fichet-Calvet E, Daffis S et al (2006) Mastomys natalensis and Lassa fever, West Africa. Emerg Infect Dis 12:1971–1974. https://doi.org/10.3201/eid1212.060812

Lembo T (2012) The blueprint for rabies prevention and control: a novel operational toolkit for rabies elimination. PLoS Negl Trop Dis. https://doi.org/10.1371/journal.pntd.0001388

Leung T, Davis SA (2017) Rabies vaccination targets for stray dog populations. Front Vet Sci 4:52. https://doi.org/10.3389/fvets.2017.00052

Lindstrom SE, Cox NJ, Klimov A (2004) Genetic analysis of human H2N2 and early H3N2 influenza viruses, 1957–1972: evidence for genetic divergence and multiple reassortment events. Virology 328:101–119

Linthicum KJ, Britch SC, Anyamba A (2016) Rift Valley fever: an emerging mosquito-borne disease. Annu Rev Entomol 61:395–415. https://doi.org/10.1146/annurev-ento-010715-023819

Lopez AD, Mathers CD (2006) Measuring the global burden of disease and epidemiological transitions: 2002–2030. Ann Trop Med Parasitol 100:481–499. https://doi.org/10.1179/1364859 06X97417

Lukashevich IS, Clegg JC, Sidibe K (1993) Lassa virus activity in Guinea: distribution of human antiviral antibody defined using enzyme-linked immunosorbent assay with recombinant antigen. J Med Virol 40:210–217. https://doi.org/10.1002/jmv.1890400308

Lumley S, Horton DL, Hernandez-Triana LLM et al (2017) Rift valley fever virus: strategies for maintenance, survival and vertical transmission in mosquitoes. J Gen Virol 98:875–888

Lutomiah J, Omondi D, Masiga D et al (2014) Blood meal analysis and virus detection in blood-fed mosquitoes collected during the 2006-2007 Rift Valley fever outbreak in Kenya. Vector Borne Zoonotic Dis 14:656–664. https://doi.org/10.1089/vbz.2013.1564

Ly HJ, Ikegami T (2016) Rift Valley fever virus NSs protein functions and the similarity to other bunyavirus NSs proteins. Virol J 13:118. https://doi.org/10.1186/s12985-016-0573-8.

Mandyata CB, Olowski LK, Mutale W (2017) Challenges of implementing the integrated disease surveillance and response strategy in Zambia: a health worker perspective. BMC Public Health 17(1):746. https://doi.org/10.1186/s12889-017-4791-9.

Mara D, Lane J, Scott B, Trouba D (2010) Sanitation and health. PLoS Med 7:e1000363. https://doi.org/10.1371/journal.pmed.1000363

Mariën J, Borremans B, Gryseels S et al (2017) No measurable adverse effects of Lassa, Morogoro and Gairo arenaviruses on their rodent reservoir host in natural conditions. Parasit Vectors. https://doi.org/10.1186/s13071-017-2146-0

Mariner J (2018) Rift Valley fever surveillance. FAO Animal Production and Health Manual No. 21. Food and Agriculture Organization of the United Nations (FAO), Rome

Marston HD, Folkers GK, Morens DM, Fauci AS (2014) Emerging Viral Diseases: Confronting Threats with New Technologies. Sci Transl Med 6(253):253ps10–253ps10

Mateer EJ, Huang C, Shehu NY, Paessler S (2018) Lassa fever-induced sensorineural hearing loss: a neglected public health and social burden. PLoS Negl Trop Dis 12:e0006187. https://doi.org/10.1371/journal.pntd.0006187

McCormick JB, King IJ, Webb PA et al (1987) A case-control study of the clinical diagnosis and course of Lassa fever. J Infect Dis. https://doi.org/10.1093/INFDIS/155.3.445

McLinden JH, Bhattarai N, Stapleton JT et al (2017) Yellow fever virus, but not Zika virus or dengue virus, inhibits T-cell receptor-mediated T-cell function by an RNA-based mechanism. J Infect Dis 216:1164–1175. https://doi.org/10.1093/infdis/jix462

Meegan JM, Hoogstraal H, Moussa MI (1979) An epizootic of Rift Valley fever in Egypt in 1977. Vet Rec 105:124–125. https://doi.org/10.1136/vr.105.6.124

Mertens PE, Patton R, Baum JJ, Monath TP (1973) Clinical presentation of Lassa fever cases during the hospital epidemic at Zorzor, Liberia, march-April 1972. Am J Trop Med Hyg 22:780–784

Meseko CA, Odaibo GN, Olaleye DO (2014) Detection and isolation of 2009 pandemic influenza a/H1N1 virus in commercial piggery, Lagos Nigeria. Vet Microbiol 168(1):197–201. https://doi.org/10.1016/j.vetmic.2013.11.003

Meseko CA, Olaleye DO, Capua I, Cattoli G (2013) Swine influenza in sub-Saharan Africa- current knowledge and emerging insights. Zoonoses Public Health:2013. https://doi.org/10.1111/zph.12068

Monath TP (2001) Yellow fever: an update. Lancet Infect Dis 1:11–20

Monath TP, Newhouse VF, Kemp GE et al (1974) Lassa virus isolation from Mastomys natalensis rodents during an epidemic in Sierra Leone. Science 185:263–265. https://doi.org/10.1126/science.185.4147.263

Monath TP, Vasconcelos PFC (2015) Yellow fever. J Clin Virol 64:160–173

Morens DM, Folkers GK, Fauci AS (2008) Emerging infections: a perpetual challenge. Lancet Infect Dis 8:710–719

Morse SS (1995) Factors in the emergence of infectious diseases. Emerg Infect Dis 1(1):7–15

Moya A, Holmes EC, Gonzalez-Candelas F (2004) The population genetics and evolutionary epidemiology of RNA viruses. Nat Rev 2:279–288

Nakiire L, Masiira B, Kihembo C, Katushabe E, Natseri N, Nabukenya I, Komakech I, Makumbi I, Charles O, Adatu F, Nanyunja M, Nsubuga P, Woldetsadik SF, Tusiime P, Yahaya AA, Fall IS, Wondimagegnehu A (2019) Healthcare workers' experiences regarding scaling up of training on integrated disease surveillance and response (IDSR) in Uganda, 2016: cross sectional qualitative study. BMC Health Serv Res 19(1):117. https://doi.org/10.1186/s12913-019-3923-6

Nash D, Mostashari F, Fine A et al (2001) The outbreak of West Nile virus infection in the new York City area in 1999. N Engl J Med 344:1807–1814. https://doi.org/10.1056/NEJM200106143442401

Ng'ang'a CM, Bukachi SA, Bett BK (2016) Lay perceptions of risk factors for Rift Valley fever in a pastoral community in northeastern Kenya. BMC Public Health 16:32. https://doi.org/10.1186/s12889-016-2707-8

Nhamo G, Nhemachena C, Nhamo S (2019) Is 2030 too soon for Africa to achieve the water and sanitation sustainable development goal? Sci Total Environ 669:129–139

Nyakarahuka L, de St. Maurice A, Purpura L et al (2018) Prevalence and risk factors of Rift Valley fever in humans and animals from Kabale district in Southwestern Uganda, 2016. PLoS Negl Trop Dis 12:e0006412. https://doi.org/10.1371/journal.pntd.0006412

O'Hearn AE, Voorhees MA, Fetterer DP et al (2016) Serosurveillance of viral pathogens circulating in West Africa. Virol J 13:163. https://doi.org/10.1186/s12985-016-0621-4

Obi RK, Shenge JA (2018) The environment and transmission of viral diseases. In: Okoye CU, Abah D (eds) Dynamics of natural resource and environment in Nigeria; theory, practice, bureaucracy, advocacy. DEBEES Printing and Publishing Company, Enugu, Nigeria. pp 66–78

Olsen B, Munster VJ, Wallensten A, Waldenstrom J, Osterhaus AD, Fouchier RA (2006) Global patterns of influenza a virus in wild birds. Science 312:384–388

Opayele AV, Odaibo GN, Olaleye OD (2018) Rift valley fever virus infection among livestock handlers in Ibadan, Nigeria. J Immunoass Immunochem 39:609–621. https://doi.org/10.1080/15321819.2018.1525739

Parish F, Sirin A, Charman D, Joosten H, Minayeva T, Silvus M, Stringer L (2008) Assessment on Peatlands, Biodiversity and Climate Change: Main Report; Global Environment Centre, Kuala Lumpur and Wetland International; Wageningen

Parish CR, Holmes EC, Morens DM, Park EC, Burke DS, Calisher CH, Laughlin CA, Saif LJ, Daszak P (2008) Cross-species virus transmission and the emergence of new epidemic diseases. Microbiol Mol Biol Rev 72:457–470

Parish IA, Marshall HD, Staron HD, Lang PA, Brüstle A, Chen JH, Cui W, Tsui YC, Perry C, Laidlaw BJ, Ohashi P, Weaver CT, Kaech SM (2014) Chronic viral infection promotes sustained Th1-derived immunoregulatory IL-10 via BLIMP-1. Journal of Clinical Investigation 124(8):3455–3468

Pepin M, Bouloy M, Bird BH et al (2010) Rift Valley fever virus (Bunyaviridae: Phlebovirus): an update on pathogenesis, molecular epidemiology, vectors, diagnostics and prevention. Vet Res 41(61). https://doi.org/10.1051/vetres/2010033

Phalkey RK, Yamamoto S, Awate P, Marx M (2013) Challenges with the implementation of an integrated disease surveillance and response (IDSR) system: systematic review of the lessons learned. Health Policy Plan 30(1):131–143

Possas C, Lourenço-de-Oliveira R, Tauil PL et al (2018) Yellow fever outbreak in Brazil: the puzzle of rapid viral spread and challenges for immunisation. Mem Inst Oswaldo Cruz 113:e180278. https://doi.org/10.1590/0074-02760180278

Richmond JK, Baglole DJ (2003) Lassa fever: epidemiology, clinical features, and social consequences. BMJ 327:1271–1275. https://doi.org/10.1136/bmj.327.7426.1271

Rosenstierne MW, McLoughlin KS, Olesen ML, Papa A, Gardner SN, Engler O, Plumet S, Mirazimi A, Weidmann M, Neidrig M, Fomsgaard A, Erlandsson L (2014) The microbial detection Array for detection of emerging viruses in clinical samples- a useful Panmicrobial diagnostic tool. PLoS One 9(6):e100813. https://doi.org/10.1371/journal.pone.0100813

Safronetz D, Lopez JE, Sogoba N et al (2010) Detection of Lassa virus, Mali. Emerg Infect Dis 16:1123–1126. https://doi.org/10.3201/eid1607.100146

Salomão C, Nacima A, Cuamba L, Gujral L, Amiel O, Baltazar C, Cliff J, Gudo ES (2017) Epidemiology, clinical features and risk factors for human rabies and animal bites during an outbreak of rabies in Maputo and Matola cities, Mozambique, 2014: implications for public health interventions for rabies control. PLoS Negl Trop Dis 11(7):e0005787. https://doi.org/10.1371/journal.pntd.0005787

Shearer FM, Moyes CL, Pigott DM et al (2017) Global yellow fever vaccination coverage from 1970 to 2016: an adjusted retrospective analysis. Lancet Infect Dis. https://doi.org/10.1016/S1473-3099(17)30419-X

Shenge JA, Odaibo GN, Olaleye DO (2018) Genetic diversity of hepatitis C virus among blood donors and patients with clinical hepatitis in Ibadan Nigeria. Arch Basic Appl Med 6(1):79–85

Smith DS, Pathirana S, Davidson F, Lawlor E, Power J, Yap PL, Simmonds P (1997) The origin of hepatitis C virus genotypes. J Gen Virol 78:321–328

Tatem AJ, Hay SI, Rogers DJ (2006) Global traffic and disease vector dispersal. Proc Natl Acad Sci U S A 103:6242–6247

Taylor LH, Latham SM, Woolhouse MEJ (2001) Risk factors for human disease emergence. Philos Trans R Soc B Biol Sci 356(1411):983–989

Ter Meulen J, Lukashevich I, Sidibe K et al (1996) Hunting of peridomestic rodents and consumption of their meat as possible risk factors for rodent-to-human transmission of Lassa virus in the Republic of Guinea. Am J Trop Med Hyg 55:661–666. https://doi.org/10.4269/ajtmh.1996.55.661

Theiler M, Smith DS (1937) The Use of Yellow Fever Virus Modified by in Vitro Cultivation for Human Immunization. J Exp Med 65(6):787–800

Tomori O, Fabiyi A, Sorungbe A et al (1988) Viral hemorrhagic fever antibodies in Nigerian populations. Am J Trop Med Hyg 38:407–410. https://doi.org/10.4269/ajtmh.1988.38.407

van Doorn RH (2014) Emerging infectious diseases. Medicine (Abingdon) 42(1):60–63

Wang Y, Zhang X, Wei H (2011) Laboratory detection and diagnosis of filoviruses. Virol Sin 26:73–80

Webby RJ, Webster RG (2001) Emergence of influenza a viruses. Philos Trans R Soc Lond B 356:1817–1828

Weidmann M, Hufert FT, Sall AA (2007) Viral load among patients infected with Maburgvirus in Angola. J Clin Virol 39:65–66

Whitehouse CA (2004) Crimean-Congo hemorrhagic fever. Antivir Res 64:145–160

WHO (2004) Report of the WHO/FAO/OIE joint consultation on emerging zoonotic diseases/in collaboration with the health Council of the Netherlands. World Health Organization, Geneva. http://www.who.int/iris/handle/10665/68899

WHO (2018) Update on Ebola Outbreak and Emergency Preparedness and Response in DRC. Accessed September 3, 2018.

Williams R, Malherbe J, Weepener H et al (2016) Anomalous high rainfall and soil saturation as combined risk indicator of rift valley fever outbreaks, South Africa, 2008–2011. Emerg Infect Dis 22:2054–2062. https://doi.org/10.3201/eid2212.151352

Woolhouse ME, Haydon DT, Antia R (2005) Emerging pathogens: the epidemiology and evolution of species jumps. Trends Ecol Evol 20:238–244

Woolhouse MEJ, Gowtage-Sequeria S (2005) Host range and emerging and Reemerging pathogens. Emerg Infect Dis 11(12):1842–1847

Wulff H, Fabiyi A, Monath TP (1975) Recent isolations of Lassa virus from Nigerian rodents. Bull World Health Organ 52:609–613

Yu H, Zhou YJ, Li GX et al (2009) Further evidence for infection of pigs with human-like H1N1 influenza viruses in China. Virus Res 140:85–90

Chapter 8
Antibiotics Use in African Aquaculture: Their Potential Risks on Fish and Human Health

S. M. Limbu

8.1 Introduction

It is globally undoubtedly accepted that antibiotics have saved many lives and eased the suffering of many millions of animals (Byarugaba 2004). However, antibiotics, of either natural or synthetic origin, are used abusively in human, livestock, agriculture, and aquaculture both to prevent proliferation and destroy bacteria (Mehdi et al. 2018). Consequently, antibiotics exist ubiquitously in the environment and are currently deemed as a global pandemic problem posing a health risk to aquatic animals and humans. Unfortunately, the risks caused by antibiotics globally are expected to continue because, between 2000 and 2015 their consumption increased 65% from 21.1 to 34.8 billion defined daily doses (DDDs), and the antibiotic consumption rate increased 39% from 11.3 to 15.7 DDDs per 1000 inhabitants per day (Klein et al. 2018). Astonishingly, the increase in consumption of antibiotics was driven by low- and middle-income countries (LMICs), a characteristic possessed by the majority of African countries.

The African continent poses peculiar features regarding antibiotics consumption. First, most African countries are generally characterized by poverty, ignorance, poor sanitation, hunger and malnutrition, poor and inadequate health care systems, civil conflicts and bad governance (Byarugaba 2004), coupled with an inappropriate prescription as well as self-medication and free sale of antibiotics (Sanou et al. 2018). Secondly, most African countries have weak regulatory agencies and absence/weak regulations concerning antibiotics usage. Accordingly, antibiotics are indiscriminately given as over-the-counter drugs at community pharmacies

S. M. Limbu (✉)
Department of Aquatic Sciences and Fisheries Technology, University of Dar es Salaam, Dar es Salaam, Tanzania

Laboratory of Aquaculture Nutrition and Environmental Health (LANEH), School of Life Sciences, East China Normal University, Shanghai, People's Republic of China

© Springer Nature Switzerland AG 2020
A. L. K. Abia, G. R. Lanza (eds.), *Current Microbiological Research in Africa*,
https://doi.org/10.1007/978-3-030-35296-7_8

(Mukonzo et al. 2013), which have been strongly correlated with antibiotic-resistant bacteria (ARB) and antibiotics resistance genes (ARGs) in aquatic animals such as fish and humans in LMICs (Alsan et al. 2015). The lack of regulatory agencies in African countries have caused the indiscriminate use of antibiotics in human for disease treatments and as therapeutic and growth promoters in livestock, agriculture, and aquaculture production. As growth promoters, antibiotics are believed to improve feed conversion, promote animal growth, and reduce mortality and morbidity rates resulting from clinical and subclinical illnesses (Foka et al. 2018).

Moreover, antibiotics applied in fish are poorly absorbed in the intestine, and subsequently are released into the aquatic environments where they selectively cause ARB and ARGs (Fu et al. 2017). Thus, antibiotic resistance in bacteria and genes that cause diseases in man is an issue of significant concern, which is expected to become the leading global cause of death by 2050 (O'Neill 2016). Although the misuse of antibiotics in human medicine is the principal cause of ARB and ARBs in Africa, the use of antibiotics in food animals and their subsequent release into the aquatic environments are contributory factors (Barton 2000; Goutard et al. 2017; Adegoke et al. 2018). Apart from ARB and ARGs, antibiotics also cause human health risk due to their residue amounts in various contaminated foods consumed in Africa (Darwish et al. 2013).

The increasing human population in Africa has led to an increase in reliance on aquaculture to supply safe, reliable, and economical food, contributing 10% of the total global population engaged in fisheries and aquaculture, second only to Asia with 84% (FAO 2018). In some African countries such as Ghana and Sierra Leone, fish contributes or exceeds, 50% of total animal protein intake (FAO 2016). Aquaculture production, mainly from catfish and tilapia, accounted for 17–18% of total fish production in Africa (Fig. 8.1), with a general increasing trend (Fig. 8.2) (FAO 2018). The per capita fish food consumption for Africa was reported as 9.9 kg/year in 2015, partly contributed by unreported data (FAO 2018).

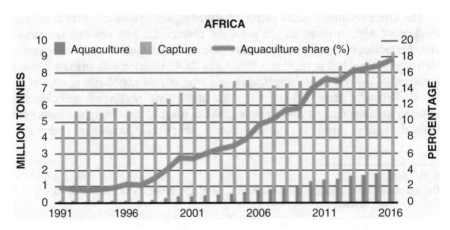

Fig. 8.1 Aquaculture contribution to total fish production (excluding aquatic plants). Source: FAO (2018)

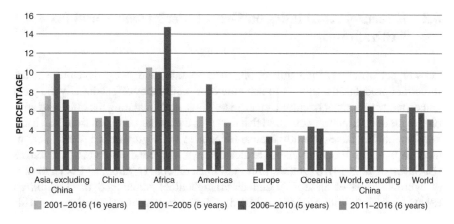

Fig. 8.2 Average annual growth rate of aquaculture production by volume (excluding aquatic plants). Source: FAO (2018)

Although the available information suggests minimal use of antibiotics in aquaculture in African countries, potential contamination of fish from fertilizers used in animals treated with antibiotics is unavoidable (Shah et al. 2012; Wamala et al. 2018), because most farmers fertilize their ponds. However, the effects of antibiotics on fish anatomy and physiology from an African perspective are currently poorly understood, and information on ARB and ARGs in fish and humans due to fish exposure is currently scattered and unfocused. Furthermore, antibiotics are increasingly used in humans and other food animals, with a concomitant prevalence of ARB and ARGs in LMICs (Bernabé et al. 2017). It has been shown that infections caused by ARB may increase health care costs due to patients' need for more diagnostic tests, more extended hospitalization periods, and poor treatment outcomes (Nyasulu et al. 2012). Despite all these, little attention has been directed towards understanding the antibiotics residues in fish, other food animals, and humans in Africa (Adegoke et al. 2018).

For the first time, this chapter organizes and synthesizes the available information in the literature on the potential risks of antibiotics on cultured fish and human health from Africa. The chapter assesses the effects of antibiotics on fish growth performance, feed utilization, hepatotoxicity and nephrotoxicity, and hematological parameters. It further evaluates the potential human health risks caused by the existence of ARB and ARGs in fish and other consumed foods, in addition to direct risks due to the consumption of fish products containing antibiotics residues. The information generated informs policies to limit the use of antibiotics in food animals by enforcing policies, which regulate their use in Africa to safeguard human health.

8.2   Effects of Antibiotics on Growth Performance and Feed Utilization

Growth performance, feed utilization, survival rate, and body development are important production attributes to fish growers because they affect directly the yield and economics of an aquaculture enterprise. Thus, understanding the effects of antibiotics on these aspects in cultured fish deserves a peculiar consideration. In Africa, very few studies have currently used antibiotics to study growth and related parameters on fish. The literature visited indicated oxytetracycline studied in *O. niloticus* (El-Sayed et al. 2014), oxytetracycline and florfenicol in *O. niloticus* ♀ × *O. aureus* ♂ hybrids (Reda et al. 2013), and chloramphenicol researched in *O. niloticus* (Shalaby et al. 2006) and African catfish, *Clarias gariepinus* (Nwani et al. 2014) were the only antibiotics used. Results from these few studies indicated improved growth of treated fish compared to controls. For example, the growth performance of *O. niloticus* increased significantly with increasing levels of chloramphenicol (Shalaby et al. 2006). Moreover, feeding diets containing oxytetracycline and florfenicol in *O. niloticus* ♀ × *O. aureus* ♂ hybrids (Reda et al. 2013) and *O. niloticus* (El-Sayed et al. 2014) resulted into faster growth performance in treated than control fish.

The precise reasons for the enhanced growth performance of fish after antibiotics administration are subject to scrutiny. Increased growth has been attributed to higher feed consumption and reduced feed conversion ratio. Indeed, the growth rate was increased in *O. niloticus* treated with chloramphenicol (Shalaby et al. 2006), oxytetracycline and florfenicol (Reda et al. 2013), and oxytetracycline (El-Sayed et al. 2014), in which feed consumption and intake were increased, while feed conversion ratio was reduced. Moreover, apparent protein, lipid, carbohydrate, and energy digestibility were increased in *O. niloticus* fed on chloramphenicol (Shalaby et al. 2006). These results should be interpreted with caution due to limited studies and the existence of contradicting results elsewhere. It has been recently shown that antibiotics, particularly oxytetracycline, do not cause growth promotion in finfish (Trushenski et al. 2018) and causes multiple effects in Nile tilapia including reduced nutrients digestibility and digestive enzymes (Limbu et al. 2018), growth performance (Limbu et al. 2019a; Limbu et al. (2019b), protein and feed efficiencies (Limbu et al. 2019b).

Like growth performance, studies conducted on the effects of antibiotics on survival rate are also limited. The results obtained in the limited studies do not show any influence of antibiotics on fish survival rate. Exposure to dietary oxytetracycline (El-Sayed et al. 2014) and chloramphenicol (Shalaby et al. 2006) both in *O. niloticus* and chloramphenicol in *C. gariepinus* (Nwani et al. 2014) did not significantly influence survival rate of treated fish relative to control. Results detailing the effects of antibiotics on body development are based on *C. gariepinus* exposed to chloramphenicol baths (Nwani et al. 2014). In this study, treated fish had abnormal behavioral changes at higher concentration of chloramphenicol. The fish swam near the water surface, lost equilibrium, swam erratically, had hyperactivity, and stayed motionless on the bottom of the culture tank. Furthermore, exposed fish had clinical toxic signs such as lightening in skin color of the body surface, erosion of fins and tails, and increased mucus secretions from the whole body. These results indicate that antibiotics application on fish leads to body malformation and damage, which may lead to physiological and metabolic dysfunctions affecting fish health.

8.3   Effects of Antibiotics on Fish Health

8.3.1   Oxidative Stress, Hepatotoxicity, and Nephrotoxicity

Antibiotics used in fish production induce oxidative stress, which affects antioxidant enzymes that protect fish body from reactive oxygen species (ROS) (Limbu et al. 2018). Changes in the activities of antioxidant enzymes indicate an imbalance in the ROS production in the body. Limited studies have been conducted in Africa to assess the antioxidant capacity of fish exposed to antibiotics. A study conducted by Olaniran et al. (2018) indicated reduced glutathione S transferase (GST) and superoxide dismutase (SOD) activities in *C. gariepinus* exposed to tetracycline. The decreased antioxidants in fish treated with antibiotics may be caused by an excess accumulation of free radicals, such as superoxide anion and hydrogen peroxide beyond the antioxidant capacity to counteract (Yonar et al. 2011; Yonar 2012; Oliveira et al. 2013; Wang et al. 2014). This may oxidize amino acids and cofactors, which may affect the general fish health. Malondialdehyde (MDA) is the main oxidative product of peroxidized polyunsaturated fatty acids and is an important index of lipid peroxidation. The extent of lipid peroxidation is measured in tissues by quantification of thiobarbituric acid reactive substances (TBARS) expressed as MDA concentration (Nunes et al. 2015). Lipid peroxidation is the initial step of cellular membrane damage caused by xenobiotics such antibiotics (Yonar et al. 2011; Yonar 2012). Limited studies have reported on lipid peroxidation using MDA. Reduced MDA level was reported in *C. gariepinus* exposed to tetracycline, indicating lack of lipid peroxidation (Olaniran et al. 2018).

The liver of fish and other vertebrates is known for its digestive, metabolism, storage, and detoxification functions. The introduction of antibiotics in fish body through medicated feeds may cause liver damage effects that might impair its functions (Dobšíková et al. 2013). Limited studies have reported on the effects of antibiotics on the hepatosomatic index (HSI) as an indicator of hepatotoxicity in fish. The chloramphenicol medicated feeds used in *O. niloticus* did not cause significant variations in HSI in experimental compared to control fish (Shalaby et al. 2006). Few studies conducted limit the ability to draw logical conclusions. Studies conducted in other parts of the world showed HSI was reduced (Refstie et al. 2006; Limbu et al. 2018) and increased (Topic Popovic et al. 2012; Nakano et al. 2018; Trushenski et al. 2018) in different fish species.

The amount of circulating proteins reflects an organism's physiology. Plasma proteins and glucose in the circulatory system transport lipids, hormones, vitamins, and minerals and regulate cellular activities, functioning of the immune system, and blood clotting. Imbalances in the plasma protein and glucose counts indicate liver damage, which interferes with its normal functions. A dose-dependent increase in plasma protein levels was reported in *O. niloticus* exposed to chloramphenicol diet, indicating osmoregulatory dysfunction, hemodilution, or tissue damage surrounding blood vessels (Shalaby et al. 2006). Moreover, plasma glucose increased significantly with increasing levels of chloramphenicol (Shalaby et al. 2006). Although limited, this study indicated that antibiotic medications in cultured fish lead to disturbances in plasma proteins and glucose, which indicate hepatotoxicity.

Except for plasma proteins and glucose, liver dysfunction is manifested by increased levels of specific serum enzymes activities, which signal cellular leakage and impaired liver cell membrane integrity and function. Alanine transaminase (ALT) and aspartate aminotransaminase (AST) are required in the metabolism of amino acids, and their change in activities reflect their leakage into the blood after cytolysis in the liver (Han et al. 2014). Thus, AST and ALT enzymes are commonly used to detect hepatotoxicity due to xenobiotics exposure (Saravanan et al. 2012). Studies from Africa assessing liver damage in cultured fish after antibiotics by using AST and ALT have reported contrasting results. The administration of florfenicol diet in *O. niloticus* did not alter ALT activity (Reda et al. 2013).

On the contrary, the activities of AST and ALT in plasma decreased significantly with increasing levels of dietary chloramphenicol in *O. niloticus* (Shalaby et al. 2006). The observed decrease in AST and ALT activities in fish is either due to insufficient detoxification mechanisms to prevent the toxicity action of antibiotics on these enzymes or failure of liver damaged cells to synthesize AST and ALT proteins (Saravanan et al. 2012). On the other hand, oxytetracycline-supplemented diets increased significantly ALT activity in *O. niloticus* (Reda et al. 2013). Increased ALT activity is due to the ability of antibiotics to accumulate or bind to different cells leading to damage and disintegration of cells, releasing ALT into blood circulation, suggesting impaired liver function.

Histopathological effects provide a quick diagnosis to detect abnormalities in various fish tissues and organs after antibiotics exposure. Antibiotics use indicate species- and antibiotic-specific histopathological effects in the liver and kidney of treated fish. Feeding dietary oxytetracycline and florfenicol in *O. niloticus* (Reda et al. 2013) induced various pathological alterations in liver and kidney of treated fish. Moreover, both dietary oxytetracycline and florfenicol decreased creatinine in the treated *O. niloticus* than the control fish (Reda et al. 2013). The existence of several histopathological damages in the liver of treated fish is due to liver degenerations (Reda et al. 2013) and inhibition of somatic cells in mitochondrial protein synthesis by antibiotics resulting in lack of oxidative ATP-generating capacity, which causes proliferation arrest of normal and malignant epithelial cells (Bakke-McKellep et al. 2007). These changes induce hepatotoxicity and nephrotoxicity.

8.3.2 Effects of Antibiotics on Hematological Parameters

Hematological parameters provide essential information on the health of cultured fish after antibiotics application. Results conducted in hematological parameters are still contrasting. Dietary chloramphenicol exposure did not affect mean corpuscular volume (MCV) and mean corpuscular hemoglobin concentration (MCHC) in *O. niloticus* (Shalaby et al. 2006) and monocytes, eosinophils, and basophils in *C. gariepinus* (Nwani et al. 2014). Moreover, florfenicol did not show significant differences in immunoglobulin M (IgM) total levels and phagocytic activity in *O. niloticus* when compared to the control fish (Reda et al. 2013).

However, a concentration- and time-dependent decrease in hemoglobin (Hb), red blood cells (RBC) counts, and MCV were detected in *C. gariepinus* exposed to chloramphenicol bath (Nwani et al. 2014). The different toxic effects of chloramphenicol bath on various organs caused the observed decrease in Hb, RBC, and MCV in fish. Chloramphenicol suppressed the production of hematological parameters caused by their toxic accumulation in lymphoid organs and pronephros (Nwani et al. 2014). The decreased RBC counts after exposure to antibiotics is due to swelling of RBC, the release of immature erythrocytes, anemia caused by tissues damage, damaged RBC, decrease in erythrocyte life span, and suppressive effects of antibiotics on erythropoietic tissues (Shalaby et al. 2006; Nwani et al. 2014). The deecreased Hb may limit the oxygen-carrying capacity of the fish blood (Nwani et al. 2014) and affect their survival rate. The inhibition of these hematological parameters may lead to sustained toxic effects caused by both dietary and bath exposure to antibiotics, resulting in tissue damage and immunity suppression with possible fatal outcomes.

Notwithstanding the above results, dietary chloramphenicol exposure in *O. niloticus* increased RBC, Hb, and hematocrit (Shalaby et al. 2006) and its bath elevated WBC, neutrophil count, and lymphocytes in *C. gariepinus* (Nwani et al. 2014). The use of oxytetracycline diet increased lysozyme activity in *O. niloticus* (Reda et al. 2013). Increased WBC count and lysozyme activity indicate a protective mechanism of the fish body to antibiotics-induced stress, a condition termed as leukocytosis, which signals a response of damaged tissues and immune system stimulation to counteract antibiotics toxicity (Ambili et al. 2013). An increase in RBC is due to a compensation mechanism for impaired oxygen uptake caused by tissue damages due to the presence of antibiotics in the fish body and high percentage of circulating immature RBC (Ambili et al. 2013). On the other hand, the increased lymphocyte count (lymphocytosis) and the formation of blood cellular components (hematopoiesis) are features of infection due to increased disease-fighting cells after antibiotics exposure in fish. In general, dietary and bath antibiotics exposure in cultured fish cause leukocytosis, hematopoiesis, and lymphocytosis, suggesting sustained toxic effects and compensatory responses to conciliate the fish body to normal health conditions.

8.4   Potential Human Health Risks from Consumption of Antibiotics-Cultured Fish

8.4.1   Antibiotic-Resistant Bacteria and Antibiotic Resistance Genes

The widespread and indiscriminate use of antibiotics in different environmental compartments including fish, agriculture, and human health have led to the development of ARB, ARGs, and transposons. Resistant bacteria and resistance genes may be horizontally or vertically transferred among bacterial communities, the environment, and finally human being via transposons (Biyela et al. 2004). The presence of ARB and ARGs in humans affects the ability of antibiotics to treat diseases and thus compromise their health. Thus, presently, the existence of ARB and ARGs in the environments, particularly those conferring resistance to antibiotics used to treat human diseases, is an issue of major global concern. Although the misuse of antibiotics in human medicine is the principal cause of the problem, ARB and ARGs originating from animals such as fish and agriculture production are also responsible (Barton 2000).

The literature shows that ARB and ARGs pose a human health risk in various African countries contributed by consuming contaminated fish, shrimp, vegetables, and various food sources as well as drinking contaminated water (Table 8.1). The human health risk posed by ARB from fish consumption appears to be widely spread because both cultured and wild fish have been shown to contain them. Various ARB have been isolated in cultured fish from Ghana (Agoba et al. 2017), Tanzania (Shah et al. 2012; Mhongole et al. 2017), and Uganda (Bosco et al. 2012; Wamala et al. 2018). Moreover, wild fish from Uganda (Wamala et al. 2018), Algeria (Dib et al. 2018), South Africa (Fri et al. 2018), and Egypt (Ramadan et al. 2018) were all shown to contain ARB. The ARB contained in fish in the different countries originate from various sources including animal-origin fertilizers (Shah et al. 2012; Omojowo and Omojasola 2013), the aquatic environment (Stenstrom et al. 2016), and possibly fish feeds. It is possible that the ARB from the different compartments are transferred to humans in Africa. Indeed, ARB have been detected in humans from Ethiopia (Kibret and Abera 2014), Ghana (Obeng-Nkrumah et al. 2013), Ivory Coast (Moroh et al. 2014), Libya (Mohammed et al. 2016), and Morocco (El Bouamri et al. 2015). This is an alarming situation because most of the bacteria isolated exhibited high resistance to common antibiotics used for treating frequently occurring diseases in humans in Africa and most of them had multiple antibiotic resistance (MAR) (Bosco et al. 2012; Omojowo and Omojasola 2013; Mohammed et al. 2016; Agoba et al. 2017; Apenteng et al. 2017; Wamala et al. 2018). Although correct and appropriate food cooking procedures may kill bacteria, contamination can occur through improper handling before cooking (Darwish et al. 2013) and possibly through bacteria-human contact because ARB are ubiquitous (Mhongole et al. 2017). Indeed, high levels of antimicrobial resistance (AMR) were obtained in food

Table 8.1 Antibiotic-resistant bacteria isolated from fish and other environments in Africa

Resistant bacterial strain	Resistance to antibiotic	Sample isolated	Country	Reference
Most isolated bacteria (≥70%)	Penicillin, ampicillin, flucloxacillin, and tetracycline	Catfish and tilapia farms	Ghana	Agoba et al. (2017)
10% of isolates	Resistant to all the nine tested antimicrobials (MAR)	Water, sediments, and fishpond	Tanzania	Shah et al. (2012)
Pseudomonas aeruginosa	77.78% were MAR	Fish ponds	Ghana	Apenteng et al. (2017)
Salmonella typhi	70% were MAR			
Escherichia coli	66.67% Resistance to more than two classes			
Salmonella spp.	82.7% Resistant to trimethoprim sulfamethoxazole	Human	Uganda	Bosco et al. (2012)
	85.3% resistant to trimethoprim-sulfamethoxazole	Animal-food origin		
Salmonella spp.	94% Sulfamethoxazole, 61% streptomycin, 22% tetracycline, 17% ciprofloxacin and nalidixic acid, 11% trimethoprim, and 6% gentamycin and chloramphenicol	Fish from pond and wastewater	Tanzania	Mhongole et al. (2017)
Aeromonas spp.	100% Penicillin and ampicillin and 23.2% cefotaxime	Fish from pond and water from wild	Uganda	Wamala et al. (2018)
Plesiomonas shigelloides	100% Penicillin and oxacillin			
Escherichia coli	100% Ampicillin, amoxicillin, cephalothin, amikacin, kanamycin, gentamicin, neomycin, and tobramycin	Wild fish and shrimp	Algeria	Dib et al. (2018)
Vibrio spp.	76.2% Amoxicillin, 67.5% ampicillin, 38.3% erythromycin, and 35.0% doxycycline	Wild fish and water from fish farms	South Africa	Fri et al. (2018)
Aeromonas hydrophila	100% Cefoxitin, 84% ampicillin, 56% ceftazidime, and 40% cefotaxime	Fish from market	Egypt	Ramadan et al. (2018)
Escherichia coli, Aeromonas hydrophila, Salmonella typhi, Staphylococcus aureus, and *Shigella dysenteriae*	100% Tetracycline, 85.6% ampicillin, 83.3% amoxicillin, 47.6% gentamicin, 66% chloramphenicol, 44.4% erythromycin, and 18.3% nalidixic acid	Cow dung fertilizer for fishponds	Nigeria	Omojowo and Omojasola (2013)

(continued)

Table 8.1 (continued)

Resistant bacterial strain	Resistance to antibiotic	Sample isolated	Country	Reference
Acinetobacter spp.	30–100% Penicillin G, ceftriaxone, nitrofurantoin, erythromycin, and augmentin, 10% oxytetracycline, and 9% minocycline	Freshwater and soil samples	South Africa	Stenstrom et al. (2016)
Escherichia coli, Klebsiella spp., and *Proteus* spp.	85.6% Erythromycin, 88.9% amoxycillin, and 76.7% tetracycline	Human	Ethiopia	Kibret and Abera (2014)
ESBL producers	92.6% Cotrimoxazole, 91.2% gentamicin, 44.8% amikacin, and 41.1% ciprofloxacin	Human	Ghana	Obeng-Nkrumah et al. (2013)
Escherichia coli, Staphylococcus aureus, Klebsiella pneumoniae, and *Enterobacter aerogenes*	78.9% Amoxicillin, 73.1% tetracycline, and 81.8% trimethoprim/sulfamethoxazole	Human	Ivory Coast	Moroh et al. (2014)
Klebsiella oxytoca	64.5% MAR	Human	Libya	Mohammed et al. (2016)
Providencia rettgeri	63.2% MAR			
Pseudomonas aeruginosa	52.1% MAR			
Acinetobacter baumannii, Citrobacter freundii, and *Enterobacter aerogenes*	47.4% MAR			
Enterobacter amnigenus biogroup 2	42.1% MAR			
Enterobacter cloacae	(40.8%) MAR			
ESBL-producing *Klebsiella pneumoniae* strains	89% Trimethoprim–sulfamethoxazole, 89% gentamicin, 84% ciprofloxacin, and 50% amikacin	Human	Morocco	El Bouamri et al. (2015)

Key: *MAR* multiple antibiotic resistance and *ESBL* Extended-Spectrum Beta-Lactamase

animals including fish intended for human consumption in Nigeria (Oloso et al. 2018).

Consistent to the existence of ARB in fish from aquaculture and wild environments, their corresponding ARGs also have been detected in cultured fish from Tanzania (Shah et al. 2012) and wild fish from Tanzania (Moremi et al. 2016), Egypt (Ramadan et al. 2018), Algeria (Brahmi et al. 2018; Dib et al. 2018), and South Africa (Fri et al. 2018) (Table 8.2). Coherent to ARB, it is possible that the ARGs detected originate from the aquatic environment (Adesoji and Ogunjobi 2016; Lyimo et al. 2016; Stenstrom et al. 2016). In general, the human health risk

Table 8.2 Antibiotic resistance genes isolated from fish and other environments in Africa

Antibiotic resistance genes	Resistance to antibiotic	Sample isolated	Country	Reference
tetA(A) and *tetA(G)*	Tetracycline	Water, sediments, and fishpond	Tanzania	Shah et al. (2012)
sul1 and *sul2*	Sulfonamides			
intl1 and *int2*	Transfer of genes[a]			
dfrA1, dfrA7, dfrA12	Trimethoprim			
strA-strB	Streptomycin			
cat-1	Chloramphenicol			
bla$_{TEM}$	β-Lactam/amoxicillin			
mefA	Erythromycin			
sul1, sul2	Sulfonamides	Wild fish and water samples	Tanzania	Moremi et al. (2016)
tet(A), tet(B)	Tetracycline			
*aac(6′)-Ib-cr, qnr*S1	Fluoroquinolones			
*aac(3)-lld, str*B, *str*A	Aminoglycosides			
*dfr*A14	Trimethoprim			
bla$_{CTX-M-15}$	β-Lactams			
bla$_{TEM}$, *bla*$_{CTX-M}$, *bla*$_{CMY}$, *bla*$_{OXA}$	β-Lactams	Fish from market	Egypt	Ramadan et al. (2018)
bla$_{CTX-M-15}$	β-Lactams	Wild fish and shrimp	Algeria	Dib et al. (2018)
bla$_{CTX-M}$	β-Lactams	Wild fish	Algeria	Brahmi et al. (2018)
oqxAB	Quinolones			
Qnr, aac(6′)-Ib-cr	Fluoroquinolones			
bla$_{OXA}$	β-Lactams	Wild fish and water from fish farms	South Africa	Fri et al. (2018)
Tet(A), tet(M)	Tetracycline			
sul1, sul2	Sulfonamides			
*dfr*1	Trimethoprim			
*erm*B	Macrolides, lincosamides, and streptogramin			
*str*A	Aminoglycosides			
nptII	Neomycin			
SXT integrase	Transfer of genes[a]			
Tet(A), tet(E) tet(B), tet(M), Tet39	Tetracycline	Treated and untreated water	Nigeria	Adesoji et al. (2015)
Tet(A), Tet(B)	Tetracycline	Drinking water sources	Tanzania	Lyimo et al. (2016)
bla$_{TEM-1}$, *bla*$_{CTX-M}$	β-Lactams			
bla$_{TEM}$, *bla*$_{SHV}$, *bla*$_{CTX}$	β-Lactams	Drinking water sources	Nigeria	Adesoji and Ogunjobi (2016)
blaAIM-1, blaGES-21	β-Lactams	Wastewater	Burkina Faso	Bougnom et al. (2019)
Enterobacteriaceae plasmid replicons	Transfer of genes[a]			

(continued)

Table 8.2 (continued)

Antibiotic resistance genes	Resistance to antibiotic	Sample isolated	Country	Reference
Tet(B), Tet(39)	Tetracycline	Freshwater and soil samples	South Africa	Stenstrom et al. (2016)
Sul 3	Sulfonamides	Rhizospheres plant	South Africa	Adegoke and Okoh (2015)
bla_{CTX-M}, bla_{TEM},	β-Lactams	Various foods	Egypt	Hammad et al. (2018)
tet(A), tet(E)	Tetracycline			
intI1	Transfer of genes[a]			
bla_{OXA-23}, bla_{OXA-51}	Carbapenems	Human	Senegal	Diene et al. (2013)

Key: *ESBL* Extended-Spectrum Beta-Lactamases
[a]Indicates transposons responsible for the transfer of ARGs

associated with ARGs is not only contributed by fish. Reasonably, ARGs have also been found in drinking water in Tanzania (Lyimo et al. 2016) and Nigeria (Adesoji et al. 2015; Adesoji and Ogunjobi 2016), Rhizospheres plants in South Africa (Adegoke and Okoh 2015), various foods in Egypt (Hammad et al. 2018), and wastewater used for urban agriculture in Burkina Faso (Bougnom et al. 2019). Accordingly, ARGs have been detected in the human body in Senegal (Diene et al. 2013).

Similar to ARB, the ARGs detected are those encoding resistance to common antibiotics used for the frequent treatment of human diseases in Africa. Thus, Africans are currently exposed to a double resistance to antibiotics due to the presence of ARB and ARGs. Indeed, transposons and plasmids for transfer of ARGs have been detected in wild fish and water from fish farms in Tanzania (Shah et al. 2012) and South Africa (Fri et al. 2018), various foods in Egypt (Hammad et al. 2018), and wastewater used for agriculture in Burkina Faso (Bougnom et al. 2019). Since antibiotics exist ubiquitously in the environment, Africans are exposed to high health risks due to their close interaction with livestock and the aquatic ecosystem (Wamala et al. 2018), which signifies increased morbidity and mortality (Gyansa-Lutterodt 2013) due to the failure of antibiotics to treat bacterial diseases. The obtained results emphasize the need for policies and mechanisms to limit the use of antibiotics in food animals production in order to protect human health. Moreover, physicians should devise some methods to change patients' treatment pattern depending on antibiotics susceptibility results. Antibiotics may also pose direct public health effects due to their residuals in different foods consumed by humans as detailed below.

8.4.2 Direct Potential Human Health Risk from Consumption of Fish

Globally, antibiotics residues in foods have attracted much attention in recent years because of growing food safety and public health concerns (Capita and Alonso-Calleja 2011; Landers et al. 2012; Berendonk et al. 2015). Their presence in food animals represent socioeconomic challenges in global trade and consumed animal products (Okocha et al. 2018). In most countries, the use of antibiotics for food animals production requires a withdrawal period before the product can be sold for human consumption. Despite this regulation, most antibiotics are used without observing such a regulation both in fish (Pham et al. 2015) and other animals (Mubito et al. 2014). Consequently, high levels of antibiotics exist in food animals intended for human consumption, which pose a direct human health risk.

Limited studies have been conducted in Africa to detect antibiotics residues in fish and other foods. In Nigeria, Olatoye and Basiru (2013) found oxytetracycline levels in cultured *C. gariepinus* in the liver and fillets exceeded the Codex Alimentarius Commission established maximum residue limit of 600 and 200 µg/kg, respectively (Table 8.3). Similarly, a study conducted by Olusola et al. (2012) in Nigeria also found tetracycline exceeded international limits of 200 µg/kg from fresh and frozen *C. gariepinus* and *O. niloticus,* while chloramphenicol, which has a zero tolerance level, was detected in Officers' Mess. It has been reported that, in

Table 8.3 Residues of antibiotics from fish and other environments in Africa

Antibiotic	Residue amount	Tissue/sample	Country	Reference
Oxytetracycline	875.32 ± 45 µg/kg	Fish liver	Nigeria	Olatoye and Basiru (2013)
	257.2 ± 133 µg/kg	Fish fillets		
Tetracycline	2185 ± 412 µg/kg	Fresh and frozen fish	Nigeria	Olusola et al. (2012)
Chloramphenicol	837 ± 165 µg/kg			
Ampicillin	0.36 ± 0.04 µg/L	WWTPs	Kenya	Kimosop et al. (2016)
	0.79 ± 0.07 µg/L	Hospital		
Sulfamethoxazole	1.8 µg/L	River water	Kenya	Ngumba et al. (2016)
Trimethoprim	0.327 µg/L			
Ciprofloxacin	0.129 µg/L			
Sulfamethoxazole	0.02–38.85 µg/L	River water	Kenya	K'Oreje et al. (2016)
Trimethoprim	0.05–6.95 µg/L			
Tetracycline	0.85 ± 0.06 µg/mL	Surface water	Nigeria	Olaniran et al. (2018)
	0.23 ± 0.01 µg/mL	Untreated effluent water		
Sulfamethoxazole	34.50 µg/L	WWTPs	South Africa	Matongo et al. (2015)
Oxytetracycline	785.58 ± 210.80 µg/L	Cow milk	Tanzania	Ridhiwani (2015)
Oxytetracycline	2604.1 ± 703.7 µg/kg	Cattle muscle	Tanzania	Kimera et al. (2015)
	3434.4 ± 606.4 µg/kg	Cattle liver		
	3533.1 ± 803.6 µg/kg	Cattle kidney		

Africa, as in other parts of the world, antibiotic residues in animal-derived foods more commonly exceed the world health organization (WHO) threshold residue levels (Darwish et al. 2013). This further highlights the high human health risk caused by antibiotics because they have also been detected in wastewater treatment plants (WWTPs) in Kenya (Kimosop et al. 2016) and South Africa (Matongo et al. 2015), surface water in Nigeria (Olaniran et al. 2018), hospitals (Kimosop et al. 2016) and river water (K'Oreje et al. 2016; Ngumba et al. 2016) in Kenya, cow's milk in Tanzania (Ridhiwani 2015) and Algeria (Layada et al. 2016), untreated effluent water from a cow market in Nigeria (Olaniran et al. 2018), and cattle muscle, liver, and kidney (Kimera et al. 2015) in Tanzania.

In practice, the human health risk resulting from antibiotics in Africa may be much higher because of multiple sources. Antibiotics are widely abused by humans for therapy, sometimes without physicians' prescription, and the quantity of antibiotics prescribed in African countries intended for the treatment of various diseases are high (Adegoke et al. 2018), contributing to elevated levels of residues. The antibiotic residues have been reported to spread rapidly, irrespective of geographical, economic, or legal differences in African countries (Darwish et al. 2013). This represents a serious concern because antibiotics, particularly chronic dietary oxytetracycline used in fish production, have been recently reported to cause direct human health risk in children (Limbu et al. 2018). In general, despite the existence of limited studies on antibiotics residues from cultured fish, the results obtained from *C. gariepinus* and *O. niloticus* suggest a widespread human health risk because the two fish species are widely consumed in African countries. Thus, there is an urgent need to control the use of antibiotics in fish intended for human consumption in order to protect human health.

8.5 Conclusion

It is clear that studies on antibiotics used in aquaculture production in Africa are still limited particularly on effects on fish anatomy and physiology. However, the existing limited data highlight toxic effects of antibiotics in the fish body and increasing prevalence of ARB and ARGs coupled with high residues of antibiotics in cultured fish, which pose a significant human health risk. The African countries require coordinated actions to tackle the indiscriminate use of antibiotics in humans, livestock, agriculture, and aquaculture at its grassroots, because currently most of them are characterized by inadequate monitoring, surveillance and weak regulatory systems. Clear policy directions for prohibiting the use of antibiotics on food animals production are urgently needed to protect human health. More studies should be conducted on the potential risks of antibiotics on fish and human health resulting from multiple exposure scenarios.

References

Adegoke AA, Okoh AI (2015) Antibiogram of *Stenotrophomonas maltophilia* isolated from Nkonkobe Municipality, Eastern Cape Province, South Africa. Jundishapur J Microbiol 8(1):e13975. https://doi.org/10.5812/jjm.13975

Adegoke AA, Faleye AC, Stenström TA (2018) Residual antibiotics, antibiotic resistant superbugs and antibiotic-resistance genes in surface water catchments: public health impact. Phys Chem Earth Parts A/B/C 105:177–183. https://doi.org/10.1016/j.pce.2018.03.004

Adesoji AT, Ogunjobi AA (2016) Detection of extended-spectrum beta-lactamases resistance genes among bacteria isolated from selected drinking water distribution channels in southwestern Nigeria. Biomed Res Int 2016:9. https://doi.org/10.1155/2016/7149295

Adesoji AT, Ogunjobi AA, Olatoye IO et al (2015) Prevalence of tetracycline resistance genes among multi-drug resistant bacteria from selected water distribution systems in southwestern Nigeria. Ann Clin Microbiol Antimicrob 14(1):35. https://doi.org/10.1186/s12941-015-0093-1

Agoba EE, Adu F, Agyare C et al (2017) Antibiotic resistance patterns of bacterial isolates from hatcheries and selected fish farms in the Ashanti region of Ghana. J Microbiol Antimicrob 9(4):35–46. https://doi.org/10.5897/JMA2017.0387

Alsan M, Schoemaker L, Eggleston K et al (2015) Out-of-pocket health expenditures and antimicrobial resistance in low-income and middle-income countries: an economic analysis. Lancet Infect Dis 15(10):1203–1210. https://doi.org/10.1016/S1473-3099(15)00149-8

Ambili TR, Saravanan M, Ramesh M et al (2013) Toxicological effects of the antibiotic oxytetracycline to an Indian major carp *Labeo rohita*. Arch Environ Contam Toxicol 64(3):494–503. https://doi.org/10.1007/s00244-012-9836-6

Apenteng JA, Osei-Asare C, Oppong EE et al (2017) Antibiotic sensitivity patterns of microbial isolates from fish ponds: a study in the Greater Accra Region of Ghana. Afr J Pharm Pharmacol 11(28):314–320. https://doi.org/10.5897/AJPP2017.4789

Bakke-McKellep AM, Penn MH, Salas PM et al (2007) Effects of dietary soyabean meal, inulin and oxytetracycline on intestinal microbiota and epithelial cell stress, apoptosis and proliferation in the teleost Atlantic salmon (*Salmo salar* L.). Br J Nutr 97(4):699–713. https://doi.org/10.1017/s0007114507381397

Barton MD (2000) Antibiotic use in animal feed and its impact on human healt. Nutr Res Rev 13(2):279–299. https://doi.org/10.1079/095442200108729106

Berendonk TU, Manaia CM, Merlin C et al (2015) Tackling antibiotic resistance: the environmental framework. Nat Rev Microbiol 13(5):310–317. https://doi.org/10.1038/nrmicro3439

Bernabé KJ, Langendorf C, Ford N et al (2017) Antimicrobial resistance in West Africa: a systematic review and meta-analysis. Int J Antimicrob Agents 50(5):629–639. https://doi.org/10.1016/j.ijantimicag.2017.07.002

Biyela PT, Lin J, Bezuidenhout CC (2004) The role of aquatic ecosystems as reservoirs of antibiotic resistant bacteria and antibiotic resistance genes. Water Sci Technol 50(1):45–50. https://doi.org/10.2166/wst.2004.0014

Bosco KJ, Kaddumulindwa DH, Asiimwe BB (2012) Antimicrobial drug resistance and plasmid profiles of isolates from humans and foods of animal origin in Uganda. Adv Infect Dis 2(4):151–155. https://doi.org/10.4236/aid.2012.24025

Bougnom BP, Zongo C, McNally A et al (2019) Wastewater used for urban agriculture in West Africa as a reservoir for antibacterial resistance dissemination. Environ Res 168:14–24. https://doi.org/10.1016/j.envres.2018.09.022

Brahmi S, Touati A, Dunyach-Remy C et al (2018) High prevalence of extended-spectrum β-lactamase-producing *Enterobacteriaceae* in wild fish from the Mediterranean Sea in Algeria. Microb Drug Resist 24(3):290–298. https://doi.org/10.1089/mdr.2017.0149

Byarugaba DK (2004) Antimicrobial resistance in developing countries and responsible risk factors. Int J Antimicrob Agents 24(2):105–110. https://doi.org/10.1016/j.ijantimicag.2004.02.015

Capita R, Alonso-Calleja C (2011) Antibiotic-resistant bacteria: a challenge for the food industry. Crit Rev Food Sci Nutr 53(1):11–48. https://doi.org/10.1080/10408398.2010.519837

Darwish WS, Eldaly EA, El-Abbasy MT et al (2013) Antibiotic residues in food: the African scenario. Jpn J Vet Res 61(Suppl):S13–S22

Dib AL, Agabou A, Chahed A et al (2018) Isolation, molecular characterization and antimicrobial resistance of Enterobacteriaceae isolated from fish and seafood. Food Control 88:54–60.
https://doi.org/10.1016/j.foodcont.2018.01.005

Diene SM, Fall B, Kempf M et al (2013) Emergence of the OXA-23 carbapenemase-encoding gene
in multidrug-resistant *Acinetobacter baumannii* clinical isolates from the principal Hospital of
Dakar, Senegal. Int J Infect Dis 17(3):e209–e210. https://doi.org/10.1016/j.ijid.2012.09.007

Dobšíková R, Blahová J, Mikulíková I et al (2013) The effect of oyster mushroom β-1.3/1.6-D-
glucan and oxytetracycline antibiotic on biometrical, haematological, biochemical, and immunological indices, and histopathological changes in common carp (*Cyprinus carpio* L.). Fish
Shellfish Immunol 35(6):1813–1823. https://doi.org/10.1016/j.fsi.2013.09.006

El Bouamri MC, Arsalane L, El Kamouni Y et al (2015) Antimicrobial susceptibility of urinary
Klebsiella pneumoniae and the emergence of carbapenem-resistant strains: a retrospective
study from a university hospital in Morocco, North Africa. Afr J Urol 21(1):36–40. https://doi.
org/10.1016/j.afju.2014.10.004

El-Sayed SAA, Ahmed SYA, Abdel-Hamid NR (2014) Immunomodulatory and growth performance effects of ginseng extracts as a natural growth promoter in comparison with oxytetracycline in the diets of Nile tilapia (*Oreochromis niloticus*). Int J Livest Res 4(1):130–142

FAO (2016) The state of world fisheries and aquaculture 2016. Contributing to food security and
nutrition for all. Food and Agriculture Organization of the United Nations (FAO), Rome, p 200

FAO (2018) The state of world fisheries and aquaculture 2018—meeting the sustainable development goals. Food and Agriculture Organization of the United Nations (FAO), Rome, p 210

Foka FET, Kumar A, Ateba CN (2018) Emergence of vancomycin-resistant enterococci in South
Africa: implications for public health. S Afr J Sci 114(9/10):1–7. https://doi.org/10.17159/
sajs.2018/4508

Fri J, Ndip RN, Njom HA et al (2018) Antibiotic susceptibility of non-cholera Vibrios isolated
from farmed and wild marine fish (*Argyrosomus japonicus*), implications for public health.
Microb Drug Resist 24(9):1296–1304. https://doi.org/10.1089/mdr.2017.0276

Fu J, Yang D, Jin M et al (2017) Aquatic animals promote antibiotic resistance gene dissemination in water via conjugation: role of different regions within the zebrafish intestinal tract,
and impact on fish intestinal microbiota. Mol Ecol 26(19):5318–5333. https://doi.org/10.1111/
mec.14255

Goutard FL, Bordier M, Calba C et al (2017) Antimicrobial policy interventions in food animal
production in South East Asia. BMJ 358:j3544. https://doi.org/10.1136/bmj.j3544

Gyansa-Lutterodt M (2013) Antibiotic resistance in Ghana. Lancet Infect Dis 13(12):1006–1007.
https://doi.org/10.1016/S1473-3099(13)70196-8

Hammad AM, Moustafa A-EH, Mansour MM et al (2018) Molecular and phenotypic analysis of hemolytic *Aeromonas* strains isolated from food in Egypt revealed clinically important multidrug resistance and virulence profiles. J Food Prot 81(6):1015–1021. https://doi.
org/10.4315/0362-028x.Jfp-17-360

Han J, Zhang L, Yang S et al (2014) Detrimental effects of metronidazole on selected innate
immunological indicators in common carp (*Cyprinus carpio* L.). Bull Environ Contam Toxicol
92(2):196–201. https://doi.org/10.1007/s00128-013-1173-4

Kibret M, Abera B (2014) Prevalence and antibiogram of bacterial isolates from urinary tract
infections at Dessie Health Research Laboratory, Ethiopia. Asian Pac J Trop Biomed 4(2):164–
168. https://doi.org/10.1016/S2221-1691(14)60226-4

Kimera ZI, Mdegela RH, Mhaiki CJN et al (2015) Determination of oxytetracycline residues in
cattle meat marketed in the Kilosa district, Tanzania. Onderstepoort J Vet Res 82:01–05. https://
doi.org/10.4102/ojvr.v82i1.911

Kimosop SJ, Getenga ZM, Orata F et al (2016) Residue levels and discharge loads of antibiotics
in wastewater treatment plants (WWTPs), hospital lagoons, and rivers within Lake Victoria
Basin, Kenya. Environ Monit Assess 188(9):532. https://doi.org/10.1007/s10661-016-5534-6

Klein EY, Van Boeckel TP, Martinez EM et al (2018) Global increase and geographic convergence in antibiotic consumption between 2000 and 2015. Proc Natl Acad Sci 115:1–8. https://doi. org/10.1073/pnas.1717295115

K'Oreje KO, Vergeynst L, Ombaka D et al (2016) Occurrence patterns of pharmaceutical residues in wastewater, surface water and groundwater of Nairobi and Kisumu city, Kenya. Chemosphere 149:238–244. https://doi.org/10.1016/j.chemosphere.2016.01.095

Landers TF, Cohen B, Wittum TE et al (2012) A review of antibiotic use in food animals: perspective, policy, and potential. Public Health Rep 127(1):4–22. https://doi. org/10.1177/003335491212700103

Layada S, Benouareth D-E, Coucke W et al (2016) Assessment of antibiotic residues in commercial and farm milk collected in the region of Guelma (Algeria). Int J Food Contam 3(1):19. https://doi.org/10.1186/s40550-016-0042-6

Limbu SM, Zhou L, Sun S-X et al (2018) Chronic exposure to low environmental concentrations and legal aquaculture doses of antibiotics cause systemic adverse effects in Nile tilapia and provoke differential human health risk. Environ Int 115:205–219. https://doi.org/10.1016/j. envint.2018.03.034

Limbu SM, Ma Q, Zhang M-L et al (2019a) High fat diet worsens the adverse effects of antibiotic on intestinal health in juvenile Nile tilapia (*Oreochromis niloticus*). Sci Total Environ 680:169–180. https://doi.org/10.1016/j.scitotenv.2019.05.067

Limbu SM, Zhang H, Luo Y et al (2019b) High carbohydrate diet partially protects Nile tilapia (*Oreochromis niloticus*) from oxytetracycline-induced side effects. Environ Pollut 113508. https://doi.org/10.1016/j.envpol.2019.113508

Lyimo B, Buza J, Subbiah M et al (2016) IncF plasmids are commonly carried by antibiotic-resistant *Escherichia coli* isolated from drinking water sources in northern Tanzania. Int J Microbiol 2016:7. https://doi.org/10.1155/2016/3103672

Matongo S, Birungi G, Moodley B et al (2015) Pharmaceutical residues in water and sediment of Msunduzi River, KwaZulu-Natal, South Africa. Chemosphere 134:133–140. https://doi. org/10.1016/j.chemosphere.2015.03.093

Mehdi Y, Létourneau-Montminy M-P, Gaucher M-L et al (2018) Use of antibiotics in broiler production: global impacts and alternatives. Anim Nutr 4(2):170–178. https://doi.org/10.1016/j. aninu.2018.03.002

Mhongole OJ, Mdegela RH, Kusiluka LJM et al (2017) Characterization of *Salmonella* spp. from wastewater used for food production in Morogoro, Tanzania. World J Microbiol Biotechnol 33(3):42. https://doi.org/10.1007/s11274-017-2209-6

Mohammed MA, Alnour TMS, Shakurfo OM et al (2016) Prevalence and antimicrobial resistance pattern of bacterial strains isolated from patients with urinary tract infection in Messalata central hospital, Libya. Asian Pac J Trop Med 9(8):771–776. https://doi.org/10.1016/j. apjtm.2016.06.011

Moremi N, Manda EV, Falgenhauer L et al (2016) Predominance of CTX-M-15 among ESBL producers from environment and fish gut from the shores of Lake Victoria in Mwanza, Tanzania. Front Microbiol 7(1862):1–11. https://doi.org/10.3389/fmicb.2016.01862

Moroh JLA, Fleury Y, Tia H et al (2014) Diversity and antibiotic resistance of uropathogenic bacteria from Abidjan. Afr J Urol 20(1):18–24. https://doi.org/10.1016/j.afju.2013.11.005

Mubito EP, Shahada F, Kimanya ME et al (2014) Antimicrobial use in the poultry industry in Dar-es-Salaam, Tanzania and public health implications. Am J Res Comm 2(4):51–63

Mukonzo JK, Namuwenge PM, Okure G et al (2013) Over-the-counter suboptimal dispensing of antibiotics in Uganda. J Multidiscip Healthc 6:303–310. https://doi.org/10.2147/JMDH. S49075

Nakano T, Hayashi S, Nagamine N (2018) Effect of excessive doses of oxytetracycline on stress-related biomarker expression in coho salmon. Environ Sci Pollut Res 25(8):7121–7128. https:// doi.org/10.1007/s11356-015-4898-4

Ngumba E, Gachanja A, Tuhkanen T (2016) Occurrence of selected antibiotics and antiretroviral drugs in Nairobi River Basin, Kenya. Sci Total Environ 539:206–213. https://doi.org/10.1016/j. scitotenv.2015.08.139

Nunes B, Antunes SC, Gomes R et al (2015) Acute effects of tetracycline exposure in the freshwater fish *Gambusia holbrooki*: antioxidant effects, neurotoxicity and histological alterations. Arch Environ Contam Toxicol 68(2):371–381. https://doi.org/10.1007/s00244-014-0101-z

Nwani CD, Mkpadobi BN, Onyishi G et al (2014) Changes in behavior and hematological parameters of freshwater African catfish *Clarias gariepinus* (Burchell 1822) following sublethal exposure to chloramphenicol. Drug Chem Toxicol 37(1):107–113. https://doi.org/10.3109/0 1480545.2013.834348

Nyasulu P, Murray J, Perovic O et al (2012) Antimicrobial resistance surveillance among nosocomial pathogens in South Africa: systematic review of published literature. J Exp Clin Med 4(1):8–13. https://doi.org/10.1016/j.jecm.2011.11.002

O'Neill J (2016) Tackling drug-resistant infections globally: final report and recommendations. Review of antimicrobial resistance. HM Government and Wellcome trust, London. https://amr-review.org/sites/default/files/160525_Final%20paper_with%20cover.pdf

Obeng-Nkrumah N, Twum-Danso K, Krogfelt KA et al (2013) High levels of extended-spectrum beta-lactamases in a major teaching hospital in Ghana: the need for regular monitoring and evaluation of antibiotic resistance. Am J Trop Med Hyg 89(5):960–964. https://doi.org/10.4269/ajtmh.12-0642

Okocha RC, Olatoye IO, Adedeji OB (2018) Food safety impacts of antimicrobial use and their residues in aquaculture. Public Health Rev 39:1–22. https://doi.org/10.1186/s40985-018-0099-2

Olaniran EI, Sogbanmu TO, Saliu JKJEM et al (2018) Biomonitoring, physico-chemical, and biomarker evaluations of abattoir effluent discharges into the Ogun River from Kara Market, Ogun State, Nigeria, using *Clarias gariepinus*. Environ Monit Assess 191(1):44. https://doi.org/10.1007/s10661-018-7168-3

Olatoye IO, Basiru A (2013) Antibiotic usage and oxytetracycline residue in African catfish (*Clarias gariepinus*) in Ibadan, Nigeria. World J Fish Mar Sci 5(3):302–309. https://doi.org/10.5829/idosi.wjfms.2013.05.03.71214

Oliveira R, McDonough S, Ladewig JCL et al (2013) Effects of oxytetracycline and amoxicillin on development and biomarkers activities of zebrafish (*Danio rerio*). Environ Toxicol Pharmacol 36(3):903–912. https://doi.org/10.1016/j.etap.2013.07.019

Oloso NO, Fagbo S, Garbati M et al (2018) Antimicrobial resistance in food animals and the environment in Nigeria: a review. Int J Environ Res Public Health 15(6):23. https://doi.org/10.3390/ijerph15061284

Olusola AV, Folashade PA, Ayoade OI (2012) Heavy metal (lead, cadmium) and antibiotic (tetracycline and chloramphenicol) residues in fresh and frozen fish types (*Clarias gariepinus*, *Oreochromis niloticus*) in Ibadan, Oyo state, Nigeria. Pak J Biol Sci 15(18):895–899. https://doi.org/10.3923/pjbs.2012.895.899

Omojowo F, Omojasola F (2013) Antibiotic resistance pattern of bacterial pathogens isolated from poultry manure used to fertilize fish ponds in new Bussa, Nigeria. Albanian J Agric Sci 12(1):81–85

Pham DK, Chu J, Do NT et al (2015) Monitoring antibiotic use and residue in freshwater aquaculture for domestic use in Vietnam. EcoHealth 12(3):480–489. https://doi.org/10.1007/s10393-014-1006-z

Ramadan H, Ibrahim N, Samir M et al (2018) *Aeromonas hydrophila* from marketed mullet (*Mugil cephalus*) in Egypt: PCR characterization of β-lactam resistance and virulence genes. J Appl Microbiol 124(6):1629–1637. https://doi.org/10.1111/jam.13734

Reda RM, Ibrahim RE, E-NG A et al (2013) Effect of oxytetracycline and florfenicol as growth promoters on the health status of cultured *Oreochromis niloticus*. Egypt J Aquat Res 39(4):241–248. https://doi.org/10.1016/j.ejar.2013.12.001

Refstie S, Bakke-McKellep AM, Penn MH et al (2006) Capacity for digestive hydrolysis and amino acid absorption in Atlantic salmon (*Salmo salar*) fed diets with soybean meal or inulin with or without addition of antibiotics. Aquaculture 261(1):392–406. https://doi.org/10.1016/j.aquaculture.2006.08.005

Ridhiwani R (2015) Assessment of antibiotic residues in raw cows' milk produced by small scale dairy farms in Bagamoyo District, Tanzania. Masters dissertation. Sokoine University of Agriculture, Morogoro, Tanzania, p 91

Sanou M, Ky/Ba A, Coulibali P et al (2018) Assessment of the prevalence of extended-spectrum β-lactamase producing gram-negative bacilli at the Charles De Gaulle Paediatric university hospital (CDG-PUH), Ouagadougou, Burkina Faso. Afr J Microbiol Res 12(13):300–306. https://doi.org/10.5897/AJMR2017.8778

Saravanan M, Devi KU, Malarvizhi A et al (2012) Effects of ibuprofen on hematological, biochemical and enzymological parameters of blood in an Indian major carp, *Cirrhinus mrigala*. Environ Toxicol Pharmacol 34(1):14–22. https://doi.org/10.1016/j.etap.2012.02.005

Shah SQA, Colquhoun DJ, Nikuli HL et al (2012) Prevalence of antibiotic resistance genes in the bacterial flora of integrated fish farming environments of Pakistan and Tanzania. Environ Sci Technol 46(16):8672–8679. https://doi.org/10.1021/es3018607

Shalaby AM, Khattab YA, Abdel Rahman AM (2006) Effects of garlic (*Allium sativum*) and chloramphenicol on growth performance, physiological parameters and survival of Nile tilapia (*Oreochromis niloticus*). J Venom Anim Tox Trop Dis 12:172–201

Stenstrom TA, Okoh AI, Adegoke AA (2016) Antibiogram of environmental isolates of *Acinetobacter calcoaceticus* from Nkonkobe Municipality, South Africa. Fresenius Environ Bull 25(8):3059–3065

Topic Popovic N, Howell T, Babish JG et al (2012) Cross-sectional study of hepatic CYP1A and CYP3A enzymes in hybrid striped bass, channel catfish and Nile tilapia following oxytetracycline treatment. Res Vet Sci 92(2):283–291. https://doi.org/10.1016/j.rvsc.2011.03.003

Trushenski JT, Aardsma MP, Barry KJ et al (2018) Oxytetracycline does not cause growth promotion in finfish1. J Anim Sci 96(5):1667–1677. https://doi.org/10.1093/jas/sky120

Wamala SP, Mugimba KK, Mutoloki S et al (2018) Occurrence and antibiotic susceptibility of fish bacteria isolated from *Oreochromis niloticus* (Nile tilapia) and *Clarias gariepinus* (African catfish) in Uganda. Fish Aquat Sci 21(1):6. https://doi.org/10.1186/s41240-017-0080-x

Wang H, Che B, Duan A et al (2014) Toxicity evaluation of β-diketone antibiotics on the development of embryo-larval zebrafish (*Danio rerio*). Environ Toxicol 29(10):1134–1146. https://doi.org/10.1002/tox.21843

Yonar ME (2012) The effect of lycopene on oxytetracycline-induced oxidative stress and immunosuppression in rainbow trout (*Oncorhynchus mykiss*, W.). Fish Shellfish Immunol 32(6):994–1001. https://doi.org/10.1016/j.fsi.2012.02.012

Yonar EM, Yonar MS, Sibel S (2011) Protective effect of propolis against oxidative stress and immunosuppression induced by oxytetracycline in rainbow trout (*Oncorhynchus mykiss*, W.). Fish Shellfish Immunol 31(2):318–325. https://doi.org/10.1016/j.fsi.2011.05.019

Chapter 9
Prospects for Developing Effective and Competitive Native Strains of Rhizobium Inoculants in Nigeria

A. I. Gabasawa

9.1 Introduction

Most legumes possess a unique ability to fix N_2 through a mutualistic relationship with root nodule rhizobia, which are unique soil- and nodule-living bacteria (Nyoki and Ndakidemi 2013). This interaction could be advantaged to enhance crop yield especially in sub-Saharan Africa (SSA) (Osei et al. 2018), where yields are below their expectation (Abaidoo et al. 2013). Farmers in this region have traditionally used chemical fertilizers in the past centuries for improved crop yields. However, they realized that such fertilizers affect soil fertility negatively by hampering many beneficial microorganisms that positively enhance the growth and yield of crops. These chemical fertilizers detrimentally affect humans as well. Biofertilizers thus became an alternative, as they were eco-friendlier to both the environment and farmers (Devi and Sumathy 2018). Biofertilizers do not contain environmentally toxic substances and readily enrich the soil, and therefore their use safeguards soil health. Microbial inoculants can play an increasingly significant role in the agricultural advancement of developing countries (Alori and Babalola 2018). Using an effective and persistent *Rhizobium* strain would reduce or eliminate the need for synthetic nitrogen-based and other chemical fertilizers (Baez-Rogelio et al. 2017).

9.2 Historical Perspective

The inoculation of plants with beneficial bacteria can be, however, traced back to antiquities (Bashan 1998). Farmers knew, from experience, that when they took soil from a previously legume-cropped area and mixed with soil in which nonlegumes

A. I. Gabasawa (✉)
Department of Soil Science, Institute for Agricultural Research, Ahmadu Bello University, Zaria, Nigeria

© Springer Nature Switzerland AG 2020
A. L. K. Abia, G. R. Lanza (eds.), *Current Microbiological Research in Africa*,
https://doi.org/10.1007/978-3-030-35296-7_9

were to be cultivated, yields often improved (Bashan 1998). The act of blending "inherently inoculated" soil with seeds turned into a prescribed method for legume inoculation in the USA by the end of the nineteenth century (Smith 1992). Since the commercialization of this soil enrichment approach, the practice of legume inoculation with rhizobia eventually has become common.

Inoculation with such non-symbiotic, associative rhizosphere bacteria, as *Azotobacter*, was utilized on a vast scale in Russia during the 1930s and 1940s, the outcomes of which were inconclusive leading to the approach being dumped later (Rubenchik 1963). An endeavor to utilize *Bacillus megaterium* for phosphate solubilization during the 1930s on a large scale also failed in Eastern Europe as reported by Macdonald (1989). Before embarking on a lengthy program of selecting inoculant strains of root nodule bacteria, it is vital to understand whether there is a need to inoculate and this can be achieved through three fundamental treatment experiments (Brockwell and Bottomley 1995; Brockwell et al. 1995; Date 2000). The lack of these experiments could have led to the failure of the large-scale historical trials, as many producers lack the appropriate background skills of adequately interpreting the results obtained (Date 2000).

In the late 1970s, two breakthroughs were experienced in plant inoculation technology. Firstly, *Azospirillum* was found to enhance nonlegume plant growth (Döbereiner and Day 1976), by a direct effect on plant metabolism (Bashan and Holguin 1997a, b). Secondly, biocontrol agents, mostly *Pseudomonas fluorescens* and *Pseudomonas putida*, were introduced and began to be intensively investigated by many researchers (Bashan 1998). Many works, including Kloepper (1994), Tang (1994), and Tang and Yang (1997), reported that different such other bacterial genera as *Bacillus*, *Flavobacterium*, *Acetobacter,* and a few *Azospirillum*-related microbes were thus additionally examined and evaluated.

Most soils used for leguminous crops production in Nigeria are nutrient-deficient, especially of total N, hence, their relatively poor productivity (Machido et al. 2011; Laditi et al. 2012). The soils are also usually low in available P and organic C, thereby making them even inherently worse in their fertility status (Yakubu et al. 2010; Machido et al. 2011). Several other biotic and abiotic factors like crops uptake and removal, denitrification, volatilization, and leaching further make the soils vulnerable to nutrient loss. In order to maintain or maximize agricultural productivity, amelioration of the depleted nutrients, primarily N and P, is paramount and is usually achieved by the application of environmentally less friendly mineral fertilizers (Udvardi et al. 2015; Song et al. 2017). Their staggering costs and a relatively lesser availability in the region (Sanginga 2003; Rurangwaa et al. 2018) pose another hitch to their judicious use by the resource-deficient farmers. These and related factors fuelled the drive towards biological nitrogen fixation (BNF), which has potentials for mitigating negative impacts associated with using mineral fertilizers (Yakubu et al. 2010). However, many soils lack adequate amounts of native rhizobia and some naturally occurring strains are lacking in terms of effectiveness or competitiveness, and fail to effectively achieve an enhanced BNF process (Westhoek et al. 2017) and hence the dire need for providing external sources of rhizobia to enable effective nodulation and consequent N_2 fixation, known as inoculation (Date 2000).

Legumes, on another hand, will generally only respond to inoculation where: (1) compatible rhizobia are absent, (2) the population of compatible rhizobia is small, and (3) the indigenous rhizobia are less effective in N_2 fixation with the intended legume than selected inoculant strains (Vanlauwe and Giller 2006). Although a commercial legume inoculants' production and use in the USA and the UK dated back to as early as 1895 (Nelson 2004), local production of this type of biofertilizer only started in the 1980s and 1990s in Africa (N2Africa 2013). There was, therefore, no regular use of rhizobial inoculants by West African mainstream farmers, including those of Nigeria (Bala 2011a). Since the introduction of these BNF inoculants, soybean production, for example, has continuously and dramatically increased in South Africa from 84,000 t, in 1987, to 1,320,000 t, in 2016. In Nigeria, an increase was also recorded from 40,000 t to 680,000 t in the same 1987 and 2016, respectively (Khojely et al. 2018).

9.3 Important Terms and Definitions

9.3.1 Bacterial Inoculant

Bacterial inoculant is a formulation that usually contains one or more beneficial bacterial strains (or species) in an easy-to-use and economical carrier material (Alori and Babalola 2018). The material can either be natural, inorganic, or derived from specific molecules. The inoculant is the means by which bacteria are transported from the industry to the living plant via the soil. The needed impacts of the inoculant on plant growth and development can be in the form of leguminous BNF, enhancement of mineral uptake, biocontrol of soil-borne diseases, weathering of soil minerals, and nutritional and hormonal impacts. Bacterial inoculants may, however, require bureaucratic and, hence, costly registration processes in several countries (O'Callaghan 2016).

9.3.2 Biofertilizer

This is a widely used term which also refers to a "bacterial inoculant." It also refers to preparations of microorganism(s) for a complete or partial substitution for chemical fertilization, for example, rhizobial inoculants. Many other effects of the bacteria on plant growth are, however, ignored. The word "fertilizer" is used in some countries to allow easier registration of the commodity for commercial use, as observed by Bashan (1998). The term biofertilizer (microbial inoculants) can generally be defined as any preparation that contains live or latent cells of efficient strains of nitrogen (N_2) fixing, phosphate solubilizing, or cellulolytic microorganisms used for application on seeds, soils, or composting materials/areas to increase the populations of such microorganisms and hence accelerate a given microbial process and compliment the level of plant-available nutrients (Mohammadi and Sohrabi 2012). The term biofertilizer may, therefore, broadly be used to mean all organic resources (manure) used for plant growth and rendered into plant-available forms, which may be through microorganisms and plant associations (Akhtar and Siddiqui 2009). Biofertilizers are essential components of integrated nutrient management. These potential biological fertilizers play a crucial role in the productivity and sustainability of soil and also protect the environment (Mohammadi and Sohrabi 2012). They are cost-effective, eco-friendly, and a renewable source of plant nutrients to supplement chemical fertilizers in sustainable agricultural systems (Malusá et al. 2012). Beneficial microorganisms in biofertilizers speed up and ameliorate plant growth and protect plants from pests and diseases (El-yazeid et al. 2007). The most common organisms used as biofertilizer component are nitrogen-fixers (N_2-fixers), potassium-solubilizers (K-solubilizer) and phosphorus-solubilizers (P-solubilizers), or in combination with molds or fungi. The bacteria used in biofertilizers mostly have a close relationship with plant roots. *Rhizobium*, for example, has a symbiotic relationship with legume roots, and rhizobacteria inhabit root surfaces or rhizosphere soil (Mohammadi and Sohrabi 2012).

9.4   Rhizobial Taxonomy

According to Bergey et al. (1923), bacteria were only included in rhizobia when they had the capacity of nodulating. However, when they had similar morphologies but could not nodulate, they were excluded from rhizobia. Nodulation, host range, and behavior on growth media were also later considered (Baldwin and Fred 1929; Fred et al. 1932) for rhizobial classification. Based on growth behavior on media, Fred et al. (1932) classified rhizobia as either fast or slow growing (Young 1996). Rhizobia, therefore, is a selected bacterial group capable of forming root nodules on legumes, and occasionally on the stems of some legumes, and can as such fix atmospheric nitrogen (N_2) to fully or partially meet the nitrogen (N) requirements of the legume host plant (Gage 2017). Frank (1889) proposed the name "rhizobia" to describe root nodule bacteria. All nodule-forming bacteria have from then been known as rhizobia. Biological nitrogen fixation, which is an N_2-fixation process via different prokaryote members (specifically diazotrophs), approximately contributes about 16% of the total N input in croplands (Ollivier et al. 2011). Rhizobia are, therefore, significant contributors to BNF, and the legume-rhizobium symbiosis can fix as much as up to 450 kg N ha^{-1} year^{-1} (Unkovich and Pate 2000).

Moulin et al. (2002) reported that as a group, rhizobia are not monophyletic and have, therefore, been classified as alpha- and beta- (α- and β-). Rhizobia currently consist of 61 species in 13 different genera, namely *Rhizobium, Mesorhizobium, Sinorhizobium, Bradyrhizobium, Azorhizobium, Allorhizobium, Methylobacterium, Burkholderia, Cupriavidus, Devosia, Herbaspirillum, Ochrobactrum,* and *Phyllobacterium*. The taxonomy of rhizobia is in constant flux (Ahmad et al. 2008).

Rhizobium, Mesorhizobium, Sinorhizobium, Bradyrhizobium, Azorhizobium, and *Allorhizobium* belong to the alpha-Proteobacterial subdivision of the purple bacteria, an incredibly diverse group (Pierre and Simon 2010). *Rhizobium* contains 33 species, 24 originating from legume nodules while *Sinorhizobium* includes nine species isolated from legume nodules. Also, *Bradyrhizobium* has seven species from legume nodules, and *Azorhizobium* has two species nodulating legumes.

The complete list of known rhizobia species is continuously updated (Khan et al. 2010). The technological advancements in morphological, biochemical, physiological, serological, and sequence analysis used for taxonomic classification could still make classification unstable (Manvika and Bhavdish 2006). Further studies on the genetic diversity of rhizobia will, however, help in understanding the evolutionary histories of the rhizobium-legume symbioses. This will, consequently, help in devising worthwhile planning strategies aimed at reaping the utmost benefit from the symbioses.

9.5 Rhizobial Ecology and Diversity

Studies have targeted to uncover the nature of rhizobial symbionts in their native environments as it has been discovered that one of the significant problems in the application of BNF technology is the establishment of the introduced inoculant strains. Nodulation and nitrogen fixation in this symbiosis require that host and microorganism are compatible, but also that the soil environment is appropriate for the exchange of signals that precede infection (Hirsch et al. 2003; Zhang et al. 2002). Earlier reviews have reported the influence of biotic and abiotic soil factors on rhizobium ecology (Amarger 2001; Sessitsch et al. 2002). A problem identified by many of the reviews adequately describing changes at the population level. Tools, like intrinsic antibiotic resistance (Beynon and Josey 1980), serology (Bohlool and Schmidt 1973; Purchase et al. 1951), and multilocus enzyme electrophoresis (Pinero et al. 1988; Eardly et al. 1990), have all facilitated the acquisition of insight into rhizobial population structure in the soil, and how this could be influenced by the host and environment. However, it is only with the development of advanced molecular (Hirsch et al. 2003; Thies et al. 2001) and computational tools that the consideration of large populations of rhizobia on a routine basis been possible. The nodule formation on the leguminous host keeps on being viewed as the essential phenotypic characteristic due to the evident agricultural significance of rhizobia. Techniques such as fatty acids methyl esters (FAME) (Leite et al. 2018), whole-cell protein analysis using sodium dodecyl sulfate-polyacrylamide gel electrophoresis (SDS-PAGE) (Dekak et al. 2018) and multilocus enzyme electrophoresis (MLEE) (Van Berkum et al. 2006), and recently whole-genome sequencing (Seshadri et al. 2015) have effectively been utilized to characterize and classify obscure strains and depict novel rhizobial species.

Traditionally, rhizobial variation has been determined using characteristics such as growth rate and colony morphology (size, shape, color, texture, and general appearance) and antibiotic resistance methods (Graham et al. 1991). However, these methods cannot sufficiently discriminate between all the variations exhibited in the target species. They cannot delineate sources of observed phenotypic variation that may be due to environmental factors or underlying genetic factors. Molecular means have now been accessible to appreciate the diversity and structure of the bacterial population. The *16S rRNA* gene sequencing is a very crucial parameter in rhizobial classification and techniques that depend on the disparity in ribosomal RNA genes have been regularly connected to the identity of species (Laguerre et al. 1994). The traditionalist idea of *16S rRNA* genes has, however, restricted its utilization due to strain level discrimination. The intergenic spacer (IGS) existing between 16S and 23S rRNA genes was depicted to be much diverse (Massol-Deya et al. 1995) and restriction fragment length polymorphism (RFLP) of the polymerase chain reaction (PCR)-intensified IGS was used in the characterization of rhizobia (Nour et al. 1994; Sessitsch et al. 1997). The advancement in PCR prompted new fingerprinting strategies. For example, techniques like Random Amplification of Polymorphic DNA (RAPD) using subjective oligonucleotide PCR primers of irregular grouping have now been used to produce strain-explicit fingerprints of rhizobia (Koskey et al. 2018).

Studies have shown that tropical rhizobia are diverse with subgroups of varied symbiotic specificity and effectiveness. Studies by Bala and Giller (2007) showed rhizobia of the same phylogenetic grouping nodulating *Calliandra calothyrsus, Gliricidia sepium,* and *Leucaena leucocephala* in some soils, but failing to nodulate at least one of the hosts in other soil, thus suggesting that rhizobial phylogeny and host range (infectiveness) were only weakly linked. Rhizobia are heterotrophic, competent bacteria that can survive as large populations for decades in the absence of host legumes (Giller 2001), but the presence of a compatible host legume confers protection to the microsymbionts against environmental stresses (Andrade et al. 2002). On the other hand, a greater diversity of rhizobia in soil populations broadens the range of legume hosts that can be nodulated in such soils. Therefore, a mutual benefit between aboveground (legume) and belowground (rhizobia) biodiversity exists.

9.6 Determinants of Host Specificity in Rhizobia

Host specificity plays a vital role in rhizobia, especially in establishing an effective symbiosis. There is a difference in the specificity of interaction between leguminous species and rhizobia. A few legume-rhizobia symbioses are more specific, for example, when a legume host specifically forms root nodules only when infected with a particular rhizobium. Some other legumes will, however, form their nodules with a variety of rhizobia (Vance et al. 2000). Broughton et al. (2000) observed that specificity encompasses the recognition of a bacterium by a host and vice versa, via signal compounds exchange, which instigates differential expression of the gene in both.

Albeit many host plants, only some symbionts can lead to the development of nitrogen-fixing nodules. Such tropical leguminous trees as Acacia, Prosopis, or Calliandra can, exceptionally, form the nodulation symbioses with diverse rhizobia from various genera. The specificity existing between symbiotic accomplices, however, limits the development of non-fixing ineffective nodules by the host legume as observed by Perret et al. (2000). The formation and development of root nodules require additional energy and nutrients from the host. Rhizobia are different in their reaction to various signal molecules that are produced by legumes. A few rhizobia have a restricted host range and therefore form nodules with a few legumes. *Azorhizobium caulinodans, Sinorhizobium saheli,* and the sesbaniae biovar of *Sinorhizobium terangae,* for example, only nodulate *Sesbania rostrata* (Boivin et al. 1997) and *Rhizobium galegae* is the main symbiont of *Galega officinalis* and *Galega orientalis* (Lindström 1989). Conversely, some rhizobia have an expansive host range and, hence, are fit to nodulate a wide range of legumes with different degrees of promiscuity. For it has recently been reported that two rhizobial strains, *Mesorhizobium japonicum* (strain Opo-235) and *M. kowhai* (strain Ach-343) could nodulate a wide range of host including species of diverse legume genera from two tribes (Galegeae and Trifolieae) (Safronova et al. 2019).

Legumes may, on the other hand, also be host to only one kind of symbiont (*Galega* spp.) or establish symbioses with a wide range of rhizobia (*Leucaena leucocephala, Calliandra calothyrsus, Phaseolus vulgaris*). Distantly related rhizobia can nodulate the same host; for example, *Sinorhizobium fredii, Bradyrhizobium japonicum,* and *Bradyrhizobium elkanii* all nodulate *Glycine max.* Members of *Rhizobium, Sinorhizobium,* and *Bradyrhizobium* are less related to each other than to other non-rhizobial genera. Stem and root-nodulating *Azorhizobium caulidonans* and root-nodulating *Sinorhizobium fredii* and *Sinorhizobium terangae* bv. sesbaniae, both symbionts of *Sesbania rostrata*, also represent two taxonomically distant genera.

9.7   Mechanisms of Biological Nitrogen Fixation

Biological nitrogen fixation (BNF), a system used only by specific prokaryotes, is catalyzed by a two-part nitrogenase complex (Yan et al. 2010). Nitrogenase catalyzes the simultaneous reduction of one N_2 and $2H^+$ into ammonia (NH_3) and a molecule of hydrogen gas, as thus:

$$N_2 + 8H_2 + 16ATP + 2NH_3 + 2NH_3 + 2H_2 + 16ADP + 16Pi$$

The immediate electron donor is ferredoxin, a potent reducing agent. The reaction is driven by the hydrolysis of 2 ATP molecules for each electron transferred (Wheelis 2008). Carvalho et al. (2011) observed that the best-known BNF system occurs between legume hosts and bacteria (rhizobia). The mutual interaction between the legume roots and given soil rhizobia accounts for the development of a specified organ, the mutual root nodule, that primarily functions in BNF. Root nodules in legumes make a vital contribution to the soil N content, which plays a significant role in agriculture (Alla et al. 2010). Legume root exudates enhance the production of Nod factor signals of rhizobia, which are readily distinguished by compatible plant receptors leading to the formation of nodules, in which are bacteroids, differentiated bacteria and N_2 (Oldroyd and Downie 2008). Maintenance of nitrogenase activity in the root nodule is subject to a fragile equilibrium. At first, a high rate of oxygen respiration is indispensable in order to supply the energy needs of the N reduction activity (Sanchez et al. 2011), but oxygen also inactivates the nitrogenase complex irreversibly. These opposing needs are reconciled by oxygen flux control via a diffusable barrier present in the nodule cortex and by leghemoglobin, an oxygen carrier of the plant which is exclusively present in the root nodules (Minchin et al. 2008).

Besides N_2 fixation, some rhizobial species are capable of growing under conditions of low oxygen using nitrate (NO_3) as an electron acceptor to support respiration in the process of denitrification in which bacteria sequentially reduce NO_3 to nitrite (NO_2) and finally to N_2 (Van Spanning et al. 2005, 2007). In this process, NO_3 is reduced to NO_2 by either a membrane-bound or a periplasmic NO_3 reductase, and NO_2 reductases catalyze the reduction of NO_2 to nitric oxide (NO). Nitric oxide is then further reduced to nitrous oxide (N_2O) by NO reductases and, finally, N_2O is converted to N_2 by the N_2O reductase enzyme. The importance of denitrification in legume-rhizobia symbiosis can best be appreciated when the oxygen concentration in soils decreases (soil hypoxia) due to environmental stresses as flooding of the roots. Following such conditions, the denitrifying process could be a mechanism of generating ATP for the survival of soil rhizobia and also to preserve the functioning of nodules (Sanchez et al. 2011).

9.8 Significance of Biological Nitrogen Fixation

The atmospheric environment is an almost homogeneous blend of gases, the amplest of which is N (78.1%) (Garrison 2006). Around 96% of the N taken by crop plants has been estimated as N derived from the atmosphere (López-Bellido et al. 2006). Biological N fixation includes the transformation of N_2 to ammonium (NH_4), which is a plant-available N form (Vessey et al. 2005). The idea of BNF is that the dinitrogenase catalyzes the response and part triple-bond idle atmospheric N (N_2) into natural ammonia molecule (Cheng 2008). The BNF is viewed as an inexhaustible asset for sustainable agriculture, as it decreases fertilizer use, and hence augments financial farmers gains (Walley et al. 2007). Also, it assumes a vital role in appraisal of rhizobial diversity, adds to global knowledge of soil microbial biodiversity and the handiness of rhizobial accumulations, and to the foundation of long-term methodologies that are aimed at expanding the contributions of biologically fixed N to agriculture. The N_2 fixed by legumes can incredibly contribute to economically buoyant and environmentally suitable agriculture, as suggested by Odair et al. (2006). It has been assessed that 80–90% of the plant-available N found in the environments is sourced from BNF (Rascio and Rocca 2008). It (BNF) also adds to the renewal of soil N, and hence circumvents the dire need for chemical N fertilizers (Larnier et al. 2005). Biological N fixation offers an economically alluring and ecologically encouraging methods for lessening external N fertilizer demand and input (Yadvinder-Singh et al. 2004). Agricultural systems sought to have gradually metamorphosed towards enhancing environmental quality and exclude its (environmental) deterioration. The use of inoculants that are composed of diazotrophic bacteria and used as N fertilizer alternatives is, therefore, one of the most vibrant methods of bypassing environmental deterioration (Roesch et al. 2007).

9.9 Rhizobial Bio-Prospecting Studies

The Agricultural Research Service (ARS) of the U.S. Department of Agriculture (USDA) has maintained a collection of nitrogen-fixing legume symbionts for most of the twentieth century (Van Berkum 2002). Although many rhizobial isolation studies appear in scientific literature, there has been little attempt to evaluate global trends across diverse strain collections. The most comprehensive studies focus on a particular rhizobial species recovered from several host legumes at multiple locations or on populations or communities of rhizobia recovered from a particular host legume over a wide geographic range (Han et al. 2008). The absence of a global synthesis can be attributed to the difficulty in comparing studies that use diverse methods for rhizobial sampling and strain typing. The use of diverse sampling strategies means that collections of isolates are rarely equivalent, except in related studies arising from individual research groups. Comparing published studies is also difficult because strain typing methods vary in their discriminatory power and are usually species-specific (Li et al. 2009) and therefore influence the number of strain types identified.

9.10 Rhizobia Identification

A typical rhizobial cell is a small- to medium-sized (0.5–0.9×1.2–3.0 μm) Gram-negative, motile rod, exhibiting the characteristic presence of copious β-hydroxybutyrate granules forming 40–50% of the cell dry weight, easily observed using metachromatic granules stains. Most strains produce sticky gum-like substances of varying composition. According to a study by Gupta et al. (2007), rhizobia are typically observed, on Yeast Mannitol Agar (YMA) media, as translucent, viscid, slimy, and individual dome-shaped colonies, having a lifted component with whole edges.

9.11 Inoculant Formulations

9.11.1 Optimal Characteristics of a Carrier for Inoculants

The carrier, a delivery vehicle of live microorganisms, helps in transporting the microorganisms from the production factory to the field where they are utilized and is the significant portion (by volume or weight) of the inoculant. There is presently no universally accepted carrier or formulation available for microbial release into the soil (El-Ramady et al. 2018). Carrier materials and formulation type, therefore, vary, and it can be a slurry or powder (Bashan et al. 2014). A suitable carrier must have the capacity to deliver the right number of viable cells in good physiological condition at the right time (Malusá et al. 2012). Other essential characteristics of a suitable inoculant carrier have been reported (Bashan 1998):

1. A carrier should be nearly sterile or easily sterilized, as chemically and physically uniform as possible with consistent quality and be suitable for as many bacterial species and strains as possible. Wet carriers should also have a high water-holding capacity.
2. It should also be easily manufactured and mixed by existing factories, allow additional nutrients, have an easily adjustable pH, be made from relatively cheap raw materials, and be in adequate supply.
3. A good carrier should also be easy to handle, provide rapid and controlled release of bacteria into the soil, and can be applied using appropriate standard machines.
4. It should be environmentally nontoxic, biodegradable, and non-polluting, thereby minimizing such environmental risks as the dispersal of cells into the groundwater or atmosphere.
5. It should have enough storage shelf life of a year or two when kept at room temperature.

No single carrier can possess all these qualities; it should, however, have as many as possible to be a good one.

9.11.2 Types of Existing Carriers for Inoculants

Peat is almost the most widely used carrier for rhizobia, the only inoculant being sold in large volume today (Ruíz-Valdiviezo et al. 2015). However, this carrier has records of several disadvantages, some of which include wide and source-dependent variability in its quality, thereby possibly presenting difficulties in inoculant dosage and clear storage conditions (Reddy and Saravanan 2013). Availability of peat is rare, and hence its exorbitant cost in most countries in Asia and Africa (Bashan et al. 2016). Some peats may release compounds toxic to the bacteria when sterilized by heat resulting in low bacterial counts (Kaljeet et al. 2011). The mining of peat has also been regarded as being unfriendly to wetland ecosystems (Margenot et al.

2018). In terms of delivery, peat powder can easily be blown away from seeds by an air-delivery system of the planter. Peat may also interfere with the seed-monitoring mechanism of the planters (Nehra and Choudhary 2015). A possible remedy is an addition of adhesives or slurries to the inoculant during its application to the seeds to improve its adhesion, but this requires additional time, labor, and cost for a process that is already labor-intensive (Nehra and Choudhary 2015). Today, inoculant carriers can generally be divided into five basic categories including soils, waste plant materials, inert materials, plain lyophilized microbial cultures, and oil-dried and liquid inoculants, as reviewed by Bashan et al. (2014)). Due to the drawbacks observed with peat, many alternatives consisting of different formulations of the basic materials have, therefore, been evaluated (Trivedi et al. 2005; Albareda et al. 2008; Nehra and Choudhary 2015).

9.11.3   Inoculant Production

There are several essential issues to be considered in inoculant production, among which include the microbial growth profile, types and optimum condition of the organism and formulation of the inoculum. The methods of inocula formulation and application and inocula product storage are all very critical for a successful biological product.

In the process of inoculant production, the target microorganism can either be introduced into a sterile or non-sterile carrier (Bashan et al. 2014). The former carrier has microbiologically significant advantages over the latter but has not been cost-effective from commercial perspectives in most cases (Bashan et al. 2016). For an inoculant to contain an effective bacterial strain and for its success or failure as a biological agent to be determined, formulation is the most critical consideration. The formulation stage is the industrial process and practice of successfully converting a promising laboratory-proven bacterium into a commercial field product. The biofertilizer formulations are, therefore, expected to conform with all the numerous characteristics mentioned earlier and also surmount two significant constraints against living organisms, that is, (1) viability loss during short storage in growers' warehouse (as most developing African countries lack appropriate refrigeration facilities), and (2) long shelf life and stability over a broad temperature range in the marketing distribution systems (O'Callaghan 2016).

The six main steps paramount in inocula production are the choice of active organisms, isolation and selection of the target microbes, selection of method/carrier material, selection of propagation method, prototype testing, and large-scale testing (Bashan et al. 2016). Active organisms must be decided based on activity objective; isolation is vital in separating target microbes from their habitation. Usually, microbes are isolates from plants' root. Best candidate isolates are selected following various stages of routine selection processes. Also, paramount is deciding the form of inoculant carrier. Selecting a befitting propagation method is mainly through understanding the optimum growth requirements of organisms, and this can be achieved by obtaining the microbial growth profile under different conditions. The prototype (usually in different forms) inoculant is made and tested at diverse environments with a view to evaluating the effectiveness and efficacy of the product.

9.12   Forms of Inoculants Dispersal

Inoculants are mostly known to come in four primary dispersal forms, as previously reviewed (Bashan et al. 2014; Alori and Babalola 2018).

1. Powders: these are often used as a pre-planting seed coating with particle size typically ranging between 0.075 and 0.25 mm to ensure a better chance for the inoculant to properly adhere to the seeds (Malusá et al. 2012).
2. Slurries are powder-based inoculants formed by suspending the base inoculant material in liquid, usually water, and applying the mixture directly to furrows. The seed can, alternatively, be dipped into the suspension just before planting (O'Callaghan 2016).
3. Granular inoculants are directly applied to furrows together with the seeds. Granular size ranges are from 0.35 to 1.18 mm (Hungria et al. 2005) and are usually used for broadacre applications (O'Callaghan 2016). There are also some bead-like forms that are synthetic variations of these granular forms, and which can be in macro (1 to 3 mm) sizes in diameter and used as a granular form. They can also be in micro size (100–200 μm) used as a powder for seed coating. These types of inoculant are, however, not suitable for developing countries as their application usually requires heavy and sophisticated machinery which in most cases is not available in such countries (Bashan et al. 2014).
4. With liquid inoculants, seeds are either evenly sprayed or dipped into the inoculant before sowing and later be sown after drying (Bashan et al. 2014). This ensures even coverage of seeds with relatively no planter-related problems or inoculum.

Despite the diverse forms of inoculant and the different ways in which they can be applied, the use of any inoculant will depend on its availability, cost, and crop/environment-specific needs. For example, although inoculants applied as seed coating may be cost-effective, their use is severely challenged by the need for proper pre- and post-application, and they are sometimes less effective than granular inoculants that are directly applied to the soil (Jones and Olson-Rutz 2018).

9.13   Potentials for Production and Use of Inoculants in Africa

Despite the high recorded rates of economic growth of more than 5 years, SSA is still by far the poorest region globally (The Economist Intelligence Unit 2014). This is not unconnected with the region's level of food insecurity, which is undermined by sporadic poverty and limited utilization of modern agricultural technologies. Poor soil fertility, pests and diseases and such low-skilled and unsustainable farming methods like continuous cropping are some of the region's direct causes of the food insecurity (Oruru and Njeru 2016). About 60–70% of mineral fertilizers applied to farms is lost, particularly, via leaching, volatilization, and erosion, for example (Hardarson et al. 2003). Only an estimated 30–40% of the mineral fertilizers applied is, therefore, utilized by plants, worldwide (Chianu et al. 2011). In such instances, biotechnology has great potential to increase the productivity in SSA (Chianu et al. 2011; Oruru and Njeru 2016), especially through the conservation and sustainable use of soil microbes (Macdonald and Singh 2014).

Many opportunities could be readily available if countries in SSA could develop a long-term approach to policies on, specifically, BNF and generally on biotechnology. Policies like these should be able to: (1) advance national biotechnology need appraisal and implementation, (2) target research on biotechnology and executing same to needs, (3) give motivating forces and conditions to commercialization of biotechnology research and endeavors, (4) advance partnerships between immediate public research for development (R4D) and multinational biotechnological industries, (5) enhance scientific limits and technological framework for the execution of an ideal biotechnology, and (6) incorporate biotechnology hazard management into existing agricultural, health, and environmental routines. The potential advantages of biotechnology, like *Rhizobium* inoculation, may, otherwise, not be tapped for the enhancement of human welfare in the SSA.

Furthermore, approaches such as BNF and *Rhizobium* inoculation should be able to circumvent the need to: (1) make agricultural and non-governmental organizations (NGOs) stronger as they diligently serve the interests of subsistent farmers as they embrace biotechnology, (2) upgrade their capability, and (3) enhance their support in adjusting and testing BNF and *Rhizobium* inoculation advancements. The SSA countries must settle on an integrated biotechnology approach instead of their much-adopted ad hoc approaches. The former, however, also needs an intervention of policy as observed by Brenner (1996). The (integrated) approach will guarantee that biotechnology research readily takes care of the grievances of resource-poor farmers. For example, Botha et al. (2004) reported that soybeans CB 1809 strain was up to 60% superior to other isolates tested in efficient BNF from Bergville and Morgenzon, in South Africa, and was almost 73% of isolates from Koedoeskop. Mpepereki et al. (2000) reported that inoculation with *Bradyrhizobium japonicum* in the SSA increased soybean (*Glycine max* (L.) Merrill) yield from 500 to 1500 kg ha^{-1}. Mugabe (1994) had earlier observed that the majority of countries in Africa could reduce much of the expenditures incurred on fertilizer imports via a full utilization of BNF. *Rhizobium* alone could provide an estimated >50% of the fertilizers required for crop production in most of the marginal environments of Kenya, Tanzania, and Zimbabwe. Unlike most African countries, the agricultural sector is overwhelmingly dominated by a detectable level of commercial farms in South Africa and Zimbabwe. This is translated into an easier adoption of commercial inoculant production and use (Bala 2011b). Chianu et al. (2011), however, observed that socio-economic and policy constraints were the most critical challenges that seriously undermined the much-needed production and use of *Rhizobium* inoculants in SSA. The limited capacity of most national agricultural research systems in SSA causes an absence of expertise in setting preferences and priorities in the use of biotechnology. This condition militates against the development and production of BNF-based technologies. Research and development (R&D) programs in the African continent are presently, more or less, isolated, with low-level monitoring and evaluation, vis-à-vis severely low funding. Inherent variability in legumes' response to inoculation (Ronner et al. 2016) and an over-dependence of BNF on such factors as legume agronomy; and edaphic and other environmental factors (van Heerwaarden et al. 2018) are some of the severe hindrances to effective inoculation programs. Also, such aspects as inoculant source, variety, and management types and level usually differ between countries. There also exist variations in climatic and edaphic conditions across various proximities. This, invariably, makes it even more difficult to draw reliable conclusions on the efficacy of inoculants derived from local trials (van Heerwaarden et al. 2018) and hence suggests that the use of inoculants and diverse varieties may need to be directed towards specific frameworks as observed by Ronner et al. (2016). This, therefore, indicates the dire need for comparative studies focussing on the efficacy of legume technologies that contain various pulses being implemented across SSA.

In early studies conducted on inoculation in various parts of Africa, there was a grain yield advantage in soybean (*Glycine max*) in tropical Africa (Sivestre 1970; Nangju 1980; Bromfield and Ayanaba 1980). Also, Sivestre (1970) observed yields, in inoculated soybean, of 1440 kg ha^{-1} compared to uninoculated ones, which had a yield of only 240 kg ha^{-1}. In a paper presented at an international workshop on *Rhizobium* inoculation, held in Tanzania, in 2008, Bala had cited work as reporting yield increase of 80–300% due to inoculation in the Democratic Republic of (DR)-Congo. Ndakidemi et al. (2006) observed in an on-farm trial conducted with rhizobial inoculants (*Rhizobium tropici* strain CIAT 899, for common bean, and *Bradyrhizobium japonicum* strain USDA 110, for soybean) at Moshi and Rombo, two districts in northern Tanzania, that at harvest, soybean and common bean (*Phaseolus vulgaris*) development was significantly higher with *rhizobial* inoculation when compared with an uninoculated control or N and P supply. Grain yields of *P. vulgaris* were also increased by 60–78% due to inoculation alone and 82–95% due to inoculation and P application at 26 kg P$_2$O$_5$ ha^{-1}, relative to the uninoculated and/or unfertilized plots. There was also a 127–139% increase in grain yield via inoculation only and 207–231% via inoculation and 26 kg P$_2$O$_5$ ha^{-1} application. Hence, the combined application of inoculants and P fertilizer to *G. max* and

P. vulgaris increased grain yield and biomass production when compared with the use of only N and P or strains of rhizobia (Ndakidemi et al. 2006).

To isolate and test the effectiveness of N_2-fixing bacteria from Africa's large-biodiversity ecosystem, vis-à-vis supervising factors that affect legumes, the rhizobia, and their symbiosis while ensuring for effective rhizobia, is paramount because it may eventually result in identifying superior inoculants that can readily improve legume growth, development, and yield. This will, later, provide for an economic boom for legume farmers. Expanded and emphasized research activities are, therefore, direly needed to promote this eco-friendly and cheap technology for the many resource-poor smallholder farmers in Africa (Simon et al. 2014). Some earlier studies conducted in South Africa, as reported by Van Rensburg and Strijdom (1969), revealed that some local soybean varieties formed a specific symbiosis with *B. japonicum*. It is, however, of paramount importance to appreciate that even promiscuous soybeans that seldom require inoculation, and that are popular in a few parts of Africa, at times respond to inoculation. Osunde et al. (2003), in a study carried out at five locations in Nigeria's moist savanna region, revealed that the promiscuous soybean cultivars (Tropical Glycine cross (TGx) 1456-2E and TGx 1660-19F) favorably responded to inoculation. *Magoye*, a Zambian, exceptionally promiscuous line released in 1981, however, readily nodulated in all tested southern African soils and seldom responded to inoculation in Zambia and Zimbabwe (Mpepereki et al. 2000). Sanginga et al. (2001) reported that the principal criterion for selection in the IITA, for more than a decade, was promiscuity without in-depth microbiological studies. It was based on these results that IITA introduced a program on soybean breeding in 1978 aimed at developing "promiscuity" in soybean cultivars that nodulate with local *Bradyrhizobia* in the soil, thereby excluding the necessity of inoculation (Kueneman et al. 1984). However, studies conducted in the early 2000s on symbiotic effective nature of local rhizobia that nodulate promiscuous soybean in 92 soils of Zimbabwe led to the identification of three isolates that were of utmost N_2-fixing potential in the *Magoye* cultivar than MAR 1491, which is a commercial strain (Musiyiwa et al. 2005). The M3 isolate was, however, later identified to be more superior to the commercial strains MAR 1491 and MAR 1495, as reported by Zengeni and Giller (2007). Okogun and Sanginga (2003) observed no statistically significant difference between the yield of inoculated and uninoculated crops (promiscuous soybean varieties—TGx 1485-1D, TGx 14562E, TGx 1448-2E, and TGx 1660-19F)) at three sites in the savanna of Nigeria, even though the native rhizobial population in soils at these sites were to a certain extent different.

Nitrogen-inoculants and BNF have had an extended history in Africa. It dates to the colonial agricultural research days when attempts were made to develop pasture legume inoculants to boost exotic cattle productivity (Odame 1997). Also, the United Nations Educational, Scientific and Cultural Organization (UNESCO) established a few Microbiological Resource Centres (MIRCENs), spread within the five continents, which were supported by the United Nations Environment Programme (UNEP) and the Food and Agriculture Organisation (FAO) of the United Nations (UN) to elevate BNF status of third world countries (Odame 1997). The main functions of MIRCENs in Africa, located in Cairo, Dakar, and Nairobi, included, among others, the collection, identification, testing, and maintenance of strains, and preparation and distribution of inoculants or their cultures that were compatible with local crop plants. Other functions were to deploy local *rhizobia* inoculant technologies, promote research, and provide advisory services, training, guidance, and counseling to institutions and individuals engaging in rhizobiology research activities. The Nairobi MIRCEN project, for example, promoted and transferred BNF technologies such as pulses' inoculants, pasture legumes, and trees to research scientists and other stakeholders of agricultural relevance in Kenya and other East African countries. It (Nairobi MIRCEN) also extended its activities into *Rhizobium* strains screening for adaptation to abiotic stresses like soil acidity, extremely high temperatures, and drought, especially while considering the various environmental stresses that hamper successful BNF and as two-thirds of agricultural land in Kenya is vulnerable to these stress types. The whole idea was aimed at gradually intensifying screening trials for rhizobia that could be adaptable to such ecological menaces. This project also used the potential of a symbiotic association with a fungal strain (mycorrhiza), on plant roots that aid the plant to extract water and P from the soil environment. The MIRCEN also developed *Biofix*, a very marketable biofertilizer (Odame 1997). Kenya Institute of Organic Farming (KIOF) and the Organic Matter Management Network (OMMN), which are Kenya-based nongovernmental organizations (NGOs), played a serious role to reckon with in the distribution of *Biofix* to farmers. Researches are also being conducted with *Biofix* in Nigeria. Over time, however, the active involvement of KIOF in promoting *Biofix* had over-stretched and, therefore, its human and financial resources waned (Chianu et al. 2011).

The FAO in the 1990s supported a project to select more promising rhizobial strains in Tanzania, as reported by Mugabe (1994). This resulted in the development of Nitrosua, a biofertilizer for profitable soybeans production by Sokoine University of Agriculture (SUA), Morogoro. The SUA, in collaboration with the Ministry of Agriculture and some NGOs, also established extension activities for the dissemination of Nitrosua to local Kenyan farmers. These activities, however, also waned over time, as stated by Bala (2008). At least two firms (Madhavani Ltd. and the BNF of Makerere University, established in 1990 with the help of the USAID) in Uganda have produced inoculants. The two firms, however, functioned until 1997 and produced 14.2 tonnes of soybean inoculants between 1995 and 1997 for the FAO. Inoculant production in Rwanda started at the Institut des Sciences Agronomiques du Rwanda (ISAR) in 1984. Production capacity reached 2.4 tonnes per annum by 1990 (Cassien and Woomer 1998). Unfortunately, activities were, halted by the civil war fought in the 1990s. At the end of the war, however, the laboratory was renovated, and BNF activities were resumed, but pre-civil war records were not yet reached by the early 2000s (Giller 2001).

First commercial quantities of inoculants were seen in the South African markets in 1952, although their quality was highly debatable until the 1970s after an independent quality control system was introduced (Strijdom 1998). All inoculants from 1967 were produced with sterilized peat and contained at least 5×10^8 rhizobial cells g^{-1} of peat (Strijdom 1998). The quality control strategies introduced ensured for comparison of South African inoculants with the best quality of inoculants produced outside the African continent (Strijdom and van Rensburg 1981). Farmers mainly growing crops like soybean, cowpea (*Vigna unguiculata*), and groundnut (*Arachis hypogaea*) enjoyed the production of the inoculants (Deneyschen et al. 1998). Khonje (1989) reported that the production of commercial quantities of inoculants in Malawi started in the 1970s, where they were made available in 50-g packets for crops like soybean and cowpea. They were produced by Chitedze Agricultural Research Station, Lilongwe. Sales rose dramatically from as little as 450 packets, in 1976, to over 1800, between 1987 and 1988. In Zimbabwe, the presence of a mega and highly established commercial soybean sector readily suggests sporadic spread and use of inoculants in that country (Mpepereki et al. 2000). Soil Productivity Research Laboratory (SPRL) controlled a BNF enhancement technology project in the 1990s in Zimbabwe, which was supported by the International Atomic Energy Agency (IAEA) as reported by Chianu et al. (2011). This project reached an inoculants mass production capacity of 120,000 packets per year, which were distributed via extension services to smallholder farmers. Mugabe (1994) reported that mycorrhizal inoculation research was also undertaken in some regions by The University of Zimbabwe.

A study in Ghanaian cowpea grown in fields have reported a nitrogen fixation of up to 402.3 mg plant^{-1}, resulting in an average of 19.5 kg N-fixed ha^{-1} (Naab et al. 2009). Values between 4 and 29 (i.e., 15% and 56% of plant N) were, in contrast, reported in the semiarid south-western region of Zimbabwe (Ncube et al. 2007). The application of rhizobial inoculants may, therefore, hold potential for increasing plant nutrition and overall soil fertility in areas such as these. Although many strains of effective rhizobia have been identified, and are now readily available, it has still been observed that the rhizobial strains often under-perform in conditions that differ from their original habitat (Zhang et al. 2003; Law et al. 2007). Also, their effective nature relies on environmental factors like soil texture (Law et al. 2007), soil pH (Botha et al. 2004), soil temperature (Zhang et al. 2003; Suzuki et al. 2014), and various types of the host plant (Pule-Meulenberg et al. 2010). This is, especially, relevant for areas like the Botswanan Okavango region given its adverse climate, nature of available local plant varieties, and its soil heterogeneity. Law et al. (2007) already reported a popular strain of inoculant not affecting Botswana-grown cowpea and peanut. Application of soybean inoculant in South Africa, furthermore, boomed seed yields at only one of three sites, as observed by Botha et al. (2004). In a study by Grönemeyer et al. (2014), the authors observed a predominance of distinct genotypes which were only found in SSA, to date, at which point, sometimes the geographic distribution may prove more local. In view of the premise that "the environment chooses," these outcomes indicate that some autochthonous species like *Bradyrhizobia* are highly fitted to particular environmental requirements and, should, as such, be favored for an inoculant formulation (Grönemeyer et al. 2014).

Ever since N2Africa project was launched in Ethiopia, the inoculant production capacity of, for example, Menagesha Bio-tech Industry (MBI), a private company based in the country, experienced a sixfold expansion (Wolde-meskel et al. 2018). This was exemplified by a surge in annual chickpea inoculant production increment from, not more than, 28,000 to up to 165,000 sachets. Distribution and sales of the inoculants also rose to 7- and 13-fold as reported by Ampadu-Boakye et al. (2017) and Wolde-meskel et al. (2018), respectively. Increased nodulation, biomass production, and N accumulation in soil-grown groundnut were achieved after inoculation with native rhizobium strains of northern Ghana (Osei et al. 2018) and the authors also reported that out of the isolates recently tested in Ghana, KNUST 1002 was observed to be highly effective. Its performance was like that of 32H1, a groundnut reference strain. Apart from only two *Rhizobium* isolates (KNUST 1003 and 1007) studied, all the strains selected in the experiment were intimately related to *B. yuanmingense*, which confirmed the species as a major groundnut microsymbiont (Osei et al. 2018).

9.14 Prospects on Inoculants Production and Use in Nigeria

There is an overwhelming opportunity to increase productivity through the use of *Rhizobium* inoculants in the Nigerian farming systems as it has been established that isolates of indigenous *Rhizobium* can produce fruitful results (Sanginga et al. 1988; Sanginga et al. 1994; Sessitsch et al. 2002). Numerous studies have also demonstrated that inoculants containing indigenous strains outperformed commercially available inoculants (Hungria et al. 2000; Ballard et al. 2003; Aliyu et al. 2014a). The recent hike in the economic recession in Nigeria, due to doom in oil price in the international market (Osalor 2016), has adversely affected many sectors within the country, including agriculture and general food security. This necessitated the inauguration of the Agriculture Promotion Policy (APP) by the Federal Government of Nigeria (FGN) through the Ministry of Agriculture and Rural Development (FMARD). This was an effort of the FGN to shift the nation from oil- to agriculture-based economy (FMARD 2016). This development is a relatively bright future for the agricultural sector and inoculant production and use. Continued significant increase in the price of mineral fertilizers, especially of the N-based, results in ever-increasing food prices (IFDC 2008; Nehring et al. 2008). This, therefore, necessitates the need for developing alternative soil fertility management strategies. The unlimited potential for developing and disseminating BNF and inoculation technologies is, therefore, revealed. Small-holder and resource-deficient farmers who cannot manage the staggering prices of mineral fertilizers may easily access and utilize *Rhizobium* inoculants (Osei et al. 2018). Another reason for the need to explore indigenous strains of *Rhizobium* inoculants is to safeguard the environment against pollution (land, water, and otherwise). Biofertilizers are more environment-friendly than their mineral counterparts (Malusá et al. 2012), even when the availability of the latter to the resource-poor farmers, in Nigeria, is not commendable. Resource-poor small-scale farmers, the primary producers of legumes in Africa, unusually apply fertilizers during legume production. The crop is, therefore, mostly dependent upon biologically fixed N by indigenous N-fixers. Rhizobia isolation for leguminous crops production has always received negligible attention in Africa.

This is, among other reasons, due to a dearth of much-needed research or lackadaisical attitude of researchers and ignorance of its significance in legume production vis-à-vis lack of proper commitment from skilled personnel to promote the technology. Assessment of the efficacy of isolated rhizobia is vital for the preparation of inoculants, the recommendation of host specificity, and symbiotic effectiveness (Simon et al. 2014).

Rhizobium strains vis-à-vis corresponding inoculation methods developed for certain conditions at a given location under a specific farming system may not perform equally well at another location practicing a different farming system (Sanginga et al. 1994). This, therefore, necessitates exploring the potentials in Nigeria's native strains of *Rhizobium*. Another area of concern is the storage conditions of the inoculants, which presents a severe constraint to the viability of rhizobia within the legume inoculants (Kaljeet et al. 2011; Abd El-Fattah et al. 2013). This will invariably give room for more research opportunities, and hence a bright future for the inoculant industry in Nigeria.

Several organizations are known for funding research activities aimed at adapting inoculant technology to the situations where it will be utilized in tropical countries. These organizations include the United Nations Development Program (UNDP) which supports International Agricultural Research Centre(s) (IARCs) through the Consultative Group on International Agricultural Research (CGIAR) and for a specific research program involving International Institute of Tropical Agriculture (IITA) and Boyce Thompson Institute (BTI)/Cornell University. The UNEP and UNESCO also support inoculant technology under the MIRCEN Project, whereas the FAO is also considering her role in the adaptation of inoculant technology for use in developing country like Nigeria. Also, the USAID, via its contracts with the University of Hawaii (Nitrogen Fixation by Tropical Agricultural Legumes NifTAL Project) and United States Department of Agriculture (USDA), is another effort. Another organization is The Beltsville Agricultural Research Centre (ARC) (World *Rhizobium* Study and Collection Center), which provides grants under Section 211(d) to the U.S. Universities' Consortium on BNF in the Tropics, and through a series of smaller grants that are administered by the USDA Science and Education Administration/Cooperative Research (SEA/CR). The USAID and several other governmental and non-governmental agencies supporting the CGIAR are also sponsoring work at Centro International de Agricultural Tropical (CIAT), IITA, International Crop Research Institute for the Semi-Arid Tropics (ICRISAT), and International Centre for Agricultural Research in Dry Areas (ICARDA) on the adaptation of inoculant technology for use in the tropics. These are among many opportunities that can boost research activities geared towards indigenous inoculant production in Nigeria.

It is evident that the BNF benefit to nonlegumes as the inclusion of legumes in a cropping system is small compared to the level of nitrogenous fertilizer used in the more intensive cereal production systems of the developed world. Thus, the principal contribution of BNF to human nutrition will continue to be via the protein in legume grains. Any suggestion of substantial replacement of nitrogen fertilization of cereals and root crops by biologically fixed nitrogen is unrealistic because these crops respond to levels of nitrogen fertilizer far more significantly than those currently supplied through BNF by legumes. Thus, there is an urgent need to devise ways to increase the contribution that BNF by legumes can make to cropping systems as a complement to nitrogen fertilizer-based production, rather than as an alternative to it. Yusuf et al. (2009) observed that many rhizobial genotypes had been identified through experiments, and these were the genotypes that significantly improved N balance in the soil. This displays the importance of inoculation, especially in N-poor tropical soils. Numerous studies, past and present, showed a promising trend in the field of inoculation technology in Nigeria. In the early 1980s, Ranga-Rao et al. (1981) reported that a series of field experiments were conducted in 1978, in Nigeria, to screen some N_2-fixation-efficient strains of *B. japonicum* that showed as high grain yield as 100% of two American soybean cultivars (Bossier and TGm 2944), whereas Asian cultivars did not indicate any significant response. Inoculation also led to encouraging grain yield increases of 40–79% in the American soybean cultivars that were grown in the Nigerian southern Guinea savannas (Nangju 1980; Pulver et al. 1982; Ranga-Rao et al. 1984). Bromfield and Ayanaba (1980) also noted that inoculation of soybean in the low pH sands of southeastern Nigeria achieved increases in grain yield of 300–500% after liming and 270–970% without liming.

In an experiment by Aliyu et al. (2014b), four indigenous strains of pasture rhizobia isolates were observed to contribute to nodulation, hence nitrogen fixation of groundnuts (Tables 9.1 and 9.2). The native strains outperformed all others, including the exotic commercial inoculant in terms of both nodule number and dry weight observed (Table 9.1).

In an earlier study, Sanginga et al. (1994) observed that about 96% of the rhizobia found in root nodules consisted of two main serotypes (IRc1045 and IRc1050). Both were confirmed as strains of indigenous rhizobia earlier isolated from Nigerian soils. This further reaffirmed the bright future for indigenous inoculant industries in Nigeria.

9.15 Studies on Rhizobia Inoculants in Nigeria from the 1990s to Date

Studies conducted within this period mostly focused on the assessment of the response of promiscuous cultivated varieties of soybean to inoculation, alongside other vital nutrients that were deficient. A few trials studied specific and promiscuous soybean cultivars. Based on vegetative parameters, the response of two soybean cultivars (SAMSOY 2 and TGX 1448-2E) to *Bradyrhizobium* inoculation (mixed with two other strains: R25B and IRj 2180A) was, for example, not affected significantly, except for root biomass in the TGX 1448-2E. This was under an on-farm researcher-managed trial condition in the northern Guinea savanna (NGS) of Nigeria. The scenario was ascribed to conceivable high populations of indigenous

Table 9.1 Influence of soil type and pasture rhizobia isolates on nodulation

Treatment	Nodule number	Nodule dry weight (mg)
Soil (S)		
CS	224.57	119.81
FS	54.29	29.38
Mean	139.43	74.6
SE	2.94	2.67
Isolates (I)		
CPI01	193.5	114.5
CPI02	141.17	80.5
MUI03	148.33	72.83
MPI04	176	78.83
Biofix	116.5	–
Control	111.33	56.83
Reference	89.17	79
Mean	139.43	39.67
SE±	5.51	5.
Soil (S) × isolates (I)	***	***

Adapted from Aliyu et al. (2014a)
SE standard error
***$p < 0.0001$

Table 9.2 Influence of soil type and pasture rhizobia isolates on nitrogen fixation

Treatment	Nitrogen fixation (mg N)
Soil (S)	
CS	101.87
FS	63.26
Mean	82.56
SE	1.93
Inoculant (I)	
CPI01	95.45
CPI02	90.94
MUI03	78.58
MPI04	71.76
Biofix	90.52
Control	68.13
Mean	82.56
SE	3.35
Soil (S) × inoculant (I)	***

***$p < 0.0001$

rhizobia satisfactory for soybean nodulation. Okogun et al. (2004), however, observed that the promiscuous cultivar outperformed SAMSOY-2 in terms of BNF and consequently the grain yield, demonstrating that varietal variations concealed the inoculation impact.

Other works included only the promiscuous cultivars of soybean; for example, diverse responses of some promiscuous soybean cultivars to inoculation, N, and P were reported from Kano state, Nigeria, as observed in a series of experiments conducted by Anne et al. (2011). Similar trials were carried out with an early maturing TGX 1485, a promiscuous cultivar, which was inoculated with a rhizobial strain at Minna, NGS of Nigeria. All the parameters observed, including grains yield, were significantly increased by the four inoculants when compared to the control.

In another study, groundnuts (*Arachis hypogaea* L.) were inoculated with indigenous strains isolated from cowpea and the rhizobia isolates were proved effective. A higher number of nodules, nodules dry weight, and consequently greater N_2 fixation were observed compared to the control and reference treatments (Aliyu et al. 2014b) (Table 9.3).

Bashan (1998) reported that combinations of microbes, blended as inoculants that synergistically interact, were being conceived. Studies conducted on microorganisms, devoid of plants, demonstrated that a few mixtures enable the bacteria to synergistically associate with one another. This provided nutrients, expelled inhibitory products, and invigorated each other through physical and/or biochemical activities that improved some beneficial aspects of their physiology like BNF. Bashan

Table 9.3 Effect of soil management and cowpea rhizobia isolates on nodulations, shoot dry weight, N uptake, and N_2 fixation of groundnut

Treatment	Nodule number (plant^{-1})	Nodule dry weight (mg plant^{-1})	Shoot dry weight (g plant^{-1})	N uptake (mg N plant^{-1})	N_2 fixation (mg N plant^{-1})
Soil (S)					
Cultivated soil	198.5	94.42	4.92	135.55	103.01
Fallowed soil	31.33	20.67	3.64	98.28	47.47
± SE	4.55	3.58	0.20	4.04	4.68
Isolates (I)					
VUI05	139.00	65.00	3.67	102.14	73.78
VUI06	119.83	68.67	4.26	121.20	77.76
Control	111.50	56.83	4.32	106.79	70.18
Reference	89.33	39.67	4.88	137.52	
Mean	114.92	57.54	4.28	116.91	73.94
SE±	6.430	5.060	0.300	5.700	5.740
$S \times I$	*	NS	NS	**	*

Adapted from Aliyu et al. (2014b)
SE standard error of difference of means
*$p < 0.05$
**$p < 0.01$

and Holguin (1997a) reported that these bacterial synergisms benefited plant growth. Also, some plant experiments indicated co-inoculation of *Azospirillum* with other microbes could metamorphose into more improved mutual impacts on plants than a single inoculation as observed by Bashan and Holguin (1997a, b). Hence, plant growth could be expanded by double inoculation with *Azospirillum* and phosphate-solubilizing bacteria (Belimov et al. 1995). This is because *Azospirillum* is also viewed as a *Rhizobium*-"aide," which stimulates plant metabolism, nodulation, and nodule activity, all of which also invigorate many plants growth factors and plant protection against unfavorable conditions (Fabbri and Del Gallo 1995; Bashan 1998). *Azospirillum* or *Azotobacter* blended with *Streptomyces* (El-Shanshoury 1995), and *Azospirillum* with the fungal biocontrol agent, *Phialophora radicola* (Flouri et al. 1995), include examples of other fruitful mixes (Bashan 1998). Blended inoculation with diazotrophic bacteria and arbuscular-mycorrhizal fungi (AMF) created synergistic interactions that resulted in a noteworthy increment in growth, P content in plants, upgraded mycorrhizal infection, and improved the uptake of mineral nutrients like N, P, copper (Cu), iron (Fe), and zinc (Zn) (Al-Nahidh and Gomah 1991; Barea 1997; Garbaye 1994; Gori and Favilli 1995; Bashan 1998).

Recently, a comparison was made, in a study by Aliyu et al. (2018), between some isolates and commercial inoculants and a control. The control was used as a benchmark for the comparison such that those isolates that statistically surpassed the control were deemed befitting candidates for a commercial inoculant production. A statistically significant difference ($p < 0.001$) was observed between the commercial inoculants and the controls regarding nodule number. Thus, the authors concluded that 70% of the isolates had records of more nodule number when compared to the control. Regarding dry nodule weight, 74% of the isolates recorded higher weights than the controls, although only 26% were statistically significant ($p < 0.01$). Based on the dry matter yield, only 18% of the isolates had a higher record of the studied parameter, and of these, only a single isolate showed a statistically significant difference ($p < 0.05$) in dry matter yield when compared to the control.

Although native *Bradyrhizobium* strains in Africa were employed to nodulate adapted soybean cultivars, thereby eliminating the need for inoculation (Abaidoo et al. 2007), Okereke et al. (2001) warned that a good establishment of effectively nodulating legumes could not be left to chance. The process, therefore, requires the introduction of effective strains of rhizobia into the soil during the planting period. This can be rightly achieved only through inoculation, and hence an opportunity for judicious use of Nigerian native strains of rhizobia. This may be achieved through their isolation from areas of a flamboyant native population and introduction of the same into relatively less populous sites. There is also a need for a holistic approach to be geared towards improving the entire cropping systems. This should include a selection of more competitive and efficient indigenous rhizobia that could serve as local inoculants (Machido et al. 2011; Sanginga 2003), and hence a bright future for indigenous biofertilizer production firms. A rigorous but systematic identification of crops suitable for diverse cropping sequences and combinations vis-à-vis reaping the potentials in N_2-fixing legumes is paramount (Machido et al. 2011). This will open a new window for more rigorous research activities on inoculant development in Nigeria. Besides, there is a dearth of information on indigenous rhizobia. Where already identified, their symbiotic properties may not fully be understood and may differ depending on their original locations. This may lead to a possible establishment of a location-specific database on the occurrence, abundance, distribution, characteristics, and composition of the indigenous populations of rhizobia strains of Nigerian soils. More promising and versatile strains of the native rhizobial population could in the process, therefore, be identified for subsequent use as registered inoculants containing indigenous strains.

Three main factors limit the effectiveness of an inoculant: (1) its poor quality accompanied with low viability; (2) its inability to compete with indigenous rhizobia; and (3) its inability to tolerate the inherent physical and chemical conditions of the soil to which it is introduced (Cummings 2005). There are, however, many current and potential approaches that may circumvent these and other problems. Chianu et al. (2011) observed some key lessons that would ensure success at farmers' field level, some of which include: (1) an ubiquitous demonstration of the inoculants to the needs of, especially, small farmers; (2) intra-national collaborations, with the involvement of mass media; (3) well-coordinated and collaborative research-for-development programs; (4) involvement of top people of the government; (5) joint efforts of related governmental and non-governmental organizations for a long time; (6) involvement of individuals and the private sector in production and dissemination of the biofertilizers, and (7) effective farmer education on inoculation. The said strategies have been found to work effectively in some pilot areas and should be scaled up to reach more smallholder farmers (Chianu et al. 2011) in order for, especially, grain legumes farmers to maximally reap the diverse dividends of using inoculants in their cropping practices. This scaling could, however, be attained only through a desirable innovation platform involving all stakeholders and appropriate incentives to entice the private sector and industries (Chianu et al. 2011).

9.16   Conclusion

The need for initiating advanced studies on inoculation to address the difficulties confronting the use of inoculants by farmers in Africa, particularly Nigeria, can never be overestimated. Various trials, to be aimed at demonstrating the need for inoculation, should, therefore, include tests for the constraint of BNF by other nutrients like boron (B) and calcium (Ca), for example. Also, there is a dire need for a deeper examination of the economic and social cost-benefit analyses of the *Rhizobium* inoculation. The need for expanding the knowledge base on BNF utilization among farmers in Nigeria should go beyond awareness and use only. It should also include more qualitative aspects of farmers' knowledge, willingness to pay, and the long-term relevance of inoculants in farm objectives. Institutions and policies promoting the development of inoculants and widespread farmer adoption campaigns for increased production of both food and cash legumes must be encouraged. This must especially be accompanied by targeted research to effectively explore the available indigenous strains of *Rhizobium* present in Nigerian soils. This is will invariably counterpoise the problems of N fertility and its consequent cost on small-holder farmers in Nigeria. Problems of poor quality, inadequate and inefficient markets, as well as inadequate extension services on inoculants and their use must also be tackled. Some successful outcomes of many on-station and farmer-condition simulating experiments with *Rhizobium* inoculants have been recorded. These records may be used as an index for the potentials of indigenous rhizobium-based inoculants in Nigeria. Specific measures such as tax motivations and exceptions will be paramount in stimulating the advancement of BNF innovation markets and the formation of nearby inoculant firms.

There is also a need for specific policy incentives to stimulate private sector involvement, at all stages of the innovation process, to install adoption. An array of studies has glaringly made it clear that the enormous diversity of *Bradyrhizobium* species specifically in Nigeria and elsewhere, in SSA, in general, is currently underestimated. Therefore, research in diversity, and characterization of nodule symbionts, in Nigeria, and SSA, should be accentuated. This is basically because numerous strains are bound to be developed into adapted inoculants for green manure and legumes. Of particular importance is tolerance for high temperature of many African *Bradyrhizobial* species, which makes them potential candidate strains to curb the problems of global climate change that foresee increases in temperature. Such future research activities should also focus on, and address, the molecular rationale and/or basis for their tolerance to the usually deleterious temperatures (Grönemeyer and Reinhold-Hurek 2018).

References

Abaidoo RC, Sanginga N, Okogun JA et al (2007) Genotypic variation of soybean for phosphorus use efficiency and their contribution of N and P to subsequent maize crops in three ecological zones of West Africa. In: Badu-Apraku B, Fakorede MAB, Lum AF et al (eds) Demand-driven technologies for sustainable maize production in West and Central Africa. Proceedings of the fifth biennial regional maize workshop, IITA-Cotonou, Benin Republic, 3–6 May 2005. WECAMAN/IITA, Nigeria, pp 194–224

Abaidoo R, Buahen S, Turner A, Dianda M (2013) Bridging the grain legume gap through agronomy. IITA R4D Review. Issue 9. Jan 2013. http://r4dreview.org/2013/01/bridging-the-grain-legume-yield-gap-through-agronomy/. Accessed 24 Sept 2013

Abd El-Fattah DA, Eweda WE, Zayed MS, Hassanein MK (2013) Effect of carrier materials, sterilization method, and storage temperature on survival and biological activities of Azotobacter chroococcum inoculant. Ann Agric Sci 58:111–118

Ahmad I, Pichtel J, Hayat S (2008) Plant-bacteria interactions: strategies and technique to promote plant growth. Wiley, Weinheim, p 166

Akhtar MS, Siddiqui ZA (2009) Effect of phosphate solubilizing microorganisms and Rhizobium sp. on the growth, nodulation, yield and root-rot disease complex of chickpea under field condition. Afr J Biotechnol 8(15):3489–3496

Albareda M, Rodríguez-Navarro DN, Camacho M, Temprano FJ (2008) Alternatives to peat as a carrier for rhizobia inoculants: solid and liquid formulations. Soil Biol Biochem 40:2771–2779

Aliyu IA, Yahaya SM, Yusuf AA (2014a) Effect of pasture rhizobia isolates on nodulation and nitrogen fixation of groundnut (*Arachis hypogaea* L.). In: Ojeniyi SO, Obi JC, Ibia TO et al (eds) Proceedings of the 38th annual conference of the Soil Science Society of Nigeria (SSSN), held at University of Uyo, Uyo, Nigeria, between 10th and 14th Mar 2014

Aliyu IA, Yahaya SM, Gabasawa AI et al (2014b) Response of groundnut to cowpea rhizobia isolates under soils of different management practice. In: Idisi PO, Okoye BC, Idu EE et al (eds) Proceedings of the 48th annual conference of the Agricultural Society of Nigeria (ASN), held by University of Abuja, Abuja Nigeria between 24th and 27th Nov 2014

Aliyu IA, Yusuf AA, Atta A (2018) Evaluation of indigenous rhizobial isolates in search for candidate strain for commercial production. Bayero J Pure Appl Sci 11:33–39

Alla S, Coba T, Ana R et al (2010) *Flavodoxin overexpression* reduces cadmium-induced damage in alfalfa root nodules. J Crop Sci Biotechnol 326(1–2):109–121

Al-Nahidh S, Gomah AHM (1991) Response of wheat to dual inoculation with VA –mycorrhiza, and *Azospirillum,* fertilized with NPK and irrigated with sewage effluent. Arid Soil Res Rehabil 5:83–96

Alori ET, Babalola OO (2018) Microbial inoculants for improving crop quality and human health in Africa. Front Microbiol 9:1–12

Amarger N (2001) Rhizobia in the field. Adv Agron 73:109–168

Ampadu-Boakye T, Stadler M, Kanampiu F (2017) N2Africa annual report 2016, p 89. www.N2Africa.org

Andrade DS, Murphy PJ, Giller KE (2002) The diversity of *Phaseolus*-nodulating rhizobial populations is altered by liming of acid soils planted with *Phaseolus vulgaris* L. in Brazil. Appl Environ Microbiol 68:4025–4034

Anne T, Bala A, Abaidoo R et al (2011) N2Africa annual country reports 2011, p 129

Baez-Rogelio A, Morales-García YE, Quintero-Hernández V, Muñoz-Rojas J (2017) Next generation of microbial inoculants for agriculture and bioremediation. Microb Biotechnol 10:19–21

Bala A (2008) Recent advances in soybean inoculum research and applications: Towards enhancing productivity in smallholder agriculture, paper presented at an International Workshop on Rhizobium Inoculation, held at Impala Hotel, Arusha Tanzania, 17–21 March, 2008

Bala A (2011a) Update on Inoculant production by cooperating laboratories. Milestone reference number: 3.4.3. N2Africa, Oct 2011, p 8

Bala A (2011b) Emerging challenges in cross-border movement of inoculants in sub-Saharan Africa. N2Africa project (Putting Nitrogen fixation to work for smallholder farmers in Africa). Podcaster 8, Aug 2011

Bala A, Giller KE (2007) Relationships between rhizobial diversity and host legume nodulation and nitrogen fixation in tropical ecosystems. Nutr Cycl Agroecosyst 76(2–3):319–330

Baldwin IL, Fred EB (1929) Root-nodulating bacteria of leguminosae. J Bacteriol 17:141–150

Ballard RA, Shepherd BR, Charman N (2003) Nodulation and growth of pasture legumes with naturalised soil rhizobia. 3. Lucerne (Medicago sativa L.). Aust J Exp Agric 43:135–140

Barea JM (1997) Mycorriza/bacteria interactions on plant growth promotion. In: Ogoshi A, Kobayashi K, Homma Y et al (eds) Plant growth-promoting Rhizobacteria-present status and future prospects. Sapporo, Faculty of Agriculture, Hokkaido Univerisity, pp 150–158

Bashan Y (1998) Inoculants of plant growth-promoting bacteria for use in agriculture. Biotechnol Adv 16(4):729–770

Bashan Y, Holguin G (1997a) Azospirillum-plant relationships: environmental and physiological advances (1990-1996). Can J Microbiol 43:103–121

Bashan Y, Holguin G (1997b) Short- and medium-term avenues for Azospirillum inoculation. In: Ogoshi A, Kobayashi K, Homma et al (eds) Plant growth-promoting Rhizobacteria-present status and future prospects. Faculty of Agriculture, Hokkaido University, Sapporo, pp 130–149

Bashan Y, de-Bashan LE, Prabhu SR, Hernandez JP (2014) Advances in plant growth-promoting bacterial inoculant technology: formulations and practical perspectives (1998-2013). Plant Soil 378:1–33

Bashan Y, De-Bashan LE, Prabhu SR (2016) Superior polymeric formulations and emerging innovative products of bacterial inoculants for sustainable agriculture and the environment. In: Agric. Important Microorg. Springer, Singapore, pp 15–46

Belimov AA, Kojemiakov AP, Chuvarliyeva CV (1995) Interaction between barley and mixed cultures of nitrogen fixing and phosphate-solubilizing bacteria. Plant Soil 173:29–37

Bergey DH, Harrison FC, Breed RS et al (1923) Bergey's manual of determinative bacteriology. Williams & Wilkins, Baltimore, MD

Beynon JL, Josey DP (1980) Demonstration of heterogeneity in a natural population of *Rhizobium phaseoli* using variation in intrinsic antibiotic resistance. J Gen Microbiol 118:437–442

Bohlool BB, Schmidt EL (1973) Persistence and competition aspects of *Rhizobium japonicum* observed in soil by immunofluorescence microscopy. Soil Sci Soc Am J 37:561–564

Boivin C, Ndoye I, Lortet G et al (1997) The Sesbania root symbionts *Sinorhizobium saheli* and *S. teranga* bv. sesbaniae can form stem nodules on *Sesbania rostrata*, although they are less adapted to stem nodulation than *Azorhizobium caulinodans*. Appl Environ Microbiol 63:1040–1047

Botha WJ, Jaftha JB, Bloem JF et al (2004) Effect of soil *Bradyrhizobia* on the success of soybean inoculant strain CB 1809. Microbiol Res 159:219–231

Brenner C (1996) Integrating biotechnology in agriculture: Incentives, constraints and country experiences. Development Centre Studies, OECD, Paris

Brockwell J, Bottomley PJ (1995) Recent advances in inoculant technology and prospects for the future. Soil Biol Biochem 27:683–697

Brockwell J, Bottomley PJ, Thies JE (1995) Manipulation of rhizobia microflora for improving crop productivity and soil fertility: a critical assessment. Plant and Soil, 174:143–180

Bromfield ESP, Ayanaba A (1980) The efficacy of soybean inoculation on acid soil in tropical Africa. Plant Soil 54:95–106

Broughton WJ, Jabbouri S, Perret X (2000) Keys to symbiotic harmony. J Bacteriol 182:5641–5652

Carvalho TL, Ferreira PC, Hemerly AS (2011) Tropical plant biology: sugarcane genetic controls involved in the association with beneficial endophytic nitrogen fixing bacteria. J Crop Sci Biotechnol 4(1):31–41

Cassien B, Woomer PL (1998) Recent history of the BNF activities at the Institut des sciences Agronomique du Rwanda. In: Dakora FD (ed) Biological nitrogen fixation in Africa: linking process to progress. African Association for Biological Nitrogen Fixation, Cape Town, pp 33–34

Cheng Q (2008) Perspectives in biological nitrogen fixation research. J Integr Plant Biol 50:786–798

Chianu JN, Nkonya EM, Mairura FS et al (2011) Biological nitrogen fixation and socioeconomic factors for legume production in sub-Saharan Africa: a review. Agron Sustain Dev 31:139–154

Cummings SP (2005) The role and future potential of nitrogen fixing bacteria to boost productivity in organic and low-input sustainable farming systems. Environ Biotechnol 1(1):1–10

Date RA (2000) Inoculated legumes in cropping systems of the tropics. Field Crop Res 65(2–3):123–136

Dekak A, Chabi R, Menasria T, Benhizia Y (2018) Phenotypic characterization of rhizobia nodulating legumes Genista microcephala and Argyrolobium uniflorum growing under arid conditions. J Adv Res 14:35–42

Deneyschen T, Strijdom BW, Law IJ (1998) Comparison of South African legume inoculants manufactured in 1978-79 and in 1995-97. In: Dakora FD (ed) Biological nitrogen fixation in africa: linking process to progress. African Association for Biological Nitrogen Fixation, Cape Town, pp 50–51

Devi V, Sumathy VJH (2018) Production of biofertilizers from agro-wastes. Int J Eng Techn 4(1):453–466

Döbereiner J, Day JM (1976) Associative symbioses in tropical grasses: characterization of microorganisms and dinitrogen-fixing sites. In: Newton WE, Nyman CJ (eds) Proceedings of the first international symposium on nitrogen fixation, vol 2. Washington State University Press, Pullman, WA, pp 518–538

Eardly BD, Materon LA, Smith NH et al (1990) Genetic structure of natural populations of the nitrogen-fixing bacterium *Rhizobium meliloti*. Appl Environ Microbiol 56:187–194

El-Ramady H, El-Ghamry A, Mosa A, Alshaal T (2018) Nanofertilizers vs. biofertilizers: new insights. Environ Biodivers Soil Secur 2:40–50. https://doi.org/10.21608/jenvbs.2018.3880.1029

El-Shanshoury AR (1995) Interactions of *Azotobacter chroococcum, Azospirillum brasilense* and *Streptomyces mutabilis*, in relation to their effect on wheat development. *J. Agron. Crop Sci.* 175 (2):119-127

El-Yazeid AA, Abou-Aly HA, Mady MA, Moussa SAM (2007) Enhancing growth, productivity and quality of squash plants using phosphate dissolving microorganisms (bio phosphor) combined with boron foliar spray. Res J Agric Biol Sci 3(4):274–286

Fabbri P, Del Gallo (1995) Specific interaction between chickpea (*Cicer arietinum* L.) and three chickpea-rhizobium strains inoculated singularly and in combination with *Azospirillum brasilense* Cd. In: *Azospirillum* VI and related microorganisms. https://doi.org/10.107/978-3-642-8_28

Flouri F, Sini K, Balis C (1995) Interactions between *Azospirillum* and *Phialophora radicicola*. NATO ASI Ser G 37:231–237

FMARD (2016) The agriculture promotion policy (2016–2020): building on the successes of the ATA, Closing Key Gaps. Policy and Strategy Document of the Federal Ministry of Agriculture and Rural Development, Nigeria

Frank B (1889) Über die Pilzsymbiose der Leguminosen, vol 19. Verlog von Paul Parey, Berlin

Fred EB, Baldani JI, McCoy E (1932) Root nodule bacteria and legume plants. University of Wisconsin, Madison, WI

Gage DJ (2017) 2004 Infection and invasion of roots by symbiotic, nitrogen-fixing rhizobia during nodulation of temperate legumes. Microbiol Mol Biol Rev 68:203

Garbaye L (1994) Helper bacteria: a new dimension to the mycorrhizal symbiosis. New Phytol 128:197–210

Garrison T (2006) Essentials of oceanography, 4th edn. Thomson Learning Academic Resource Center, Belmont, p 368

Giller KE (2001) Nitrogen fixation in tropical cropping systems, 2nd edn. CAB International, Wallingford, p 423

Gori A, Favilli F (1995) First results on individual and dual inoculation with *Azospirillum-Glomus* on wheat. In: Fendrik I, Del Gallo M, Vanderleyden J, de Zamaroczy M (eds) *Azospirillum* VI and related microorganisms, genetics-physiology-ecology, NATO ASI series, series G: ecological sciences, vol G37. Springer, Berlin, pp 245–249

Graham PH, Sadowsky MJ, Keyser HH (1991) Proposed minimum standards for the description of new genera and species of root-and stem-nodulating bacteria. Int J Syst Bacteriol 41:582–587

Grönemeyer JL, Reinhold-Hurek B (2018) Diversity of *bradyrhizobia* in Subsahara Africa: a rich resource. Front Microbiol 9:2194. https://doi.org/10.3389/fmicb.2018.02194

Grönemeyer JL, Kulkarni A, Berkelmann D et al (2014) Rhizobia indigenous to the Okavango region in sub-Saharan Africa: diversity, adaptations, and host specificity. Appl Environ Microbiol 80(23):7244–7257

Gupta RA, Kalia A, Kapoor S (2007) In: Jain SP (ed) Bioinoculants: a step towards sustainable agriculture. New India Publishing Agency, New Delhi, pp 4–5

Han TX, Wang ET, Han LL et al (2008) Molecular diversity and phylogeny of rhizobia associated with wild legumes native to Xinjiang, China. Syst Appl Microbiol 31:287–301

Hardarson G, Bunning S, Montanez A et al (2003) The value of symbiotic nitrogen fixation by grain legumes in comparison to the cost of nitrogen fertiliser used in developing countries. In: Hardarson G, Broughton W (eds) Maximizing the use of biological nitrogen fixation in agriculture. Kluwer, Dordrecht, pp 213–220

Hirsch AM, Bauer WD, Bird DM et al (2003) Molecular signals and receptors controlling rhizosphere interactions between plants and other organisms. Ecology 84:858–868

Hungria M, Andrade DS, Chueire LMO (2000) Isolation and characterisation of new efficient and competitive bean (*Phaseolus vulgaris* L.) rhizobia from Brazil. Soil Biol Biochem 32:1515–1528

Hungria M, Loureiro MF, Mendes IC et al (2005) Inoculant preparation, production and application. In: Nitrogen fixat. agric. for. ecol. environ. Springer, Dordrecht, pp 223–253

IFDC (International Fertilizer Development Center) (2008) Soaring fertilizer prices threaten world's poorest farmers. http://ifdc.org/New_Layout/News_PressReleases/index.html

Jones C, Olson-Rutz K (2018) Inoculation and nitrogen management to optimize pulse crop yield and protein. Crop Soils 51:12

Kaljeet S, Keyeo F, Amir HG (2011) Influence of carrier materials and storage temperature on survivability of rhizobial inoculant. Asian J Plant Sci 10:331–337

Khan MS, Zaidi A, Musarral J (2010) Microbes for legume improvement. Springer, Wein

Khojely DM, Ibrahim SE, Sapey E, Han T (2018) History, current status, and prospects of soybean production and research in sub-Saharan Africa. Crop J 6:226–235

Khonje DJ (1989) Adoption of *Rhizobium* inoculation technology for pasture improvement in sub-Saharan Africa. Department of Agricultural Research, Chitedze Agricultural Research Station, Lilongwe, p 14

Kloepper JW (1994) Plant growth promoting rhizobacteria: other systems. In: Okon Y (ed) *Azospirillum*/plant associations. CRC Press, Boca Raton, FL, pp 137–166

Koskey G, Mburu SW, Kimiti JM et al (2018) Genetic characterization and diversity of Rhizobium isolated from root nodules of mid-altitude climbing bean (Phaseolus vulgaris L.) varieties. Front Microbiol 9:1–12. https://doi.org/10.3389/fmicb.2018.00968

Kueneman EA, Root WR, Dashiel KE, Hohenberg J (1984) Breeding soybeans for the tropics capable of nodulating effectively with indigenous *Rhizobium* spp. Plant Soil 8:387–396

Laditi MA, Nwoke OC, Jemo M et al (2012) Evaluation of microbial inoculants as biofertilizers for the improvement of growth and yield of soybean and maize crops in savanna soils. Afr J Agri Res 7(3):405–413

Laguerre GP, Allard MR, Revoy F, Amarger N (1994) Rapid identification of rhizobia by restriction fragment length polymorphism analysis of PCR-amplified 16S rRNA genes. Appl Environ Microbiol 60:56–63

Larnier JE, Jordan DL, Speras FJ et al (2005) Peanut response to inoculation and nitrogen fertilizer. Agron J 97:79–84

Law IJ, Botha WJ, Majaule UC, Phalane FL (2007) Symbiotic and genomic diversity of 'cowpea' bradyrhizobia from soils in Botswana and South Africa. Biol Fertil Soils 43:653–663

Leite J, Passos SR, Simões-Araújo JL et al (2018) Genomic identification and characterization of the elite strains Bradyrhizobium yuanmingense BR 3267 and Bradyrhizobium pachyrhizi BR 3262 recommended for cowpea inoculation in Brazil. Brazilian J Microbiol 49:703–713

Li W, Raoult D, Fournier P (2009) Bacterial strain typing in the genomic era. FEMS Microbiol Rev 33:892–916

Lindström K (1989) *Rhizobium galegae,* a new species of legume root nodule bacteria. Int J Syst Bacteriol 39(3):365–367

López-Bellido L, López-Bellido RJ, Redondo R, Benítez J (2006) Faba bean nitrogen fixation in a wheat-based rotation under rain fed Mediterranean conditions: effect of tillage system. Field Crops Res 98:253–260

Macdonald RM (1989) An overview of crop inoculation. In: Campbell R, Macdonald RM (eds) Microbial inoculation of crop plants (special publication of the society of general microbiology). IRL Press, New York, pp 1–9

Macdonald C, Singh B (2014) Harnessing plant-microbe interactions for enhancing farm productivity. Bioengineered 5(1):5–9

Machido DA, Olufajo OO, Yakubu SE, Yusufu SS (2011) Enhancing the contribution of the legumes to the N fertility of soils of the semi-arid zone of Nigeria. Afr J Biotechnol 10(10):1848–1853

Malusá E, Sas-Paszt L, Ciesielska J (2012) Technologies for beneficial microorganisms inocula used as biofertilizers. Sci World J 2012:1–12. https://doi.org/10.1100/2012/491206

Manvika S, Bhavdish NJ (2006) Taxonomy of rhizobia: current status. Curr Sci 90(4):486–487

Margenot AJ, Griffin DE, Alves BSQ et al (2018) Substitution of peat moss with softwood biochar for soil-free marigold growth. Ind Crop Prod 112:160–169

Massol-Deya AA, Odelson DA, Hickey RF, Tiedje JM (1995) Bacterial community fingerprinting of amplified 16S and 16-23S ribosomal DNA gene sequences and restriction endonuclease analysis (ARDRA). In: ADL A, van Elsas JD, de Bruijn FJ (eds) Molecular Microbial Ecology Manual. Kluwer Academic Publishers, Dortrecht

Minchin FR, James EK, Becana M (2008) Oxygen diffusion, production of reactive oxygen and nitrogen species, and antioxidants in legume nodules. In: Dilworth MJ, James EK, Sprent JI, Newton WE (eds) Nitrogen-fixing Leguminous symbioses. Springer Science, Berlin, pp 321–362

Mohammadi K, Sohrabi Y (2012) Bacterial biofertilizers for sustainable crop production: a review. J Agri Biol Sci 7(5):307–316

Moulin L, Chen WM, Béna G et al (2002) Rhizobia: the family is expanding. In: Nitrogen fixation. Global perspectives. CABI Publishing, New York, pp 61–65

Mugabe J (1994) Research on biofertilizers: Kenya, Zimbabwe and Tanzania. Biotechnol Dev Monit 18:9–10

Musiyiwa K, Mpepereki S, Giller KE (2005) Physiological diversity of rhizobia nodulating promiscuous soybean in Zimbabwean soils. Symbiosis 40:97–107

Mpepereki S, Javaheri F, Davis P, Giller KE (2000) Soyabeans and sustainable agriculture. Prosmicuous soybean in South Africa. Field Crops Research 65(2–3):137–149

N2Africa (Putting Nitrogen fixation to work for smallholder farmers in Africa) (2013) Objective 4. Deliver legume and inoculant technologies to farmers throughout sub-Saharan Africa. http://www.n2africa.org/objectives/objective_4. Accessed 14 Sept 2013

Naab JB, Chimphango SMB, Dakora FD (2009) N2 fixation in cowpea plants grown in farmers' fields in the Upper West Region of Ghana, measured using15N natural abundance. Symbiosis 48:37–46

Nangju D (1980) Soybean response to indigenous rhizobia as influenced by cultivar origin. Agronomy 72:403–406

Ncube B, Twomlow SJ, van Wijk MT et al (2007) Productivity and residual benefits of grain legumes to sorghum under semi-arid conditions in south western Zimbabwe. Plant Soil 299:1–15. https://doi.org/10.1007/s11104-007-9330-5

Ndakidemi PA, Dakora FD, Nkonya EM et al (2006) Yield and economic benefits of common bean (*Phaseolus vulgaris*) and soybean (*Glycine max*) inoculation in northern Tanzania. Aust J Exp Agric 46:571–577

Nehra V, Choudhary M (2015) A review on plant growth promoting rhizobacteria acting as bio-inoculants and their biological approach towards the production of sustainable agriculture. J Appl Nat Sci 7:540–556

Nehring R, Vialou A, Erickson K, Sandretto C (2008) Assessing economic and environmental impacts of ethanol production on fertilizer use in corn. In: Annual meeting, Feb 2–6, Dallas, Texas 6736, Southern Agricultural Economics Association

Nelson LM (2004) Plant growth promoting rhizobacteria (PGPR): prospects for new inoculants. Crop Manag. https://doi.org/10.1094/CM-2004-0301-05-RV

Nour SM, Cleyet-Marel JC, Beck D et al (1994) Genotypic and phenotypic diversity of Rhizobium isolated from chickpea (*Cicer arietinum* L.). Can J Microbiol 40:345–354

Nyoki D, Ndakidemi PA (2013) Economic benefits of *Bradyrhizobium japonicum* inoculation and phosphorus supplementation in cowpea (*Vigna unguiculata* (L.) Walp) grown in northern Tanzania. Am J Res Com 1(11):321–332

O'Callaghan M (2016) Microbial inoculation of seed for improved crop performance: issues and opportunities. Appl Microbiol Biotechnol 100:5729–5746

Odair A, Glaciela K, Mariangela H (2006) Sampling effects on the assessment of genetic diversity of rhizobia associated with soybean and common bean. J Soil Biol Biochem 38:1298–1307

Odame H (1997) Biofertilizer in Kenya: research, production and extension dilemmas. Biotechnol Dev Monit 30:2023

Okereke GU, Onoci CC, Onyeagba E (2001) Effectiveness of foreign *Bradyrhizobium* in enhancing nodulation, dry matter and seed yield of soybean cultivars in Nigeria. Biol Fertil Soils 33:3–9

Okogun JA, Sanginga N (2003) Can introduced and indigenous rhizobial strains compete for nodule formation by promiscuous soybean in the moist savanna agroecological zone of Nigeria? Biol Fertil Soils 38:26–31

Okogun JA, Otuyemi BT, Sanginga N (2004) Soybean yield determinants and response to rhizobial inoculation in on-farm trial in the northern Guinea savanna of Nigeria. West Afr J App Ecol 6:30–39

Oldroyd GE, Downie JA (2008) Coordinating nodule morphogenesis with rhizobial infection in legumes. Annu Rev Plant Biol 59:519–546

Ollivier J, We ST, Bannert A et al (2011) Nitrogen turnover in soil and global change. FEMS Microbiol Ecol 78:3–16

Oruru MB, Njeru EM (2016) Upscaling arbuscular mycorrhizal symbiosis and related agroecosystems services in smallholder farming systems. BioMed Res Int 4376240:12

Osalor P (2016) Nigerian economic recession and entrepreneurial revolution. Vanguard Online Newspaper of 26th Sept 2016

Osei O, Abaidoo RC, Ahiabor BDK et al (2018) Bacteria related to *Bradyrhizobium yuanmingense* from Ghana are effective groundnut micro-symbionts. Appl Soil Ecol 127:41–50

Osunde AO, Gwam S, Bala A et al (2003) Responses to rhizobial inoculation by two promiscuous soybean cultivars in soils of the southern Guinea savanna of Nigeria. Biol Fertil Soils 37:274–279

Perret X, Jabbouri S, Broughton WJ (2000) Keys to symbiotic harmony. J Bacteriol 182:5641–5652

Pierre C, Simon CA (2010) Iron uptake and homeostasis in microorganisms. Caister Academic Press, Norfolk

Pinero D, Martinez E, Selander RK (1988) Genetic diversity and relationships among isolates of *Rhizobium leguminosarum biovar phaseoli*. Appl Environ Microbiol 54:2825–2832

Pule-Meulenberg F, Belane AK, Krasova-Wade T, Dakora FD (2010) Symbiotic functioning and bradyrhizobial biodiversity of cowpea (*Vigna unguiculata* L. Walp.) in Africa. BMC Microbiol 10:89. https://doi.org/10.1186/1471-2180-10-89

Pulver EL, Brockman F, Wien HC (1982) Nodulation of soybean cultivars with *Rhizobium* spp. and their response to inoculation with *R. japonicum*. Crop Sci 22:1065–1070

Purchase HF, Vincent JM, Ward LM (1951) The field distribution of strains of nodule bacteria from species of *Medicago*. Aust J Agric Res 2:261–272

Ranga-Rao V, Thottapilly G, Ayanaba A (1981) Studies on the persistence of introduced strains of *Rhizobium japonicum* in soil during fallow and effects on soybean growth and yield. In: BNF technology for tropical agriculture, pp 309–315

Ranga-Rao V, Ayanaba A, Eaglesham ARJ, Thottappilly G (1984) Effects of Rhizobium inoculation on field-grown soybeans in Western Nigeria and assessment of inoculum persistence during a two-year fallow. Trop Agric 62:125–130

Rascio N, Rocca NL (2008) Biological nitrogen fixation. In: Encyclopedia of ecology. University of Padua, Padua, pp 412–419

Reddy CA, Saravanan RS (2013) Polymicrobial multi-functional approach for enhancement of crop productivity. Adv Appl Microbiol 82:53–113. https://doi.org/10.1016/B978-0-12-407679-2.00003-X

Roesch LF, Quadros PD, Camargo FA, Triplett EW (2007) Screening of diazotrophic bacteria *Azopirillum* spp. for nitrogen fixation and auxin production in multiple field sites in southern Brazil. J Microbiol Biotechnol 23(10):1377–1383

Ronner E, Franke AC, Vanlauwe B et al (2016) Understanding variability in soybean yield and response to P-fertilizer and rhizobium inoculants on farmers' fields in northern Nigeria. Field Crop Res 186:133–145

Rubenchik LI (1963) *Azotobacter* and its use in agriculture. In: Academy of sciences of the Ukrainian SSR, Microbiological Institute D.K. Zabolotnyi. Israel programme for scientific translations, Jerusalem. 65 p. (from a Russian text published in 1960)

Ruíz-Valdiviezo VM, Canseco LMCV, Suárez LAC et al (2015) Symbiotic potential and survival of native rhizobia kept on different carriers. Brazilian J Microbiol 46:735–742

Rurangwaa E, Bernard V, Giller KE (2018) Benefits of inoculation, P fertilizer and manure on yields of common bean and soybean also increase yield of subsequent maize. Agric Ecosyst Environ 261:219–229

Safronova V, Belimov A, Sazanova A et al (2019) Two broad host range rhizobial strains isolated from relict legumes have various complementary effects on symbiotic parameters of co-inoculated plants. Front Microbiol 10:1–14. https://doi.org/10.3389/fmicb.2019.00514

Sanchez C, Tortosa G, Granados A et al (2011) Involvement of *Bradyrhizobium japonicum* denitrification in symbiotic nitrogen fixation by soybean plants subjected to flooding. Appl Soil Ecol 43:212–217

Sanginga N (2003) Role of biological nitrogen fixation in legume-based cropping systems; a case study of West Africa farming systems. Plant Soil 252:25–39

Sanginga N, Mulongoy K, Ayanaba A (1988) Nitrogen contribution of *Leucaena/Rhizobium* symbiosis to soil and a subsequent maize crop. Plant Soil 112:137–141

Sanginga N, Danso SKA, Mulongoy K, Ojeifo AA (1994) Persistence and recovery of introduced rhizobium 10 years after inoculation on *Leucaena Leucocephala* grown on Alfisol in southwestern Nigeria. Plant Soil 159:199–204

Sanginga N, Okogun JA, Vanlauwe B, Diels J, Dashiell K (2001) Contribution of nitrogen fixation to the maintenance of soil fertility with emphasis on promiscuous soybean maize-based cropping systems in the moist savanna of West Africa. In: Tian G, Ishida F, Keatinge JDH (Eds.) Sustaining soil fertility in West Africa. American Society of Agron, Madison, pp 157–178

Seshadri R, Reeve WG, Ardley JK et al (2015) Discovery of novel plant interaction determinants from the genomes of 163 root nodule bacteria. Sci Rep 5:1–9. https://doi.org/10.1038/srep16825

Sessitsch A, Hardarson G, Akkermans ADL, de Vos WM (1997) Characterization of *Rhizobium etli* and other *Rhizobium* spp. that nodulate *Phaseolus vulgaris* L. in an Austrian soil. Mol Ecol 6:601–608

Sessitsch A, Howieson JG, Perret X et al (2002) Advances in rhizobium research. Crit Rev Plant Sci 21:323–378

Simon Z, Mtei K, Gessesse A, Ndakidemi PA (2014) Isolation and characterization of nitrogen fixing rhizobia from cultivated and uncultivated soils of northern Tanzania. Am J Plant Sci 5:4050–4067

Sivestre P (1970) Travaux de l'IRAT sur le soya, Ford Foundation Grain Legume Seminar, Jun 22–26, Ibadan, Nigeria

Smith RS (1992) Legume inoculant formulation and application. Can J Microbiol 38:485–492

Song K, Xue Y, Zheng X et al (2017) Effects of the continuous use of organic manure and chemical fertilizer on soil inorganic phosphorus fractions in calcareous soil. Sci Rep 7:1–9. https://doi.org/10.1038/s41598-017-01232-2

Strijdom BW (1998) South African studies on biological nitrogen-fixing systems and the exploitation of the nodule bacterium-legume symbiosis. S Afr J Sc 94:11–23

Strijdom BW, van Rensburg J (1981) Effect of steam sterilization and gamma irradiation of peat on the quality of rhizobium inoculants. Appl Environ Microbiol 41, 1344–1347

Suzuki Y, Adhikari D, Itoh K, Suyama K (2014) Effects of temperature on competition and relative dominance of *Bradyrhizobium japonicum* and *Bradyrhizobium elkanii* in the process of soybean nodulation. Plant Soil 374:915–924

Tang WH (1994) Yield-increasing bacteria (YIB) and biocontrol of sheath blight of rice. In: Ryder MH, Stephens PM, Bowen GD (eds) Improving plant productivity with rhizosphere bacteria. Division of Soils CSIRO, Adalaide, pp 267–273

Tang WH, Yang H (1997) Research and application of biocontrol of plant diseases and PGPR in China. In: Ogoshi A, Kobayashi K, Homma Y et al (eds) Plant growth-promoting rhizobacteria-present status and future prospects. Faculty of Agriculture, Hokkaido University, Sapporo, pp 4–9

The Economist Intelligence Unit (2014) Food security in focus: Sub-Saharan Africa. http://food-securityindex.eiu.com/

Thies J, Holmes EM, Vachot A (2001) Application of molecular techniques to studies in rhizobium ecology: a review. Aust J Exp Agric 41:299–319

Trivedi P, Pandey A, Palni LMS (2005) Carrier-based preparations of plant growth-promoting bacterial inoculants suitable for use in cooler regions. World J Microbiol Biotechnol 21:941–945

Udvardi M, Brodie EL, Riley W et al (2015) Impacts of agricultural nitrogen on the environment and strategies to reduce these impacts. Procedia Environ Sci 29:303

Unkovich MJ, Pate JS (2000) An appraisal of recent field measurements of symbiotic N_2 fixation by annual legumes. Field Crops Res 65:211–228

Van Berkum PB (2002) USDA-ARS National rhizobium germplasm collection. http://www.ars. usda.gov/is/np/systematics/rhizobium.htm. Accessed 24 May 2013

Van Berkum P, Elia P, Eardly BD (2006) Multilocus sequence typing as an approach for population analysis of Medicago-nodulating rhizobia. J Bacteriol 188:5570–5577

van Heerwaarden J, Baijukya F, Kyei-Boahen et al (2018) Soya bean response to rhizobium inoculation across sub-Saharan Africa: patterns of variation and the role of promiscuity. Agric Ecosyst Environ 261:211–218

Van Rensburg H, Strijdom BW (1969) Strains of *Rhizobium japnicum* and inoculant production in South Africa, Phytophylactica 1, 201-204. Fertil Soils 38:26–31

Van Spanning RJ, Delgado MJ, Richardson DJ (2005) The nitrogen cycle: de-nitrification and its relationship to N_2 fixation. In: Werner D, Newton WE (eds) Nitrogen fixation in agriculture, forestry, ecology and the environment. Springer, Dordrecht, pp 277–342

Van Spanning RJ, Richardson DJ, Ferguson SJ (2007) Introduction to the biochemistry and molecular biology of denitrification. In: Bothe H, Ferguson SJ, Newton WE (eds) Biology of the nitrogen cycle. Elsevier, Amsterdam, pp 3–20

Vance CP, Graham PH, Allan DL (2000) Biological nitrogen fixation. Phosphorus: a critical future need. In: Pedrosa FO, Hungria M, Yates MG, Newton WE (eds) Nitrogen fixation: from molecules to crop productivity. Kluwer Academic Publishers, Dordrecht, pp 506–514

Vanlauwe B, Giller KE (2006) Popular myths around soil fertility management in sub-Saharan Africa. Agric Ecosyst Environ 116:34–46

Vessey JK, Pawlowski K, Bergman B (2005) Root-based N_2-fixing symbioses: legumes, actinorhizal plants, *Parasponia* sp. and cycads. J Crop Sci Biotechnol 274:51–78

Walley FL, Clayton GW, Miller PR et al (2007) Nitrogen economy of pulse crop production in the Northern Great Plains. J Agron 99:1710–1718

Westhoek A, Field E, Rehling F et al (2017) Policing the legume-Rhizobium symbiosis: a critical test of partner choice. Sci Rep 7:1–10. https://doi.org/10.1038/s41598-017-01634-2

Wheelis M (2008) Principles of modern microbiology. Jones and Bartlett Publishers, Sudbury, MA, pp 187–188

Wolde-meskel E, van Heerwaarden J, Abdulkadir B et al (2018) Additive yield response of chickpea (*Cicer arietinum* L.) to rhizobium inoculation and phosphorus fertilizer across smallholder farms in Ethiopia. Agric Ecosyst Environ 261:144–152

Yadvinder-Singh BS, Ladha JK, Khind CS et al (2004) Long-term effects of organic inputs on yield and soil fertility in the rice-wheat rotation. Soil Sci Soc Am J 68:845–853

Yakubu H, Kwari JD, Ngala AL (2010) N_2 fixation by grain legume varieties as affected by rhizobia inoculation in the sandy loam soil of Sudano-Sahelian zone of northeastern Nigeria. Nigerian J Basic Appl Sci 18(2):229–236

Yan Y, Ping S, Peng J et al (2010) Global transcriptional analysis of nitrogen fixation and ammonium repression in root-associated *Pseudomonas stutzeri*. J Crop Sci Biotechnol 11:11–18

Young JPW (1996) Phylogeny and taxonomy of rhizobia. Plant Soil 186:45–52

Yusuf AA, Iwuafor ENO, Abaidoo RC et al (2009) Grain legume rotation benefits to maize in the northern Guinea savanna of Nigeria: fixed-nitrogen versus other rotation effects. Nutr Cycl Agroecosyst 84:129–139

Zengeni R, Giller KE (2007) Effectiveness of indigenous soybean rhizobial isolates to fix nitrogen under field conditions of Zimbabwe. Symbiosis 43:129–135

Zhang H, Daoust F, Charles TC et al (2002) *Bradyrhizobium japonicum* mutants allowing improved nodulation and nitrogen fixation of field-grown soybean in a short season area. J Agric Sci 138:293–300

Zhang H, Prithiviraj B, Charles TC et al (2003) Low-temperature tolerant *Bradyrhizobium japonicum* strains allowing improved nodulation and nitrogen fixation of soybean in a short season (cool spring) area. Eur J Agron 19:205–213

Chapter 10
Determination of the Ecotoxicity Changes in Biologically Treated *Cyanobacteria Oscillatoria* and *Microcystis* Using Indicator Organisms

L. L. Ndlela, P. J. Oberholster, T. E. Madlala, J. H. Van Wyk, and P. H. Cheng

10.1 Introduction

The prevalence of cyanobacterial blooms and their impact on the environment has been recorded in numerous studies (Ndlela et al. 2016; Oberemm et al. 1997; Oberholster et al. 2009a, b; Paerl et al. 2014; Preece et al. 2017). Control measures of these blooms and toxins have been well researched, and among these is the use of biological control (Ndlela et al. 2018). In the present study, the control agents were microorganisms, particularly bacterial isolates collected from natural bloom waters. Studies on biological control of cyanobacteria have been applied at laboratory scale, with a successful outcome being the lysis or stress response of the targeted cyanobacteria (Nakamura et al. 2003; Yang et al. 2012; Zhang et al. 2016). The lysis or stress of these targeted cyanobacteria may also indicate a subsequent release of intracellular toxins (Ndlela et al. 2018; Paerl et al. 2016), thereby causing the water body in which these cells lyse to have an increased concentration of cyanotoxin (Westrick et al. 2010). Of these biological control studies, a few have assessed the resulting ecotoxicity through the application of indicator organisms.

L. L. Ndlela (✉) · P. J. Oberholster
Council for Scientific and Industrial Research, Stellenbosch, South Africa

Faculty of Science, Department of Botany and Zoology, Stellenbosch University, Matieland, South Africa

T. E. Madlala
Council for Scientific and Industrial Research, Stellenbosch, South Africa

Department of Earth Sciences, University of the Western Cape, Bellville, South Africa

J. H. Van Wyk
Faculty of Science, Department of Botany and Zoology, Stellenbosch University, Matieland, South Africa

P. H. Cheng
Council for Scientific and Industrial Research, Stellenbosch, South Africa

© Springer Nature Switzerland AG 2020   257
A. L. K. Abia, G. R. Lanza (eds.), *Current Microbiological Research in Africa*,
https://doi.org/10.1007/978-3-030-35296-7_10

An example of this is a study by Keijola et al. (1988), which assessed the effectiveness of different drinking water treatment methods using toxic cyanobacteria isolates (*Anabaena*, *Microcystis* and *Oscillatoria*). Toxicity was determined through a mouse bioassay and gas chromatography measurements of cyanotoxins. Another study by Pool et al. (2003) investigated the use of hormone interleukin 6 (IL-6) as an indicator of inflammatory agents in water as an in vitro study. A review of cyanobacteria and cyanotoxin removal in drinking water by Westrick et al. (2010) found that the biological filtration treatment of cyanobacteria resulted in reduced microcystin removal efficiency by 30% in autumn months, compared to summer months. This reduced efficiency was attributed to temperature changes that affected microbial metabolism (Grützmacher et al. 2002). In the same review, a study on the removal of saxitoxins through biological filters indicated a shift from less toxic to more toxic variants increase (Kayal et al. 2008). These fluctuations are due to the response of the live organisms in the biological filters. This is an example of biological control of live cell toxins by other live cells, which is the case in the present study.

Ecotoxicity is the assessment of how an organism reacts to specific chemicals and pollutants. This area of study has been conducted in assessing the pollutants in freshwater, marine and soil environments, using organisms, commonly termed bioindicators. The bioindicator organisms are from different trophic levels and indicate the negative and positive changes in a given environment. The bioindicator categories are plant, animal or microorganisms (Parmar et al. 2016). When these indicator organisms are used to determine the change in response to a specific chemical or environmental parameter, they enable an excellent means to gauge the environmental impact of treatment or change, based on their response. For example, a study by Mohamed and Hussein (2006) assessed the response of Tilapia fish to microcystins, finding that the fish were able to survive and depurate the toxins. A later study by Mohamed et al. (2014) assessed the inhibition and toxin reduction in *Microcystis aeruginosa* by a fungus *Trichoderma citrinoviride*, where toxin concentrations were reduced to undetectable levels within 5 days. A similar approach was applied in the present study by assessing the bioassay responses compared with the cyanotoxin concentration changes. Some of the bacterial isolates employed in the present study have given an indication of microcystin reducing capacity. The changes in cyanotoxin were determined in biologically treated and untreated cyanobacteria. Aquatic and agricultural bioindicator organisms were used to confirm whether the reduction or increase in toxicity brought about by the proposed control agents impacted the bioindicator organisms. These ecotoxicity assays were performed to indicate whether the selected bacterial control agents could be viable at a larger scale or for freshwater body applications. The use of bioindicator response would help to determine whether the use of these biological control agents is environmentally friendly and could lead to improved protection of these organisms.

10.1.1   Cyanobacterial Research Context in South Africa

South Africa has significantly contributed to research in the area of cyanobacteria from the ecology of the blooms to the understanding of the toxins associated with cyanobacteria (Harding 1995; Harding et al. 2009; Van Halderen et al. 1995). Identification, monitoring and some work on biological control have indicated the continued research in this area in South Africa. At present, a greater focus has been on the monitoring of cyanobacterial bloom occurrence through remote sensing methods (Matthews and Bernard 2015), development of management strategies and research beyond the 2000s (van Ginkel 2011). The research in cyanotoxin reduction has thus far been limited to research on biological control studies conducted by Gumbo et al. (Gumbo et al. 2014, 2010; Gumbo and Cloete 2011), as well as the further classification of toxic and non-toxic bloom and species occurrences as well as screening of the response to the toxin (Magonono et al. 2018; Pool et al. 2003; van Ginkel 2012). The latter focus ties in with the assessment of biological control of cyanobacteria and the combination of ecotoxicity screening. The current study merges the applied control methods and how this affects the various indicator organisms in an aquatic ecosystem.

10.2   Materials and Methods

10.2.1   Cyanobacterial Collection Site Descriptions

Two collection sites were selected. The first site, the Brandwacht WWTW, is in the K10D quaternary catchment roughly 500 m upper east of the town of Brandwacht, between the towns of Mossel Bay and Oudtshoorn in the Western Cape province, South Africa. The Brandwacht River, found around 650 m west of the site, is the main surface water feature. Brandwacht is situated at the foot of a small hill (Die Erwe) and has an elevation of 38–54 m above sea level. The second collection site, the Klippoortjie Coal Mine, is situated within the Emalahleni Local Municipality (2678 km^2) and Nkangala District Municipality, approximately 20 km east of the town of Ogies in the Mpumalanga province, South Africa.

10.2.2   Sample Collection and Isolation of Cyanobacteria and Heterotrophic Bacteria

Water containing a bloom of *Microcystis* sp. was collected from Brandwacht wastewater treatment works, in Mossel Bay, Western Cape, South Africa (34° 3′ 3.6″ S, 22° 3′ 28.8″ E). Filamentous cyanobacteria were collected from the Klippoortjie wastewater treatment works (26° 07′ 00″ S; 29° 08′ 00″ E), near the town of Ogies,

Mpumalanga, South Africa. Samples were collected in sterile water bottles and kept on ice during commutation. Cyanobacterial isolates were identified by light microscopy at 400× magnification (Zeiss Axioskop), using the procedures mentioned by Oberholster et al. (2009a, b) and stored as non-axenic cultures at 4 °C for the duration of the study and checked monthly for dominance of the cyanobacteria of interest.

Heterotrophic bacteria showing predatory activity against cyanobacteria were isolated from the water containing cyanobacterial isolates through the use of the plaque assay as described by Gumbo et al. (2010). Three isolates were randomly selected from the plaque assay, namely, isolates 1, 3w and 3y. A culture of *Bacillus* (Isolate B) was generously donated by the Microbiology Department of the Stellenbosch University and used as a reference, based on earlier research done on this genus as an algicidal isolate (Gumbo and Cloete 2011). Bacterial isolates were then grown as pure cultures in nutrient agar and nutrient broth (Merck, Germany) at 25 °C.

10.2.3   Growth Measurements

Cyanobacteria were grown in BG-11 broth medium in 100 mL volumes (Merck, Germany) over 28 days at 25 °C in a 12 h:12 h light-dark cycle, with light illumination of approximately 60 mmol photons (PAR) $m^{-2} s^{-1}$. Chlorophyll *a* and confirmatory wet weight measurements were taken every 3 days. Chlorophyll *a* was extracted with methanol and measured according to the methods of Porra et al. (1989).

Bacterial isolates 1, 3w, 3y and B were grown in 100 mL volumes of nutrient broth (Merck, Germany) and Tween 80 broth and agar medium overnight (5 g peptone, 3 g meat extract, 10 mL Tween 80, 100 mg $CaCl_2.2H_2O$, 15 g agar per litre, pH 7.2). Master cultures were prepared with 80% of culture medium and 20% glycerol (Merck, Germany) and stored at −80 °C.

10.2.4   Exposure Experiments

10.2.4.1   Pre-growth of Cyanobacterial and Bacterial Isolates

Oscillatoria sp. and *Microcystis* sp. were cultured in 1× BG-11 broth (Sigma-Aldrich) at 25 °C in a 12 h:12 h light-dark cycle, with light illumination of approximately 60 mmol photons (PAR) $m^{-2} s^{-1}$, for 3–7 days which is when they reached the exponential growth phase, based on chlorophyll *a* measurements, based on at least three experimental repeats.

Bacterial isolates 1, 3w, 3y and *Bacillus* were grown in Tween 80 broth for 8 h at 25 °C.

Non-axenic cultures of 0.1 g (wet weight) of filamentous cyanobacteria and 1×10^6 cells of *Microcystis* sp. were added into 100 mL of sterile BG-11 medium and grown at 25 °C, respectively. After 2–3 days, 8-hour-old bacterial cells grown in Tween 80 broth were counted with a bacterial counting chamber (Helber, Marienfeld, Germany) at 400× magnification using a light microscope (Zeiss Axioskop). Cells were harvested by centrifugation at 10000 × g for 10 min centrifuged for 10 min at 10000 × g (Thermo Scientific SL 16R) and washed twice with 1× phosphate-buffered saline (PBS) (Lonza).

10.2.4.2   Addition of Bacterial Isolates to Cyanobacterial Cultures

Washed bacterial cells were re-suspended in 1 mL of PBS and added to cyanobacterial cultures. Based on preliminary chlorophyll a measurements, 0.1 g of *Oscillatoria* yielded approximately 10× more chlorophyll a, compared to one million cells of *Microcystis* (wet weight of 0.1 g). Therefore, 10× more cells were added to the filamentous cultures. Cells were added in 1:2 ratios (based on cell counts) of heterotrophic bacteria:cyanobacteria and shaken briefly after addition. The flasks were left at 25 °C for 4 days under static conditions with a 12 h light and 12 h dark cycle. After 4 days, culture samples were vacuum filtered using the 0.22 μm 250 mL Steri-cup Express filters (Merck, Germany) to separate the cells from the culture medium. The residual cells were analysed for phycocyanin; the filtrate water was analysed for total microcystin concentration changes and ecotoxicity assays. Water chemistry, alkaline phosphatase activity and microscopic analyses of the samples were also conducted.

10.2.4.3   Phycocyanin Measurements

After 4 days, culture samples were vacuum filtered using 0.22 μm 250 mL Steri-cup Express filters (Merck, Germany). The residual cells were ground to a fine powder in liquid nitrogen and re-suspended in 1 mL phosphate buffer according to the method of Moraes et al. (2011).

10.2.4.4   Cyanotoxin Detection

Total microcystins were measured in the filtrate water samples after 4 days using the Envirologix microcystin detection kit (Stargate Scientific, South Africa) according to the manufacturer's instructions.

10.2.4.5   Water Chemistry Analyses

Water chemistry of filtrate water samples was conducted using the Hach DR 3900 (Agua Africa) and powder pillows (Agua Africa) to measure the following parameters: Potassium, Nitrates, Nitrites, Zinc, Copper, Iron, Phosphates, Ammonia,

Aluminium and Sulphates. The pH of samples was also measured using the Hanna HI 991300 multi-meter (Hanna, USA). These parameters were selected based on the indication of their impact on cyanobacterial growth.

10.2.5 Ecotoxicity Assays

Cell-free filtered water samples from exposure experiments that had been analysed for changes in cyanotoxins, water chemistry and chlorophyll *a* were used for eco-toxicity assays. The selected indicators for these assays have relevance in agriculture and as ecosystem filter feeders. The water in some dam reservoirs is used for irrigation, and the occurrence of toxic blooms has potential impacts on the produce irrigated as well as the end users of the produce (Dabrowski et al. 2013). As a result, plants and seeds agriculturally produced and sensitive to pollutants were selected in this study. In the case of animal indicators, the use of crustacean filter feeders is a good indication of the ecosystem pollutants, and since they play a critical role in the food web, their sensitivity to pollutants makes them suitable bioindicators (Le et al. 2016; Sánchez et al. 2016).

10.2.5.1 *Lactuca sativa* Bioassay

The lettuce seed bioassay was conducted according to methods previously described by EPA (1996) and Bagur-González et al. (2011) with slight modifications. Briefly, 20 lettuce seeds (Starke Ayres, South Africa) were laid out on No. 1 Whatman filter paper (Sigma Aldrich) in 90 mm Petri dishes (Lasec, South Africa). Then, 3 mL of 0.22 μm filtered water from exposure experiments was added to the seeds on the filter paper and left in the dark for 120 h at ambient temperature. Each water sample was tested in triplicate. The seeds were exposed to tap water (pH 7.2, 5.3 mS/m) as a control and to treated and untreated *Oscillatoria* and *Microcystis* water samples. The number of seeds that hatched was calculated to determine the percentage of seed germination relative to the control.

10.2.5.2 *Allium cepa* Root Tip Assay

Onions weighing between 190 and 230 g were pre-grown in tap water as previously described by Barberio (2013). The ring of the root primordia at the bottom of the onion bulb was scraped with a surgical blade and rinsed with distilled water. The cleaned onions were then immersed into beakers of clean tap water, approximately 2–5 cm deep, allowing for the growth of new roots from the bulbs, over 48 h. After that, onions were exposed to tap water as a control and to bacterially treated and untreated *Oscillatoria* and *Microcystis* water samples for 48 h. At the end of the exposure, the onion roots were cut 3 cm from the tip and placed in Carnoy's fixative (three parts glacial acetic acid to one-part absolute ethanol (Merck, Germany) for 6 h. For calculation of mitotic indices, the roots were stained with 0.5% Hoechst nucleic stain (0.05 mg/mL) (Sigma-Aldrich) diluted in 1× phosphate-buffered saline (Lonza). Roots of 2–3 cm lengths were placed on clean microscope slides, and 100 μL of 0.5% Hoechst stain was added onto the root. The roots were subsequently squashed with the coverslip by applying pressure with the thumb. The onion roots were thereafter imaged after 10 min of staining, on a Carl Zeiss laser scanning confocal microscope 780 (Germany), with the following parameters: magnification-Alpha Plan-Apochromat 100×/1.46 Oil DIC M27 Elyra, laser: 2.0%, master gain: 450, pin-hole: 90 μm, beam splitters: Invis: MBS and a scan speed of 6.30 μs at the Stellenbosch University Central Analytical Facilities (CAF) fluorescence microscopy unit. The mitotic index was calculated from the number of cell nuclei actively dividing after 48 h of exposure.

10.2.5.3 *Daphnia magna* Bioassay

The Daphtox F kit was purchased from Tox Solutions Kits and Services (South Africa) and used according to the manufacturer's instructions. Eppiphia were hatched in a freshwater medium over 72 h at 22 °C. Upon neonate hatching, they were fed ground spirulina 2 h before exposure to the cyanobacteria filtrate water described in Sect. 2.4. Mortality of the neonates was observed over 24 and 48 h at room temperature in the dark. The EC_{50} was calculated based on the mortality at 24- and 48-h observations.

10.2.5.4 *Thamnocephalus platyurus* Bioassay

The Thamnotox kit was purchased from Tox Solutions Kits and Services (South Africa) and used according to the manufacturer's instructions. Cysts were hatched in a freshwater medium over 24 h at 25 °C. Upon neonate hatching, they were exposed to the cyanobacteria filtrate water described in Sect. 2.4. Mortality of the neonates was observed over 24 h. The EC_{50} was calculated based on the mortality at 24-h observations. At the end of the 24-h exposure period, the neonates were harvested and stored at 4 °C.

10.2.6 DNA Fragmentation Assay

After the conclusion of the bioassays, harvested *Thamnocephalus platyurus* crustacean samples were ground to a fine powder in liquid nitrogen with a cooled mortar and pestle. The ground samples were used to test for DNA apoptosis.

The ApoTarget™ Quick Apoptotic DNA Ladder Detection Kit (Invitrogen) was used according to the manufacturer's instruction. The samples were imaged on a 1.2% gel, stained with Gel Red nucleic acid stain (Thermo-Fischer Scientific) after electrophoresis for 2 h at 5 v/cm.

10.2.7 Statistical Analyses

Water chemistry correlation and principal component analyses of the bioassay findings (Spearman's correlation for non-parametric data) were conducted using Microsoft XLSTAT™ (2010). Visualization of the relationship between the biological indicators was conducted through a bubble plot using JMP™ (version 14) software. One-way ANOVA was conducted on data from *Lactuca sativa* to determine any statistically significant differences between the treatment groups.

10.3   Results

10.3.1   Phycocyanin Estimation

Exposures of *Oscillatoria* and *Microcystis* sp. cultures were conducted over 4 days. The measurement of phycocyanin showed that treatments with different isolates either enhanced or reduced the pigment formation relative to the untreated controls (Fig. 10.1). It is also interesting to note that at the same wet weight, phycocyanin concentrations in *Microcystis* were almost a 100-fold lower than in the case of *Oscillatoria*. This provides an interesting comparison to the chlorophyll *a* measurement, with a similar trend noted as per the chlorophyll reductions.

While isolate 1 had an inhibitory effect on *Oscillatoria* compared to the control, it appeared to have a beneficial effect on *Microcystis* based on the pigment concentrations relative to the control samples. This was observed with isolate B as well. Isolate 3w and 3y had similar effects on the phycocyanin production of both cyanobacteria.

10.3.2   Cyanotoxin Detection

After the addition of bacterial isolates to *Microcystis* cultures over 4 days, changes in cyanotoxin reduction were observed. There was a reduction in cyanotoxin concentrations in treated cyanobacteria, relative to the control sample (Fig. 10.2). Isolates 1 and B exhibited a more significant reduction (27 and 30%, respectively) in the water samples of *Microcystis* compared to isolate 3w and 3y (16 and 4%, respectively). The monitoring of cyanotoxins in *Oscillatoria* water samples showed a minimal decrease in toxins from treatment with isolate 3y (2.4%), while the other isolates had slight increases in toxicity, up to 15% by isolate 3y. These results

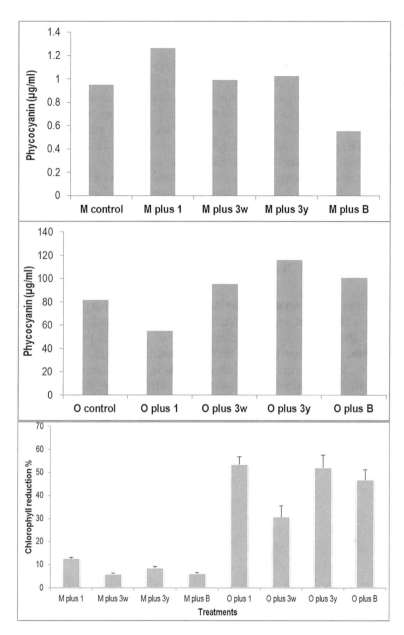

Fig. 10.1 Phycocyanin group comparisons in *Microcystis* (M) and *Oscillatoria* (O) treated with isolates 1, 3w, 3y and B

indicated a variation in bacterial performance. The data captured in Fig. 10.2 are comparable to an average of toxin reduction that indicated an overall cyanotoxin average reduction of up to 16% in *Oscillatoria* sp. water and up to 63% reduction in *Microcystis* sp. water.

10.3.3   Water Chemistry Changes

Of the parameters measured in the water chemistry of exposed and unexposed cyanobacteria (Table 10.1), phosphates had the greatest variations, while the other
parameters were similar, with less than 5% fluctuations. The changes in phosphorous concentrations were linked to the change in TN:TP ratios, which, when fluctu-
ated, created more favourable conditions for cyanobacterial growth. This is mainly
at low TN:TP ratios, around 10 or less. Due to the parameters measured, there could
be no accurate estimation of the TN:TP ratios, except to note that the ammonia
concentrations were lowest in the untreated control samples. A correlation analysis
(Table 10.2) of the data indicated that measured parameters, apart from copper and
orthophosphates, had a strong correlation value (>0.750) to each other.

10.3.4   Lactuca sativa *Bioassay Findings*

Findings from the germination of the lettuce seeds indicate that there were minimal
variations in seed germination percentages between the treated and untreated samples after exposure to water containing *Microcystis*, with less than 10% difference
in germination percentage between the treated and untreated samples. Similarly, for
Oscillatoria, the differences were less than 20% in the treated and untreated samples (Fig. 10.3). One-way ANOVA showed no statistically significant difference in
Microcystis treatment groups ($p \leq 0.81$). However, in the *Oscillatoria* treatment
group, there was a statistically significant difference ($p \leq 0.03$) between the groups.
Microcystis sp. treatment with isolate 3w resulted in a toxin reduction of 16%; however, the seed germination (45%) was 8% lower than the untreated control sample

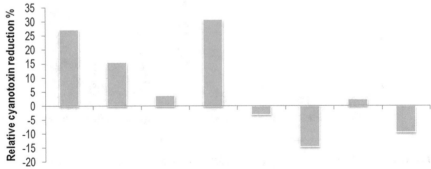

Fig. 10.2 Cyanotoxin reductions in filtrate water samples of *Microcystis* (M) and *Oscillatoria* (O)
treated with bacterial isolates 1, 3w, 3y and B, relative to the control untreated samples

Table 10.1 Water chemistry parameter measurements of treated and untreated *Microcystis* and *Oscillatoria* sample filtrates after 4 days of exposure

Treatment	Ammonia (mg/L)	Copper (mg/L)	Nitrates (mg/L)	Orthophosphate (mg/L)	Potassium (mg/L)	Sulphates (mg/L)
M control	0.77	0.04	1.08	8.7	122	18
M plus1	0.84	0.1	1.08	10.3	123	19
M plus 3w	1.07	0.03	1.08	10.2	124	19
M plus 3y	1.43	0.02	1.08	5.3	126	18
M plus B	0.94	0.02	1.08	6.22	126	18
O control	0.01	0.02	1.02	6.01	114	14
O plus 1	0.05	0.03	1.02	8.1	113	15
O plus 3w	0.24	0.02	1.02	8.7	115	14
O plus 3y	0.71	0	1.02	9.1	117	14
O plus B	0.18	0.03	1.02	9.9	117	15

Table 10.2 Spearman's correlation values of the variables measured in treated and untreated cyanobacteria *Microcystis* and *Oscillatoria*, after 4 days of exposure to heterotrophic bacteria

Variables	Ammonia	Copper	Nitrates	Orthophosphate	Potassium	Sulphates
Ammonia	1	0.063	**0.870**	0.079	**0.945**	0.750
Copper	0.063	1	0.400	0.524	0.057	0.654
Nitrates	**0.870**	0.400	1	0.070	**0.876**	**0.898**
Orthophosphate	0.079	0.524	0.070	1	0.031	0.345
Potassium	**0.945**	0.057	**0.876**	0.031	1	0.739
Sulphates	0.750	0.654	**0.898**	0.345	0.739	1

Values in bold show a strong correlation

(53%). The control sample of tap water had 70% seed germination, while tap water with 1% compost added to it had 100% seed germination. With *Oscillatoria*, the changes in toxicity of only isolate B treated samples were not reflected by the changes in seedling germination. These observed inconsistencies may be due to the slight changes in toxicity not being well indicated by the lettuce seeds, thereby indicating that their sensitivity as bioindicators is not well suited to the slight changes observed in *Oscillatoria* toxicity.

10.3.5 Allium cepa *Root Assay*

The onion root assay was done through the observation of mitotic indices in onions exposed to treated and untreated samples of cyanobacteria filtrate water. Tap water was used as a control. Nuclei in state of cell division were counted out of a total number of 100 cells to generate the mitotic index (Table 10.3). Only *Oscillatoria* treated with isolate 3y had a low mitotic index (41% vs. 85%).

The images of the onion root cells are shown in Fig. 10.4. These were taken as representative samples to indicate the differences in the mitotic stages of the onion roots exposed to samples with greater toxicity. M control cells (a) showed chromosomal bridges in anaphase and some nucleic disintegration, indicated by the arrow in Fig. 10.4a. Cells treated with isolate 1 (Fig. 10.4b) showed an intact mitotic nucleus with normal anaphase occurring at the bottom right corner of the image, indicated by the arrow.

10.3.6 Daphnia magna *Bioassay*

Testing of the response of daphnids as freshwater crustaceans to the treated and untreated water samples (Table 10.4) indicated that the *Microcystis* samples with reduced toxicity had a higher survival of neonates, compared to the control

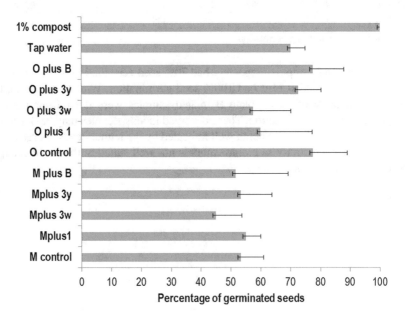

Fig. 10.3 Mean percentage germination of lettuce seeds after 120 h of incubation in different water. The bars indicate standard deviation. The 1% compost sample had no standard deviations

(untreated) sample, which had higher toxicity. There was statistical significance in the observations between the treated and untreated samples ($p \leq 0.048$) of *Microcystis*. For *Oscillatoria*, only the control and 3w sample had a lower neonate survival, with only 79% survival ($p \leq 0.13$). The freshwater control, which is the recommended control by the manufacturer, had no adverse effects on neonate survival, with a 100% survival. The inconsistencies in *Oscillatoria* samples could not be accounted for; however, the trend indicates that samples with toxicity changes greater than 15% were well detected by the bioindicators. The EC_{50} calculations of water samples indicated that toxicity is reduced in treated samples, with *Oscillatoria* filtrate water samples having an impact after 48 h compared to the 24-h effective concentrations, which remained the same for *Microcystis* samples.

10.3.7   Thamnocephalus platyurus *Bioassay*

One of the recommended bioassays for microcystin detection is the use of the crustacean *Thamnocephalus platyurus*, which is also a crucial and sensitive filter feeder in the food web. This crustacean response in 24 h gives an indication of the water quality and particularly the microcystin concentrations. The trend in *Thamnocephalus* was like that observed in *Daphnia*, with an overall 6% higher survival, however, which may indicate the slight difference in sensitivity of the isolates (Table 10.5). The difference in the EC_{50} indicates that the *Daphnia* were more sensitive to the changes in cyanotoxin as opposed to *Thamnocephalus*, with much lower EC_{50} concentrations required in *Daphnia* neonates. The longer exposure time of *Daphnia* yielded more information as opposed to the shorter exposure of *Thamnocephalus*, which may be more meaningful under a chronic exposure assay. Statistical

Table 10.3 Mean mitotic index (% occurrence in 100 cells) of treated and untreated *Microcystis* and *Oscillatoria* onion root cells after 48-h exposures

Treatment	Mitotic index
Tap water	93.33 ± 4.08
M control	85.08 ± 20.24
M plus 1	89.43 ± 15.00
M plus 3w	**35.84 ± 10.66**
M plus 3y	**67.07 ± 9.82**
M plus B	85.56 ± 12.68
O control	99.68 ± 19.55
O plus 1	90.79 ± 24.60
O plus 3w	57.83 ± 16.96
O plus 3y	**40.97 ± 7.66**
O plus B	87.55 ± 15.98

Bold figures indicate significantly lower mitotic indexes in comparison to the control samples

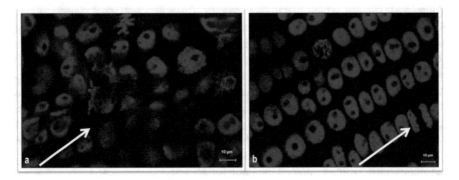

Fig. 10.4 Nucleic acid stain of *Allium cepa* roots exposed to water from untreated *Microcystis* (**a**) and *Microcystis* treated with isolate 1 (**b**). Nuclei appear intact and normal mitosis occurs in (**b**), while there is nuclear disintegration in (**a**)

Table 10.4 Neonate survival of *Daphnia magna* after 24- and 48-h exposure to biologically treated and untreated *Microcystis* and *Oscillatoria* water samples

Treatment	Neonate survival (%)	24 h EC$_{50}$	48 h EC$_{50}$
M control	50 ± 8.66	1.45	1.45
M plus1	80 ± 8.66	2.64	2.64
M plus 3w	75 ± 4.79	2.45	2.45
M plus 3y	100 ± 0		
M plus B	90 ± 2.89	5.02	5.02
O control	79 ± 5.00	6.85	1.39
O plus 1	100 ± 5.77		2.35
O plus 3w	79 ± 4.79	7.84	1.57
O plus 3y	95 ± 2.5		3.34
O plus B	95 ± 2.5		3.75

Blank blocks indicate treatments where the EC$_{50}$ was not quantified due to the low toxicity of the water sample. EC$_{50}$ values represent µg/mL of cyanotoxin

significant differences were observed in the *Microcystis* treatment group ($p \leq 0.038$), while no statistically significant difference was observed in the *Oscillatoria* treatment group ($p \leq 0.72$).

10.3.8   Overall Variation and Response Patterns

A representation of the *Allium cepa*, *Daphnia* and *Thamnocephalus* assays (Fig. 10.5) indicated similar response trends of the *Microcystis* and *Oscillatoria* filtrate water. The assays indicate that the changes in toxicity were observed in the bioindicator responses, meaning that more toxic samples had lower survival.

Table 10.5 Neonate survival of *Thamnocephalus platyurus* after 24 h of exposure to biologically treated and untreated *Microcystis* and *Oscillatoria* water samples

Treatment	Neonate survival (%)	24 h EC_{50}
M control	62 ± 30.82	2.42
Mplus1	100 ± 48.39	17.57
Mplus 3w	90 ± 44	6.12
Mplus 3y	83 ± 40.20	3.49
M plus B	93 ± 45.18	5.02
O control	100 ± 48.39	11.41
O plus 1	97 ± 46.90	5.04
O plus 3w	93 ± 45.18	3.92
O plus 3y	100 ± 48.39	11.13
O plus B	90 ± 43.78	2.88
Freshwater control	100 ± 48.39	

Blank areas indicate treatments where the EC_{50} was not quantified due to the low toxicity of the water sample. EC_{50} values represent μg/mL of cyanotoxin

Figure 10.5 is reflective of the toxicity changes in the water samples. Treatment numbers 1–5 on the y-axis indicate the control untreated cyanobacteria (1), cyanobacteria treated with isolate 1 (2), treatment with isolate 3w (3), treatment with isolate 3y (4) and treatment with isolate B (5). The x-axis indicates the percentages of survival for *Daphnia* and *Thamnocephalus* as well as the mitotic index percentages for *Allium cepa*. To obtain a more precise assessment of the findings, a principal component analysis (PCA) was conducted on the water chemistry variables that did not show any correlation to the other variables in the correlation test performed. This was done to observe whether it had any relation to the ecotoxicity assay findings (Fig. 10.6). The factor loading data from the analysis (Table 10.6) indicate that factor 1 was the in vivo bioindicator responses, while factor 2 was the agricultural bioindicator response to copper. Factor 3 indicates a relationship between the crustaceans and plants to copper and ammonia. Therefore, these were the main components influencing the data in the observations.

The bioindicator crustaceans were closely related to each other (Fig. 10.6), while the plant indicator organisms were also in the same quadrant. This indicated similarities in plant (lettuce and onion) responses and similarities in the crustacean (*Daphnia* and *Thamnocephalus*) responses, which can be expected as they are more closely related organisms. Copper concentrations, however, showed a close relation to *Microcystis* treated with isolate 1. *Oscillatoria* samples showed a close correlation to the survival of the crustaceans and the mitotic index of the onions. The *Microcystis* water samples were closely grouped to the changes in ammonia, with similarities in the 3w and 3y sample, while the M plus B and M control samples had less relation to any of the parameters.

A summary of all the findings is represented in Table 10.7. The "+" sign represents the sensitivity of the bioindicator to the changes in the different water samples. The "−" sign indicates no sensitivity. All these findings, except apoptosis, were indicated relative to the toxicity in the control sample. Therefore a "+" would mean

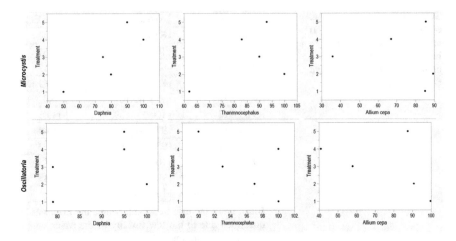

Fig. 10.5 A bubble plot representation of the bioassays which showed sensitivity to the changes in toxicity with treatments (1–5) against the percentage survival/mitotic index. Similar trends are observed for *Microcystis* and *Oscillatoria*, indicating a reduced survival with treatment from *Oscillatoria*, while an increased survival is seen for *Microcystis*

the bioindicator reflected the increase or decrease in toxicity relative to the untreated control sample.

10.3.9   *DNA Fragmentation of* Thamnocephalus platyurus *Exposed to Treated and Untreated Cyanobacteria*

Further analysis of apoptosis in *Thamnocephalus platyurus* indicated that isolates with higher survival or higher EC_{50} still showed apoptotic damage of DNA (Fig. 10.7). There was no laddering of DNA observed.

10.4   Discussion

This study assessed the use of potential algicidal bacteria in treating the extracellular toxins in mixed cultures dominated by cyanobacteria, *Oscillatoria* and *Microcystis*. Moreover, the changes in toxicity linked to variation in microcystin concentration (determined by ELISA procedure) were verified through the response of various bioindicators (in vivo exposures). A battery of bioindicator species was used, including agricultural produce (*Lactuca sativa* and *Allium cepa*) and freshwater crustaceans (*Daphnia magna* and *Thamnocephalus platyurus*).

An assessment of the viability of a biological control is a crucial step in determining whether further research is to be conducted in up-scaling or optimization.

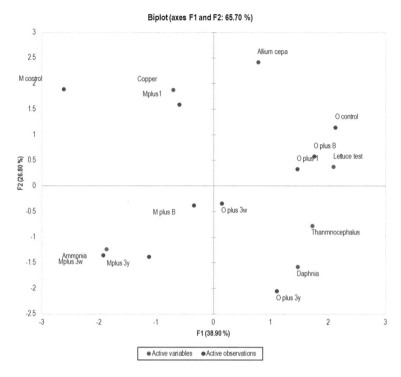

Fig. 10.6 Principal component analysis of the observations in this study and their relation to each other. Total variation accounted for by factor 1 and 2 is 66%

Table 10.6 Factor loadings from principal component analysis

	F1	F2	F3	F4	F5	F6
Daphnia	0.599	−0.533	0.400	0.406	−0.043	−0.174
Thamnocephalus	0.702	−0.262	0.542	−0.321	−0.061	0.195
Allium cepa	0.319	0.814	0.112	0.436	−0.106	0.146
Lactuca sativa	0.851	0.127	−0.299	0.001	0.414	0.003
Ammonia	−0.758	−0.418	0.281	0.273	0.266	0.163
Copper	−0.286	0.634	0.662	−0.172	0.168	−0.147

This is critical in ensuring there are no adverse environmental impacts that arise from a biological control intervention. On a higher order organism scale, the historical failures of some classical biological control interventions indicate the need for environmental response testing to gauge the viability of a control agent (Stiling 1993). The advantage in lower order organisms is host specificity. Most of the research reported on biological control of cyanobacteria has not been well conducted in terms of full-proofing (ensuring no secondary or indirect adverse impacts) biological treatments of cyanotoxins, with most studies focusing on the lytic or toxin reducing effects of bacterial isolates (Kim et al. 2008; Nakamura et al. 2003;

Table 10.7 A summary of all the findings from the toxicity and biotoxicity assays from treated and untreated *Microcystis* and *Oscillatoria* filtrate water samples

Treatment	Microcystins (µg/mL)	Sensitivity to toxin changes				Apoptosis
		L. sativa	*A. cepa*	*D. magna*	*T. platyurus*	
M. untreated	1.45					−
M plus 1	1.05[a]	+	+	+	+	−
M plus 3w	1.22[a]	−	−	+	+	−
M plus 3y	1.40[a]	+	−	+	+	+
M plus B	1.00[a]	−	+	+	+	+
O untreated	0.68					+
O plus 1	0.71	+	+	−	+	+
O plus 3w	0.78	+	+	+	+	+
O plus 3y	0.67[a]	+	−	+	+	+
O plus B	0.75	−	+	−	+	−

The positive and negative sign is based on the sensitivity to cyanotoxin changes
[a]Toxin reduction relative to the control untreated water sample

Ren et al. 2010; Su et al. 2016a, b). In the present study, the changes in the toxicity of treatment were first assessed from the cyanobacterial response but also the subsequent response of organisms exposed to the treated water.

Microcystin toxin reduction of over 80% has been reported from the use of bacteria (Su et al. 2016b) and treatment technologies such as ozonation (Liu et al. 2010). This study indicates reductions of up to 30%. In a system where the toxicity is reduced, the use of an environmental indicator can indicate the improvement of water quality (reduced toxicity) after the treatment. This is particularly relevant in cyanobacteria, where more than one toxin variant is present at a given time. Therefore the reduction of one toxin does not necessarily deem the water in question safe. Furthermore, it indicates whether there could be any secondary adverse effects from the treatment and whether a mixed response could be expected when multiple co-dominant cyanobacterial species exposures occur.

10.4.1 Cyanobacterial Response

The reduction in cyanobacterial photosynthetic pigments has been linked to cell stress or a reduction in cell abundance in water (Kasinak et al. 2015). The changes in phycocyanin, which is an accessory photosynthetic pigment, indicated that the cells exposed to isolates 3w and B were more stressed in the case of *Microcystis*, while exposure to isolate 1 and B resulted in greater stress for *Oscillatoria*. These findings might show that different isolates have different relationships with the two types of cyanobacteria. This might mean that while one isolate may be predatory to *Microcystis*, it may be beneficial to *Oscillatoria*. Phycocyanin, in this case, *c*-phycocyanin, is an accessory photosynthetic pigment with antioxidant properties,

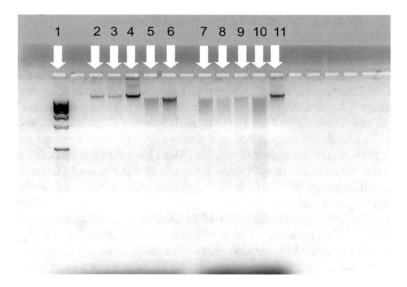

Fig. 10.7 DNA apoptosis gel of *Thamnocephalus platyurus* exposed to treated and untreated *Microcystis* and *Oscillatoria* water samples. Lanes 2–4 (M control, M plus 1 and M plus 3w) show minimal smearing. Lanes 5 and 6 show smears of DNA indicative of apoptosis, which are treatments of *Microcystis* with isolate 3y (lane 5) and B (lane 6). In the case of *Oscillatoria*, the control sample (Lane 7) and water samples treated with isolate 1 (lane 8), 3w (lane 9) and 3y (lane 10) all had apoptosis. Only the *Oscillatoria* sample treated with isolate B (lane 11) showed minimal smearing of the DNA after a 24-h exposure

and its reduction is usually indicative of cell damage or lysis (Zhang et al. 2011). However, the extraction of this pigment has been found unreliable in small samples (Horváth et al. 2013), which may have been the case in this study, where 0.1 g of cyanobacterial cells were used. Chlorophyll *a* concentrations also showed a similar trend to the phycocyanin measurements in the present study. When assessing the changes in toxicity, the most significant reduction was in water treated with isolate 1 and B for *Microcystis*, and isolate 3y in the case of *Oscillatoria*. This indicated that the changes in phycocyanin were not directly related to the changes in toxicity observed or any cell stress (Fig. 6.1). This would indicate that perhaps no cell stress can be linked to phycocyanin, unlike chlorophyll, which is another more widely used indicator of cell growth.

In the present study, the changes in the microcystin toxicity were determined through the ELISA antibody assay; this has a relative wide cross-reactive range and has also been known to detect even cleaved ADDA portions of microcystins (Samdal et al. 2014). This method, therefore, requires confirmatory testing through high-performance liquid chromatography (HPLC) to determine which toxin variants are present and also whether they are reduced or not. However, as it is a commonly applied and robust method, the study aims were to determine whether the changes in microcystin toxicity as indicated by ELISA were reflected in the biological indicator response.

10.4.2   Ecotoxicity Assays: Crustacean Response

Historically, most of the biological indicator tests have been applied mainly to test the effects of chemicals, with most of the works using plants as bioindicators. Only 2% of the literature published at the time (2006) were attributed to ecosystems and amphibians (Burger 2006). However, research linked to *Daphnia* response to the changes in the environment has been reported (Neves et al. 2015), showing it can reflect the prevalence of cyanobacterial toxins and other environmental stressors (Lürling 2003). Another study has, however, found low concentrations of microcystins to not affect the mortality or stress indicators of this crustacean under chronic exposure (Chen et al. 2005). The findings from our research indicated the ability of *Daphnia magna* to respond to toxicity changes. However, the *Daphnia* mortality did not show correspondence to slight (10–15%) fluctuations in microcystin concentration. This was seen in different isolate treatments such as isolates 3w and 3y in the case of *Microcystis*. The same observation was made in the *Oscillatoria* sample treatments, where the 3% fluctuation in toxicity between water treated with isolate 1 and the control sample was not well indicated. The 7% difference in toxicity between isolate 3y and B treatments was also not well indicated by the mortality of *Daphnia*. Only isolate 3w, which resulted in 15% more microcystins, corresponded to a lower survival rate.

When assessing the response of another freshwater crustacean, *Thamnocephalus platyurus*, the findings indicated a greater sensitivity to toxins, confirming the findings of earlier studies (Kim et al. 2009). Variation in microcystin concentration in the different isolate exposure groups showed slight mortality changes following the slight changes in toxicity. This indicated the ability of *Thamnocephalus* to detect these changes in treatments that occurred in mixed population cultures. Similar findings were reported by Bober and Bialczyk (2017) and Maršálek and Bláha, (2000), who found *Thamnocephalus* to be more sensitive than *Daphnia* to cyanobacteria. The longer-term exposure of *Daphnia* indicates a higher toxicity response over time, as opposed to the acute response of *Thamnocephalus*.

10.4.3   Seed and Plant Bioindicator Response

The impacts of microcystins in water on plants may also have significant implications for agriculture and human consumption, with a requirement for assessment of bioaccumulation of these toxins in fresh produce irrigated with toxin-contaminated water (Gutiérrez-Praena et al. 2014). Gutiérrez-Praena et al. (2014) further reported microcystins in fruit and vegetables irrigated with contaminated groundwater for irrigation and the presence of toxins in water indicated for agricultural purposes. This raises the relevance of the study of agricultural produce response to these toxins.

Other studies related to lettuce and other crops in cyanobacteria bloom prone areas indicated the presence of trace amounts of microcystin R-R in most vegetables tested (Li et al. 2014). Maisanaba et al. (2018) reported that there was increased consumption of certain raw leafy vegetables despite a higher bioavailability of the cyanobacterial toxin cylindrospermopsin in uncooked spinach compared to spinach boiled for 2 min (Maisanaba et al. 2018). Being a vegetable that is consumed primarily raw, the lettuce plant is also likely to be ingested with trace amounts of microcystins. Testing for the sensitivity of *Lactuca sativa* to wastewater polluted with heavy metals indicated its suitability as a bioindicator, with reduced germination reflecting the toxicity of the water (Charles et al. 2011). Lettuce seedlings exposed to microcystin L-R showed that germination did not differ significantly to the control in terms of fresh weight and elongation in concentrations lower than 6 µg/mL (Wang et al. 2011). A similar trend was observed in the present study. This means that for lettuce, the germination and other physical traits were not sensitive enough endpoints to indicate microcystin effects at lower concentrations (Fig. 10.3). More biochemical or molecular analyses may be useful in the case of *Lactuca sativa* assays where toxin concentrations fluctuate slightly.

The *Allium cepa* assay has been described as an efficient and reliable method for testing environmental pollutants among other materials, having a response similar to other higher order organisms such as rodents. The mitotic index has also been applied as a measure of the organism response, with increases or reductions in the mitotic index relative to the control sample being an indication of adverse impacts on the cell root (Leme and Marin-Morales 2009). In the present study, variation in the mitotic index as a biomarker did not show complete agreement with the corresponding changes in microcystin toxicity. Also, the root lengths were similar, with no morphological differences. A study of differences in tenfold concentrations of microcystin and aeruginosin indicated a good correlation between toxin concentration and the mitotic index in onion roots cells (Laughinghouse et al. 2012). Similar to the lettuce and *Daphnia* responses, minor variation in toxicity was not reflected in the mitotic index response. The most sensitive indicator to toxicity changes according to the findings of the present study was shrimp, *Thamnocephalus platyurus*.

Nonetheless, the changes in water chemistry also need to be considered as contributing factors and therefore controlled for. The changes in ammonia and orthophosphate in the treated water samples may have influenced the respective TN:TP (total nitrogen: total phosphorous) ratios, which are essential factors in the growth of cyanobacteria (Ndlela et al. 2016). Also, the potential of the cyanobacteria to produce aeruginosin (*Microcystis*) and nodularin (*Oscillatoria*) suggests that other additional toxins that may have had added impacts on the bioindicators, not accounted for in the scope of this study, may have played a significant role. Moreover, the cross-reactivity of the ELISA does not informatively indicate which toxin analogue is increasing or decreasing in the overall toxicity findings, although the extensive cross-reactivity may include a variety of toxins. Regarding isolate 3w, the lowest survival, germination and mitotic index were observed in *Microcystis* treated with this isolate, characterized by a simultaneous 15% reduction in microcystin toxicity according to the ELISA approach. This may suggest that other factors, apart from the cyanotoxins, may have caused the adverse effects of this treatment. This cannot be concluded from the present study but may be a point for consideration in future works. In terms of cyanotoxin reduction, the bacterial isolates were more effective against extracellular cyanotoxins from *Microcystis* compared to *Oscillatoria*.

The principal component analysis (PCA) indicates that the plants and crustaceans had similar response patterns. From these findings, isolate 1 and B appeared to have a positive relation to bioindicator survival across the assays conducted. When assessing the genetic apoptosis in *Thamnocephalus platyurus* (the most sensitive), all the treatments resulted in DNA apoptosis except for isolate B, which showed more intact DNA. This suggests that the reduction in cyanotoxins needs to be fully validated through biological indicators and more importantly, that specific biological control isolates may have more favourable environmental impacts, despite similar indications of microcystin reduction.

Another point to consider in this type of research is the selection of economically feasible bioindicators. *Daphnia magna* is culturable in the lab and not expensive to grow; the *Lactuca sativa* bioassay is the most affordable of the bioassays conducted in this research, although not as sensitive. *Allium cepa* experimental setup is also a reasonably feasible experiment, although the imaging requires good microscopy. *Thamnocephalus* acute exposure is also a feasible assay to conduct. Analysis beyond the fundamental mortality and visual screening is, however, required for more informative data.

10.5 Conclusion

This present study confirmed that toxicity changes in biological control systems could be reflected through bioindicators, although the sensitivity of the indicators towards microcystins may vary among organisms. The organisms used were able to indicate that toxicity was not well reduced in *Oscillatoria* treatments and toxicity changes greater than 15% could be well reflected in all the treatments. More importantly, if the biological control agent performance can be optimized (greater toxin reduction), it will be easier to predict the potential environmental impacts.

References

Bagur-González MG, Estepa-Molina C, Martín-Peinado F, Morales-Ruano S (2011) Toxicity assessment using *Lactuca sativa* L. bioassay of the metal(loid)s As, Cu, Mn, Pb and Zn in soluble-in-water saturated soil extracts from an abandoned mining site. J Soils Sediments 11:281–289. https://doi.org/10.1007/s11368-010-0285-4

Barberio A (2013) Bioassays with plants in the monitoring of water quality. https://doi.org/10.5772/50546

Bober B, Bialczyk J (2017) Determination of the toxicity of the freshwater cyanobacterium Woronichinia naegeliana (Unger) Elenkin. J Appl Phycol 29:1355–1362. https://doi.org/10.1007/s10811-017-1062-1

Burger J (2006) Bioindicators: a review of their use in the environmental literature 1970–2005. Environ Bioindic 1:136–144. https://doi.org/10.1080/15555270600701540

Charles J, Sancey B, Morin-Crini N et al (2011) Evaluation of the phytotoxicity of polycontaminated industrial effluents using the lettuce plant (Lactuca sativa) as a bioindicator. Ecotoxicol Environ Saf 74:2057–2064. https://doi.org/10.1016/j.ecoenv.2011.07.025

Chen W, Song L, Ou D, Gan N (2005) Chronic toxicity and responses of several important enzymes in Daphnia magna on exposure to sublethal microcystin-LR. Environ Toxicol 20:323–330. https://doi.org/10.1002/tox.20108

Dabrowski J, Oberholster PJ, Dabrowski JM et al (2013) Chemical characteristics and limnology of Loskop Dam on the Olifants River (South Africa), in light of recent fish and crocodile mortalities. Water SA

EPA US (1996) Ecological effects test guidelines. Gammarid acute Toxic. test OPPTS 850

Grützmacher G, Böttcher G, Chorus I, Bartel H (2002) Removal of microcystins by slow sand filtration. Environ Toxicol An Int J 17:386–394

Gumbo JR, Cloete TE (2011) The mechanism of Microcystis aeruginosa death upon exposure to Bacillus mycoides. Phys Chem Earth 36:881–886. https://doi.org/10.1016/j.pce.2011.07.050

Gumbo JR, Cloete TE, van Zyl GJ, Sommerville JEM (2014) The viability assessment of Microcystis aeruginosa cells after co-culturing with Bacillus mycoides B16 using flow cytometry. Phys Chem Earth 72:24–33. https://doi.org/10.1016/j.pce.2014.09.004

Gumbo JR, Ross G, Cloete TE (2010) The isolation and identification of predatory bacteria from a Microcystis algal bloom. Afr J Biotechnol 9:663–671. https://doi.org/10.4314/ajb.v9i5

Gutiérrez-Praena D, Campos A, Azevedo J et al (2014) Exposure of Lycopersicon esculentum to microcystin-LR: effects in the leaf proteome and toxin translocation from water to leaves and fruits. Toxins 6:1837–1854. https://doi.org/10.3390/toxins6061837

Harding GA (1995) Death of a dog attributed to the cyanobacterial (blue-green algal) hepatotoxin nodularin in South Africa. J South Afr Vet Assoc 66:256–259

Harding WR, Downing TG, van Ginkel CE, Moolman APM (2009) An overview of cyanobacterial research and management in South Africa post-2000. Water SA

Horváth H, Kovács AW, Riddick C, Présing M (2013) Extraction methods for phycocyanin determination in freshwater filamentous cyanobacteria and their application in a shallow lake. Eur J Phycol 48:278–286. https://doi.org/10.1080/09670262.2013.821525

Kasinak J-ME, Holt BM, Chislock MF, Wilson AE (2015) Benchtop fluorometry of phycocyanin as a rapid approach for estimating cyanobacterial biovolume. J Plankton Res 37:248–257

Kayal N, Newcombe G, Ho L (2008) Investigating the fate of saxitoxins in biologically active water treatment plant filters. Environ Toxicol An Int J 23:751–755

Keijola AM, Himberg K, Esala AL et al (1988) Removal of cyanobacterial toxins in water treatment processes: laboratory and pilot-scale experiments. Toxic Assess 3:643–656. https://doi.org/10.1002/tox.2540030516

Kim B-H, Sang M, Hwang S-J, Han M-S (2008) In situ bacterial mitigation of the toxic cyanobacterium Microcystis aeruginosa: implications for biological bloom control. Limnol Oceanogr Method 6:513–522. https://doi.org/10.4319/lom.2008.6.513

Kim J-W, Ishibashi H, Yamauchi R et al (2009) Acute toxicity of pharmaceutical and personal care products on freshwater crustacean (Thamnocephalus platyurus) and fish (Oryzias latipes). J Toxicol Sci 34(2):227–232

Laughinghouse HD, Prá D, Silva-Stenico ME et al (2012) Biomonitoring genotoxicity and cytotoxicity of Microcystis aeruginosa (Chroococcales, Cyanobacteria) using the Allium cepa test. Sci Tot Environ 432:180–188. https://doi.org/10.1016/j.scitotenv.2012.05.093

Le Q-AV, Sekhon SS, Lee L et al (2016) Daphnia in water quality biomonitoring - "omic" approaches. Toxicol Environ Health Sci 8:1–6. https://doi.org/10.1007/s13530-016-0255-3

Leme DM, Marin-Morales MA (2009) Allium cepa test in environmental monitoring: a review on its application. Mutat Res 682:71–81. https://doi.org/10.1016/j.mrrev.2009.06.002

Li Y-W, Zhan X-J, Xiang L et al (2014) Analysis of trace microcystins in vegetables using solid-phase extraction followed by high-performance liquid chromatography triple-quadrupole mass spectrometry. J Agric Food Chem 62:11831–11839. https://doi.org/10.1021/jf5033075

Liu X, Chen Z, Zhou N et al (2010) Degradation and detoxification of microcystin-LR in drinking water by sequential use of UV and ozone. J Environ Sci (China) 22:1897–1902

Lürling M (2003) Effects of microcystin-free and microcystin-containing strains of the cyanobacterium Microcystis aeruginosa on growth of the grazer Daphnia magna. Environ Toxicol An Int J 18:202–210

Magonono M, Oberholster JP, Shonhai A et al (2018) The presence of toxic and non-toxic cyanobacteria in the sediments of the Limpopo river basin: implications for human health. Toxins. https://doi.org/10.3390/toxins10070269

Maisanaba S, Guzmán-Guillén R, Valderrama R et al (2018) Bioaccessibility and decomposition of cylindrospermopsin in vegetable matrices after the application of an in vitro digestion model. Food Chem Toxicol 120:164–171. https://doi.org/10.1016/j.fct.2018.07.013

Matthews MW, Bernard S (2015) Eutrophication and cyanobacteria in South Africa's standing water bodies: a view from space. South Afr J Sci 111:77–85. https://doi.org/10.17159/sajs.2015/20140193

Mohamed ZA, Hashem M, Alamri SA (2014) Growth inhibition of the cyanobacterium Microcystis aeruginosa and degradation of its microcystin toxins by the fungus Trichoderma citrinoviride. Toxicon 86:51–58. https://doi.org/10.1016/j.toxicon.2014.05.008

Mohamed ZA, Hussein AA (2006) Depuration of microcystins in tilapia fish exposed to natural populations of toxic cyanobacteria: a laboratory study. Ecotoxicol Environ Saf 63:424–429. https://doi.org/10.1016/j.ecoenv.2005.02.006

Moraes CC, Sala L, Cerveira GP, Kalil SJ (2011) C-phycocyanin extraction from Spirulina platensis wet biomass. Braz J Chem Eng 28(1):45–49. https://doi.org/10.1590/S0104-66322011000100006

Nakamura N, Nakano K, Sugiura N, Matsumura M (2003) A novel cyanobacteriolytic bacterium, Bacillus cereus, isolated from a eutrophic lake. J Biosci Bioeng 95:179–184

Ndlela LL, Oberholster PJ, Van Wyk JH, Cheng PH (2016) An overview of cyanobacterial bloom occurrences and research in Africa over the last decade. Harmful Algae 60. https://doi.org/10.1016/j.hal.2016.10.001

Ndlela LL, Oberholster PJ, Van Wyk JH, Cheng PH (2018) Bacteria as biological control agents of freshwater cyanobacteria: is it feasible beyond the laboratory? Appl Microbiol Biotechnol. https://doi.org/10.1007/s00253-018-9391-9

Neves M, Castro BB, Vidal T et al (2015) Biochemical and populational responses of an aquatic bioindicator species, Daphnia longispina, to a commercial formulation of a herbicide (Primextra® Gold TZ) and its active ingredient (S-metolachlor). Ecol Indic 53:220–230

Oberemm A, Fastner J, Steinberg CEW (1997) Effects of microcystin-LR and cyanobacterial crude extracts on embryo-larval development of zebrafish (Danio rerio). Water Res 31:2918–2921. https://doi.org/10.1016/S0043-1354(97)00120-6

Oberholster PJ, Botha AM, Myburgh JG (2009a) Linking climate change and progressive eutrophication to incidents of clustered animal mortalities in different geographical regions of South Africa. Afr J Biotechnol 8(21):5825–5832. https://doi.org/10.5897/AJB09.1060

Oberholster PJ, Myburgh JG, Govender D et al (2009b) Identification of toxigenic Microcystis strains after incidents of wild animal mortalities in the Kruger National Park, South Africa. Ecotoxicol Environ Saf 72:1177–1182. https://doi.org/10.1016/j.ecoenv.2008.12.014

Paerl HW, Gardner WS, Havens KE et al (2016) Mitigating cyanobacterial harmful algal blooms in aquatic ecosystems impacted by climate change and anthropogenic nutrients. Harmful Algae 54:213–222. https://doi.org/10.1016/j.hal.2015.09.009

Paerl HW, Xu H, Hall NS et al (2014) Controlling cyanobacterial blooms in hypertrophic Lake Taihu, China: will nitrogen reductions cause replacement of non-N2 fixing by N2 fixing taxa? PLoS One 9. https://doi.org/10.1371/journal.pone.0113123

Parmar TK, Rawtani D, Agrawal YK (2016) Bioindicators: the natural indicator of environmental pollution. Front Life Sci 9:110–118. https://doi.org/10.1080/21553769.2016.1162753

Pool EJ, Jagals C, van Wyk JH, Jagals P (2003) The use of IL-6 induction as a human biomarker for inflammatory agents in water. Water Sci Technol 47:71–75

Porra RJ, Thompson WA, Kriedemann PE (1989) Determination of accurate extinction coefficients and simultaneous equations for assaying chlorophylls a and b extracted with four different solvents: verification of the concentration of chlorophyll standards by atomic absorption spectroscopy. Biochim Biophys Acta Bioenerg 975:384–394. https://doi.org/10.1016/S0005-2728(89)80347-0

Preece EP, Hardy FJ, Moore BC, Bryan M (2017) A review of microcystin detections in estuarine and marine waters: environmental implications and human health risk. Harmful Algae 61:31–45. https://doi.org/10.1016/j.hal.2016.11.006

Ren H, Zhang P, Liu C et al (2010) The potential use of bacterium strain R219 for controlling of the bloom-forming cyanobacteria in freshwater lake. World J Microbiol Biotechnol 26:465–472. https://doi.org/10.1007/s11274-009-0192-2

Samdal IA, Ballot A, Løvberg KE, Miles CO (2014) Multihapten approach leading to a sensitive ELISA with broad cross-reactivity to microcystins and nodularin. Environ Sci Technol 48:8035–8043. https://doi.org/10.1021/es5012675

Sánchez MI, Paredes I, Lebouvier M, Green AJ (2016) Functional role of native and invasive filter-feeders, and the effect of parasites: learning from hypersaline ecosystems. PLoS One 11:e0161478. https://doi.org/10.1371/journal.pone.0161478

Stiling P (1993) Why do natural enemies fail in classical biological control programs? Am Entomol 39:31–37

Su JF, Ma M, Wei L et al (2016a) Algicidal and denitrification characterization of Acinetobacter sp. J25 against Microcystis aeruginosa and microbial community in eutrophic landscape water. Mar Pollut Bull 107:233–239. https://doi.org/10.1016/j.marpolbul.2016.03.066

Su JF, Shao SC, Ma F et al (2016b) Bacteriological control by Raoultella sp. R11 on growth and toxins production of Microcystis aeruginosa. Chem Eng J 293:139–150. https://doi.org/10.1016/j.cej.2016.02.044

van Ginkel CE (2011) Eutrophication: present reality and future challenges for South Africa. Water SA

van Ginkel CE (2012) Algae, phytoplankton and eutrophication research and management in South Africa: past, present and future. Afr J Aquat Sci 37:17–25. https://doi.org/10.2989/16085914.2012.665432

Van Halderen A, Harding WR, Wessels JC et al (1995) Cyanobacterial (blue-green algae) poisoning of livestock in the Western Cape Province of South Africa. J South Afr Vet Assoc 66:260–264

Wang Z, Xiao B, Song L et al (2011) Effects of microcystin-LR, linear alkylbenzene sulfonate and their mixture on lettuce (Lactuca sativa L.) seeds and seedlings. Ecotoxicology 20:803–814. https://doi.org/10.1007/s10646-011-0632-2

Westrick JA, Szlag DC, Southwell BJ, Sinclair J (2010) A review of cyanobacteria and cyanotoxins removal/inactivation in drinking water treatment. Anal Bioanal Chem 397:1705–1714. https://doi.org/10.1007/s00216-010-3709-5

Yang L, Maeda H, Yoshikawa T, Zhou G (2012) Algicidal effect of bacterial isolates of Pedobacter sp. against cyanobacterium Microcystis aeruginosa. Water Sci Eng 5:375–382. https://doi.org/10.3882/j.issn.1674-2370.2012.04.002

Zhang B-H, Ding Z-G, Li H-Q et al (2016) Algicidal activity of streptomyces eurocidicus JXJ-0089 metabolites and their effects on microcystis physiology. Appl Environ Microbiol 82:5132–5143. https://doi.org/10.1128/AEM.01198-16

Zhang H, Yu Z, Huang Q et al (2011) Isolation, identification and characterization of phytoplankton-lytic bacterium CH-22 against Microcystis aeruginosa. Limnologica 41(1):70–77. https://doi.org/10.1016/j.limno.2010.08.001

Chapter 11
Options for Microbiological Quality Improvement in African Households

**Phumudzo Budeli, Resoketswe Charlotte Moropeng,
Mutshiene Deogratias Ekwanzala, and Maggy Ndombo Benteke Momba**

11.1 Introduction

Access to safe drinking water sources, appropriate sanitation facilities and good hygiene are fundamental not only to the health and survival of people but also to the economic growth and development of a country. These necessities are still a luxury for many people in the developing world, especially in rural areas (Peter 2010). The United Nations' Sustainable Development Goal 6 stipulates the need to "*invest in adequate infrastructure, provide sanitation facilities and encourage hygiene at every level*" to ensure universal access to safe and affordable drinking water for all by 2030. In addition to this, water-related ecosystems such as forests, mountains, wetlands and rivers must be protected as a water scarcity mitigation measure. Also, cooperation between different bodies at an international level is crucial to encourage water efficiency and support treatment technologies, especially in developing countries (Bain et al. 2014). An estimate of 663 million people rely on "unimproved" water supplies (as defined by the WHO/UNICEF Joint Monitoring Programme for Water and Sanitation), which are thought to harbour a high concentration of pathogenic contaminants, and more than 2.5 billion people are still deprived of access to improved sanitation facilities, with 946 million people practising open defecation (WHO/UNICEF 2013). Numerous water sources classified as improved are still not safe for human consumption (Bain et al. 2014). A WHO report indicates disparities between regions, with 61% of the population in sub-Saharan Africa having improved water supply sources, versus 90% or more in Latin America, the Caribbean, Northern Africa and large parts of Asia. Sub-Saharan Africa accounts for more than 40% of the global population without access to safe drinking water, followed by Southern Asia (21%) and Eastern Asia (16%) (WHO/UNICEF 2013).

P. Budeli (✉) · R. C. Moropeng · M. D. Ekwanzala · M. N. B. Momba
Department of Environmental, Water and Earth Sciences, Tshwane University of Technology,
Pretoria, South Africa

© Springer Nature Switzerland AG 2020
A. L. K. Abia, G. R. Lanza (eds.), *Current Microbiological Research in Africa*,
https://doi.org/10.1007/978-3-030-35296-7_11

Numerous studies have indicated that water is a potential source of waterborne infectious diseases, affecting many communities negatively, particularly those inhabiting rural and indigenous areas. As a result, it is estimated that water-related diseases are accountable for five million deaths annually (Baumgartner et al. 2007; Pritchard et al. 2009). Several gastrointestinal infections such as diarrhoea, dysentery, typhoid shigellosis and human enteritis are known to be caused by bacterial pathogens in water (Sobsey et al. 2002; Murcott 2006; Lantagne and Clasen 2012). The primary cause of illness and deaths in low-income countries is watery diarrhoea, called cholera, caused by bacteria known as *Vibrio cholera* (Clasen and Boisson 2006). Consumption of unsafe water, poor sanitation and hygiene have been reported to be the major causes of infectious diarrhoea, which is claimed to be responsible for 1.7 million deaths per year (3.1% of all annual deaths), 90% of which are children, virtually all in developing countries. Furthermore, 3.7% of the annual health burden worldwide (54.2 million disability-adjusted life years) is attributed to unsafe water, sanitation and hygiene (Mwabi et al. 2012).

Several studies have suggested that the key to reducing or even eradicating the burden of waterborne disease is appropriate sanitation facilities and piped water systems (Clasen and Boisson 2006; Sobsey et al. 2002; Murcott 2006; Lantagne and Clasen 2012; van Halem et al. 2009; Mwabi et al. 2012, 2013). According to Mwabi and co-workers, the establishment of this type of infrastructure could take decades, especially in impoverished rural communities of African countries (Mwabi et al. 2013). It is, therefore, important for water authorities around the world to make sure that the water that reaches households is safe for consumption and does not contain any substances that may have hazardous effects on human health (Suthar 2011). Although there are cases where good quality water is accessible in rural communities at a common access point, a problem that arises is that there is a high possibility of the water being contaminated during either transportation or inappropriate storage in the household (Moher et al. 2010; Moropeng et al. 2018). Consequently, HWT and safe storage systems have been considered as low-cost and efficient measures to decrease waterborne diarrhoeal diseases and improve the availability of safe water supply (Clasen and Boisson 2006; Fewtrell et al. 2005). These water treatment devices could be of great use during seasonal floods, heavy rainfall or other natural disasters when water sources are susceptible to faecal contamination (Mwabi et al. 2013; Wang et al. 2014).

Various water treatment methods, such as the use of disinfectants (chlorine and iodine), filtration, distillation, reverse osmosis, solar disinfectants and water purifiers, have been reported to serve in reducing endemic diarrhoeal diseases caused by waterborne pathogens and also to improve the microbial and chemical quality of drinking water (Sobsey et al. 2002; Murcott 2006; Stauber et al. 2012; Mwabi et al. 2011). Because of the low cost of manufacturing the filters using locally available materials and the simplicity and ease of construction and maintenance, these HWT systems allow the users to have access to safe potable water immediately after installation (Mol 2001). Factors that play a pivotal role in the acceptance and adoption of appropriate POU water treatment systems in rural communities include the accessibility and availability of necessary materials, quality of the intake water, duration of storage, tradition, customs, religion, educational level of the end user and availability of a skilled personnel to provide training on the maintenance of water treatment device before implementation (Murcott 2006; Mwabi et al. 2013). Another problem reported by (Mwabi et al. 2013) and Mwabi et al. (2011) concerning HWT devices is that consumers often lack a proper understanding of how to use these devices effectively and appropriately, and managers selling these products are unable to provide an appropriate solution to their customers when problems occur.

Numerous systems and devices have been widely reported in the literature; because of this diversity, it is critical to decide which device or devices would be most suitable for communities living in African rural areas. The criteria listed in Table 11.1 were considered by Mwabi et al. 2012 in order to help the communities to select a suitable device for their PoU water treatment system.

This chapter, therefore, informs communities on the selection of HWT methods/devices/systems that can produce microbiologically high-quality drinking water. It draws together evidence from published studies on home or HWT from the African continent. This chapter seeks to provide a comprehensive comparison between different water treatment systems developed for low-resource settings based on cost, water quantity and microbial removal. This is done in order to aid African households to choose a water system that fits their needs. Moreover, reviewed articles highlighted key research gaps and perceptions of future HWT options. Although the removal of other contaminants is essential as well, this chapter focuses solely on microbial removal, with the emphasis on reducing the diarrhoeal rate. This chapter on options for microbiological quality improvement in African household was compiled using the preferred reporting system based on the PRISMA guidelines (Moher et al. 2010).

Table 11.1 Selection criteria for HWTS and criteria for evaluation (Mwabi et al. 2012)

Selection criteria—to choose devices to evaluate in the lab/field	Evaluation criteria—characteristics to be tested during lab/field work
1. Can members of rural communities afford to obtain the unit? Construction and operation costs must not exceed earnings	1. Cost (capital/running)
2. A representative of a number of similar systems	2. Final water quality must comply with SANS 241
3. Systems already extensively evaluated	3. The turbidity of treated water must comply with SANS 241/WHO, <1NTU
4. Pressure requirement, maximum two metres	4. Ease of operation
5. Power requirement does not exceed equitable share	5. Storage ability and ability to deliver enough water
6. Robustness durability of the filter	6. Robustness (test)
7. Minimum required volume for basic human needs, 25 ℓ/p/d	7. Social acceptance

11.2   Search Strategy and Inclusion Criteria

The strategy used to retrieve relevant studies is summarised in Fig. 11.1.

Terms such as "household" OR "home" AND "water treatment" AND "Africa" were used to retrieve relevant information. Literature searches run on PubMed, Web of Science and African Journal returned 439, 19 and 3 results, respectively. Other sources retrieved 13 additional articles that could be included in the final analysis. When all databases were combined, 356 duplicate articles were removed. Based on the title and abstract screening for inclusion, 118 records were accessed, of which 41 were excluded. The remaining 77 records were assessed for eligibility after a full text read. Seven records were deemed ineligible for inclusion because they comprised (1) records of studies outside the African continent ($n = 4$) and (2) assessment studies without a HWT system ($n = 3$).

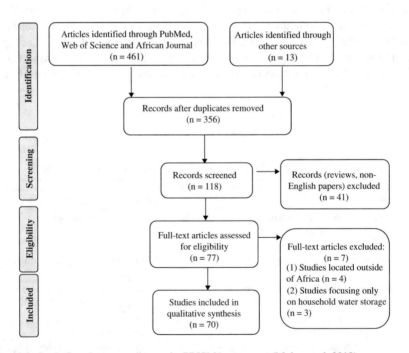

Fig. 11.1   Study flowchart according to the PRISMA statement (Moher et al. 2010)

11.3 Quality Assessment of the Included Studies and Data Analysis

The quality assessment of included studies was carried out as per the checklist provided by the Joanna Briggs Institute (Moher et al. 2010). Thus, studies that scored between 6 and 10 were included.

Data were analysed in Microsoft Excel® 2016 (Microsoft Corporation, Redmond, WA, USA) while figures were produced using PowerPoint® 2016 (Microsoft Corporation, Redmond, WA, USA) and BioVinci software (BioTuring, California, USA). In total, 474 papers were found by searching the databases. After the removal of duplicates, the remaining 118 were screened using the titles and abstracts. After reading the titles and abstracts, 77 studies were classified to be read in full; however, seven studies did not fit the criteria for inclusion. Finally, 70 full-text records were deemed eligible for inclusion in the qualitative synthesis. The studies that were included reported a home or HWT system implemented on the African continent capable of reducing or removing microbial cell count. The removal rate was expressed in percentage for most studies.

Out of 70 assessed studies, chemical methods using disinfection were the most commonly used methods (44.28%), followed by filtration (37.14%) and lastly physical methods, which accounted for 18.57% of all studies. Thirty-six household treatment methods were found across all retrieved studies. Filtration methods were found in 44.4% ($n = 16$), chemical methods in 38.9% ($n = 14$) and lastly physical methods, which accounted for only 16.7% ($n = 6$). Although a plethora of HWTS exists, only a few have been implemented and reported on the African continent. As presented in Fig. 11.2, only eight African countries have successfully implemented different HWTS between 2008 and 2018, with plausible diarrhoeal reduction outcomes. There has been a decrease in the number of implementation studies over the last decade, which may be because not all intervention findings are published.

The best implemented system, which showed a drastic reduction in diarrhoeal rate, was a biosand-zeolite silver-impregnated clay granular filter (BSZ-SICG) and silver-impregnated porous pot filter (SIPP). In microbial removal terms, not all systems were assessed to remove bacteria, protozoa and viruses. Systems that were assessed to remove all microbial contamination were barrel filters, biosand filters, BSZ-SICG, bucket filters, ceramic pot filters, colloidal silver-impregnated ceramic filters (CSF), SIPP, solar disinfection (SODIS) and solar cookers. All had impressive microbial removal capacities. As shown in Fig. 11.3a, high-performing methods of bacterial removal were BSZ-SICG, cloth, SIPP, fibreglass, solar cookers, SOPAS, iron oxide, NaDCC tablets, sodium hypochlorite, titanium dioxide and zinc oxide.

As for protozoan removal (Fig. 11.3b), high performers were the barrel filter, biosand filter, BSZ-SICG, bucket filter, ceramic candle filter, ceramic pot filter, Indian cloth, CSF, diatomaceous earth water filter, SIPP, SODIS and solar cooker. Figure 11.3c shows all systems that can remove viral particles. The best viral removal methods were the ceramic candle filter, ceramic pot filter, CSF, SIPP,

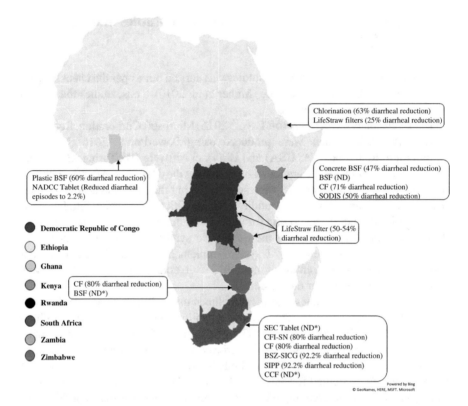

Fig. 11.2 Countries in which HWT methods were implemented. *CF* ceramic filters, BSF biosand filters, *NADCC* sodium dichloroisocyanurate, *SODIS* solar disinfection systems, *SEC* silver-embedded ceramic, *CFI-SN* Ceramic water filters impregnated with silver nanoparticles, *BSZ-SICG* biosand-zeolite silver-impregnated clay granular filter, *SIPP* silver-impregnated porous pot filter and *CCF* ceramic candle filter. *ND not determined

fibreglass, SODIS, solar cooker and iron oxide. We are expanding on these evaluated HWTs methods in detail in the following sections.

11.4 Appropriate HWT Methods

Considering the lack of skills, costs linked to operations and maintenance, and especially the distance between scattered villages and farms in rural areas, new approaches to treat and deliver microbiologically safe drinking water to rural communities at household level must be considered to prevent waterborne diseases in developing countries (Mwabi et al. 2013). Safe household water management refers to the maintenance or improvement of the microbiological quality of potable water through collection, distribution, transportation and storage in the home (Murcott 2006). Several studies have demonstrated that simple, low-cost HWTS could result

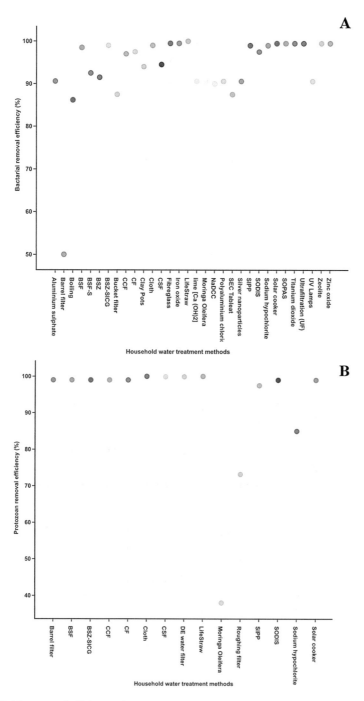

Fig. 11.3 The removal efficiency of HWT methods. (**a**) Bacterial removal efficiency. (**b**) Protozoan removal efficiency. (**c**) Viral removal efficiency

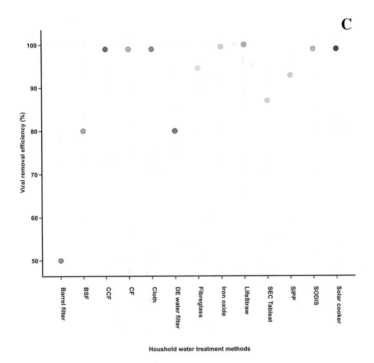

Fig. 11.3 (continued)

in substantial improvement of the microbiological quality of drinking water and thus help to reduce the risk of illness and death (Baumgartner et al. 2007; Fewtrell et al. 2005; Moropeng et al. 2018). A significant reduction in diarrhoeal disease ranging between 6 and 95% has been reported in areas where HWT has been implemented. The reduction in diarrhoeal diseases is thought to depend on the type of HWTS and how efficiently or appropriately it is used (Clasen and Boisson 2006; Moropeng et al. 2018). The systems not only provide a positive barrier against pathogenic infection in homes but can also be more cost-effective than large-scale projects, allowing individuals a choice between multiple technologies to meet their specific needs. This section focuses on various HWTS that have been developed over the years, ranging from physical removal of pathogens by adsorption, sedimentation and filtration to the destruction and control of pathogens using chemical disinfectants, ion exchange, heat inactivation (solar energy) and natural coagulants. A list of HWT devices/systems that indicates HWTS type, mechanisms, flow rate, cost and removal percentage of each system is provided. The subsequent sections review some of the HWT technologies that are available for possible adoption in African rural communities.

11.5 Filtration Methods for Treating Water at the Household Level

Filtration is a simple water treatment process capable of removing abiotic (colloids and suspended solids) and biological (pathogens) pollutants from drinking water sources mainly through a size exclusion mechanism. A well-designed filtration system can generate clean water for drinking purposes. A list of available options for filtration systems that have been developed for the production of safe, potable drinking water is summarised in Table 11.2.

Filtration technologies implemented in PoU systems are mainly biosand filtration (BSF), ceramic filtration and membrane filtration. Most of the available filters are regarded as cost-effective, as the construction procedure is simple, no degree of expertise is required for their construction, operation and maintenance, and they produce water at low cost. These systems are highly efficacious in removing pathogenic bacteria (80–100%), compared to protozoan parasites (50–100%) and viruses (50–100%). The cost of filtration systems included in this study ranged from USD 10–64. The most common filtration system employed in rural communities is BSF, with more than 500,000 people using it worldwide.

The principle of BSF is similar to that of a conventional slow sand filter. However, BSF experiences a varying flow rate and intermittent filtration through the sand layer. The spigot of this system is located higher than the filtration system medium; this allows the intake water to saturate the layers of sand throughout the operation. This also provides a conducive environment for biofilm growth which ultimately removes pollutants such as larger microorganisms and colloids (Stauber et al. 2006; Wang et al. 2014).

A new BSF system could only remove 63% of *Escherichia coli* cells, against 98% in a mature filter. However, the treated water is not within the WHO drinking water guideline. Thus, this system is not appropriate for use in households. Also, a BSF demonstrates lower removal of viruses (>80%) and offers no post-filtration residual protection, as water filtered into open or unclean storage containers has the risk of being recontaminated. Among the filtration systems summarised, the ceramic pot filter, ceramic candle filter, SIPP, LifeStraw, BSZ-SICG, ultrafiltration (UF) and ROAMPlus are the most promising systems available for adoption; however, they have significant drawbacks.

For instance, although ceramic filtration systems included in this study demonstrate higher efficiency in removing pathogenic microorganisms from water, most of their flow rate ranges between 1 and 4 L/h, which is lower than the WHO recommended limit of 25 L/P/day. Studies have shown that rural users might prefer higher flow rate devices mainly because of the large quantity of water produced, although efficient and effective pathogen removal should be the priority. Thus, it is imperative that further investigations focus on improving the flow rate of the technologies/ methods that demonstrate high effectiveness in removal of waterborne pathogens. The ability to remove pathogenic microorganisms at a high flow rate is an indication that the technology/method has greater potential to provide the required volumes of

Table 11.2 A list of filtration methods for treating water at the household level

HHWTS type	Mechanisms	Flow rate ($\ell.h^{-1}$)	Cost	Removal (%)			References
				Bacteria	Protozoan parasites	Viruses	
Barrel filter	Filtration	14.1		>50	>99	>50	Sobsey et al. (2002)
Biosand filter with zeolite	Filtration	14.1	<USD 20	85–98			Elliott et al. (2008) Mwabi et al. (2011)
BSF	Filtration	30	40 USD	98.5	>99	>80	Buzunis (2019) Duke et al. (2006) Earwaker (2006) Mwabi et al. (2013) CAWST (2017)
BSF-S	Filtration	15	USD 16	90 to 95			Sobsey et al. (2008) Barnes et al. (2009) Danley-Thomson et al. (2018)
BSZ-SICG	Filtration			>99	>99	>99	Budeli et al. (2018) Moropeng et al. (2018)
Bucket filter	Filtration	167	19 USD	80–95			Sobsey et al. (2002) Mwabi et al. (2013)
Ceramic candle filter	Filtration	1.85	64 USD	97	>99	>99	Laurent (2005) Lantagne et al. (2009) Adeyemo et al. (2015)
Ceramic pot filter	Filtration	1–3	45 USD	>95–100	>99	>99	Oyanedel-Craver and Smith (2008) Van Halem (2006) Gupta et al. (2012)
Clay pots	Filtration	–	15–20 USD	93–95			Varkey and Dlamini (2012)
Cloth	Filtration	N/A	10 USD	>99	100		Colwell et al. (2003) Huq et al. (2010)
Colloidal silver-impregnated ceramic filter (CSF)	Filtration	3.1	0.31–1.25 USD/month	90–99	99.9	>90–99	Mattelet (2006) Sobsey et al. (2008) CAWST (2017)

Diatomaceous earth water filter	Filtration	–	–	–	99.9	>80%	Farrah et al. (1991) Fulton (2000)
LifeStraw	Filtration	12–15	R400.00 per straw	a log removal of 6.9 for *E. coli*	3.6 log reduction for *Cryptosporidium* oocysts	4.7 for MS2 bacterial phage	Fritter et al. (2003) Clasen et al. (2009)
Roughing filter	Filtration	1.5			57.3–89.0		Sobsey et al. (2002)
Silver-impregnated porous pot filter (SIPP)	Filtration	3.31	USD 30	>99	>96–100	85.7–100	Mwabi et al. (2011) Mwabi et al. (2012) Adeyemo et al. (2015) Moropeng et al. (2018)
Ultrafiltration (UF)	Filtration	–		99–100	ND	ND	Molelekwa et al. (2014)

ND not determined

water needed by rural communities for drinking and other domestic purposes (Mwabi et al. 2013; Momba 2013). Against this background, it is clear that most of the ceramic filtration systems will not be the best option for adoption in rural settings for production of microbiologically safe water as a stand-alone intervention.

Filtration systems such as BSF that do not meet the selection criteria postulated by Mwabi et al. 2012, but produce a higher flow rate, may be used as a pre-filtration step for ceramic filters in order to produce consistently safe, clean water for cooking and drinking. However, this has negative cost implications for consumers, as PoUs are meant for low-income earners in developing countries. A recent study by Moropeng et al. (2018) reported an improved SIPP that produces a flow rate of 27.5 L/h and consistent removal efficiency of pathogenic bacteria and protozoan parasites during implementation. In addition to a higher flow rate and microbial removal efficacy, SIPP also offers post-filtration residual protection, which minimises the possibility of recontamination and is available at a low cost of USD 20.67. Thus, this system is one of the best HWTS options available for safe potable water production in rural households.

Although other filtration systems such as UF and ROAM plus have demonstrated complete removal of pathogenic microorganisms from water, the initial cost of these systems may have a negative impact on their adoption in rural settings. Moreover, there is a paucity of information on their performance in the removal of protozoan parasites and viruses. There is also a lack of evidence regarding the long-term effectiveness of these technologies, particularly in a programmatic, scalable context.

The LifeStraw filtration system can remove pathogenic microorganisms to a recommended level. This system is a portable device; thus, it will not be ideal for an entire household. An improved biosand filtration system (BSZ-SICG) holds a greater promise for the production of microbiologically safe water for rural householders, as it counters the drawbacks of a conventional biosand filter. A conventional biosand filter was improved by the addition of zeolite and silver-impregnated clay granules to form a BSZ-SICG filter. A diffusion plate was also incorporated between fine sand and silver granules, which resulted in an increased flow rate from 19 L/h to 38.6 L/h; this is available at a cost of USD 17.31. Unlike conventional biosand filters that do not offer post-filtration residual protection, BSZ-SICG-embedded silver leaches out to treated water, which inhibits bacterial regrowth and biofilm development in a water storage container. The adhesion of bacteria to the inner surface of the storage containers is affected by factors such as turbidity of the intake water, silver concentration impregnated in the clay pores during manufacturing and the duration of the storage period.

In addition to these benefits, BSZ-SICG has been shown to produce microbiologically safe water consistently in rural households for 8 months (Moropeng et al. 2019), thus making it a more favourable option for adoption for HWT. Moreover, Moropeng et al. (Clasen et al. 2008b) reveal that SIPP and BSZ-SICG, with proper maintenance, could produce similar water quality results over extended periods. Although this research suggests that SIPP and BSZ-SICG are the best options available for adoption under PoU filtration methods, the implementation of these systems should always be accompanied by a hygiene and sanitation programme in order to mitigate the possibilities of reversing health benefits offered by these systems through recontamination of treated water.

11.6   Physical Methods for Treating Water at the Household Level

Table 11.3 shows the list of physical methods that have been developed for HWT systems for use in rural communities. These methods are preferred because they are easy to use and are cost-effective. Among the physical methods included in this study, boiling is arguably the oldest and most commonly practised HWT method and has been widely promoted for decades (Sobsey et al. 2002). Although the recommend boiling time varies significantly from 0 to 20 min, waterborne microbes that are pathogenic to humans are killed or inactivated even before the water reaches 100 °C. Boiling is effective at inactivating all enteropathogenic bacteria, viruses, and protozoa that cause diarrhoeal disease (Clasen et al. 2008a). It is reported that pasteurisation at 60 °C kills pathogenic bacteria in 30 min, while boiling is much more rapid (Clasen et al. 2008a, b). Although boiling has been proven to be effective in inactivation of waterborne pathogens, studies of the effectiveness of boiling in actual practice in developing countries have shown mixed results. In Vietnam and India, boiling reduced the level of thermotolerant coliforms by 99% and 97%, respectively, compared to source water samples in households; householders reported always or almost always boiling their water (Clasen et al. 2008a, b). This disparity may be due to inconsistent boiling, recontamination of boiled water in storage, differences in cultural practices in the study populations, level of education and the impact of an emergency. Thus, it is recommended that water storage be stored in the same vessel in which it was boiled, then carefully handled, and preferably consumed within 24 h to minimise recontamination. To date, no studies have assessed the health impact associated with boiling water.

The solar cooker is another physical method included in this study. This technology demonstrated the highest efficiency in inactivation of bacteria (99–100%), protozoan parasites (>99%) and viruses (>99%). While it is evident that the physical methods are effective in the microbial reduction and are traditionally and widely used, they may not always be the optimal solution, especially in areas where wood and other biomass fuels or fossil fuels are in limited supply and must be purchased, and the costs of boiling water are unaffordable. Moreover, using the aforementioned methods for treating water at household level may be linked to burn injuries and respiratory infections risks from indoor stoves or fires, potential taste objections and the possibility for incomplete water treatment if the users do not bring water to full boiling temperature (Lantagne et al. 2009). These factors negatively impact the adoption rate of these methods for use in HWT.

Other physical methods such as the use of fibreglass and SOPAS have been developed; however, in addition to the aforementioned drawbacks, these methods

Table 11.3 A list of physical methods for treating water at the household level

HHWTS type	Mechanisms	Cost	Removal (%)			References
			Bacteria	Protozoan parasites	Viruses	
Boiling	Pasteurization	Variable	86.2	ND	ND	Clasen et al. (2008a)
Fibreglass	Pasteurization	–	99–100	ND	>90–99	Medina-Valtierra et al. (2004) Nangmenyi et al. (2011)
SODIS	Pasteurization	0.04 USD a month/Variable	95–100	>99	>99	Lantagne et al. (2006) Méndez-Hermida et al. (2007) Gómez-Couso et al. (2012)
Solar cooker	Pasteurization	Variable	99–100	>99	>99	Conroy et al. (2001) Sobsey et al. (2002) Gómez-Couso et al. (2012) Poonia et al. (2017)
SOPAS	Pasteurization	–	99–100	ND	ND	Reyneke et al. (2018)
UV lamps	Disinfection	–	86.2–95	ND	ND	Locas et al. (2008) Guo et al. (2012)

ND not determined

have only been tested for their removal efficacy in a laboratory setting. Studies to validate their effectiveness in inactivation of protozoan parasites and viruses are needed before they can be considered for implementation in rural communities.

Among the physiochemical methods listed in this study, SODIS would be the preferred option for the production of microbiologically safe water at the household level. SODIS is a low-cost, effective disinfection POU method that involves the prolonged exposure of water in polyethylene terephthalate or glass bottles to sunlight (Wegelin et al. 1994), allowing ultraviolet (UV) emissions to transmit into the water. The temperature in the water bottle tends to increase due to prolonged exposure; this inevitably inactivates the pathogens when coupled with UV. Over the past two decades, SODIS has been a research hotspot in HWT systems. A study by Berney et al. (Ratnayaka et al. 2009) revealed that inactivation of *Vibrio cholerae* starts once the water temperature rises beyond 40 °C. *Escherichia coli* was found to be slightly more heat resistant. Reduction in *E. coli* occurred when the water temperature rose beyond 45 °C (Berney et al. 2006; Mcguigan et al. 1998). In four randomised, controlled trials, SODIS reduced diarrhoeal disease from 9% to 86% (Conroy et al. 1996, 1999). One of these studies documented a reduction in the risk of cholera transmission in children in six households using SODIS. This technique has been deployed in different places around the world, serving more than 30 countries worldwide. However, SODIS has its disadvantages, namely the need for pretreatment (filtration or flocculation) of water of higher turbidity, user acceptability concerns because of the limited volume of water that can be treated at one time, the lack of visual improvement in water aesthetics to reinforce the benefits of treatment, the length of time required to treat water and the large supply of intact, clean, suitable plastic bottles required. Furthermore, it is heavily dependent on weather conditions and requires an exposure period of more than 48 h on cloudy days (Lantagne et al. 2009; Pooi and Ng 2018). Against this background, SODIS will not be the best option available for adoption in rural African communities, especially if used as a stand-alone intervention. However, SODIS can be coupled with other cost-effective filtration systems such as BSF for pre-treatment and chemical disinfection methods (silver tablets, PUR, chlorine tablets, etc.) to protect treated water from regrowth of microorganisms during storage. This step will among others dramatically increase the cost and deteriorate the quality of water further in terms of taste and odour, as well as being more laborious and time-consuming. These factors make SODIS a less desirable option for the production of microbiologically safe water in rural settings.

11.7 Chemical Disinfection at a Household Level

The disinfection of water is the final and most crucial step in water treatment before the distribution of the water to consumers. It necessitates the addition of a specific amount of physical or chemical agent (disinfectant) to the water. Contact between the water and the disinfectant is required for a pre-determined period to guarantee the success of the disinfection process with regard to the removal, deactivation or killing of any remaining pathogens resulting from the filtration process (Schutte and Focke 2006). The choice of disinfectant depends on several factors, which include: (1) efficacy against pathogens, (2) the ability to monitor and control the methods during the disinfection process accurately, (3) the ability to maintain a disinfectant residual within the distribution system, and (4) the ability to avoid a compromise in terms of the aesthetic quality of the drinking water (Långmark 2004). Unlike the technologies above, where microorganisms are removed from the water, disinfection results in the inactivation or death of microorganisms (Ratnayaka et al. 2009). The type of disinfectant used and dosage/exposure time differ from one microorganism to another. The inactivation/destruction mechanisms of disinfectants also differ depending on the target microorganisms.

The list of chemical disinfection/flocculation methods is summarised in Table 11.4. Among the disinfection methods, chlorination, ozonation and UV radiation are commonly used worldwide. Chemical methods are cost-effective, with prices ranging from USD 0.83 to USD 1.38/L, and are effective at inactivating most bacteria, the success rate ranging from 75 to 100%. However, these methods demonstrate lower efficiency in the removal of protozoan parasites (38–85%) and viruses (50–99%). In addition to the lower removal efficacy, most of these chemical methods have not been evaluated for their efficiency against protozoan parasites and viruses. Among the chemical methods included in this study, only silver-embedded ceramic tablets, iron oxide and hypochlorite were evaluated for their removal efficacy against viruses. The most common chemical disinfection method is chlorination.

The first disinfection employing chlorination in public water supply dated back to the early 1900s and helped reduce waterborne disease drastically in cities in Europe and the United States (Cutler and Miller 2005). Although there had been small trials of PoU chlorination previously (Mintz 1995), more extensive trials began in the 1990s as part of the activities of the Pan American Health Organisation and the CDC response to epidemic cholera in Latin America (Mintz 1995). A hypochlorite solution is effective at inactivating most bacteria and viruses that cause diarrhoeal disease (CDC 2008). In six randomised, controlled trials, the flocculant-disinfectant methods have resulted in diarrhoeal disease reductions in users ranging from 22 to 84% (Crump et al. 2005; Lule et al. 2005; Quick et al. 2002, 1996; Reller et al. 2003; Semenza et al. 1998). Also, a hypochlorite solution offers residual protection that protects against the regrowth of microorganisms in water storage containers. However, it is ineffective at inactivating some protozoan oocysts, such as *Cryptosporidium*. Other drawbacks associated with chlorination include relatively low protection against parasitic cysts, lower disinfection effectiveness in turbid waters contaminated with organic and some inorganic compounds, potential user taste and odour objections and the necessity of ensuring quality control of the solution (Lantagne et al. 2009). There is also scant information regarding the health effects associated with chlorine by-products. Although chemical disinfection methods such as **lime [Ca (OH)$_2$], silver-embedded ceramic tablets, iron** and **titanium oxide** have been developed and have demonstrated higher efficiency in bacterial inactivation, the cost of these methods remains a significant drawback that discourages adoption in rural communities.

Moringa oleifera is available at low cost and grows naturally in most parts of Africa. Its inactivation efficiency in relation to pathogenic bacteria in water ranges between 90% and 99%. However, just like many other disinfection methods, *Moringa oleifera* is ineffective at inactivation of protozoan parasites (38%). Moreover, there is little information on its ability to reduce diarrhoeal infections among rural householders during implementation.

The best option among the chemical/flocculation methods is **PUR**. The PUR product is a small sachet containing powdered ferric sulphate (a flocculant) and calcium hypochlorite (a disinfectant). To treat water with PUR, users open the sachet, add the contents to an open bucket containing 10 L of water, stir for 5 min and allow the debris to settle to the base of the bucket, and then a cotton cloth is used to strain the water into a second container. The hypochlorite content of the PUR inactivates microorganisms after 20 min. PUR has been documented to remove the vast majority of bacteria, viruses and protozoa, even in highly turbid waters (Crump et al. 2005; Souter et al. 2003). PUR has also been documented to reduce diarrhoeal disease from 16% to more than 90% in five randomised, controlled health intervention studies (Crump et al. 2005; Chiller et al. 2006; Doocy and Burnham 2006).

11.8 Implementation of HWT Methods in AFRICA

It has been estimated that 39% of Africa's population live without acceptable access to drinking water and the majority of those who are most affected live in sub-Saharan Africa (Waldman et al. 2013; WHO/UNICEF 2014). Microbiologically contaminated water leads to diarrhoeal illnesses, which account for 8% of deaths of children younger than the age of five annually in Africa (Wolf et al. 2014). This burden of illness can be addressed by interventions such as the implementation of PoU treatment methods in homes of vulnerable communities. In addition, it has been evidenced that improved water quality at the point of consumption can protect children from diarrhoeal diseases. One of the meta-analysis reviews by Wolf and co-authors (Figueroa and Kincaid 2010) suggested that water interventions could reduce diarrhoea by 34%. However, the success of such interventions depends on the user's interest and preferences.

In most cases, efforts to increase the demand for HWTS focus on the efficiency of the system in removing microbial contaminants and the health benefits, looking at diarrhoea reduction. In addition, other studies have shown that promotion of HWTS based on microbial removal efficiency and health benefits is unlikely to generate sustainable demand, as consumers select devices based on the convenience of the practice and design appeal of the product (Figueroa and Kincaid 2010; Albert et al. 2010; Luoto et al. 2011). User's preference is the key factor in accelerating the adoption rate of HWTS in the communities. Therefore, if users' preference is not taken into consideration when constructing HWTS, adoption/acceptance of HWTS

Table 11.4 A list of chemical/flocculation methods for treating water at a household level

HHWTS type	Mechanisms	Cost	Removal (%)			Reference
			Bacteria	Protozoan parasites	Viruses	
Aluminium sulphate	Disinfection/coagulation	$0.21–0.23/Kg	90–99	ND	ND	Sobsey et al. (2002) Wrigley (2007) Clarke et al. (2017)
Copper oxide	Disinfection	–	ND	ND	ND	Ren et al. (2009)
Iron oxide	Disinfection	–	99–100	ND	99–100	Chu et al. (2001) Brown and Sobsey (2009) Li (2010) Nangmenyi et al. (2011)
Lime [Ca(OH)$_2$]	Disinfection	US$9,95 a month	90–99	ND	ND	Sobsey et al. (2002) Wrigley (2007)
Moringa oleifera	Disinfection	–	90–99	38	ND	Narasiah et al. (2002) Lea (2010) Petersen et al. (2016)
NaDCC	Disinfection	0.30USD a month	>99	ND	ND	Clasen et al. (2007) Mohamed et al. (2015) Légaré-Julien et al. (2018)
Polyaluminium chloride	Disinfection	$0.2–0.8/Kg	90–99	ND	ND	Sobsey et al. (2002) Wrigley (2007) Tian et al. (2017)
Resins	Disinfection	–	ND	ND	ND	Gottlieb (2005) Rohm and Haas (2008)
Silver nanoparticles	Disinfection	–	90–99			Sondi and Salopek-Sondi (2004)
Silver-embedded ceramic tablet	Disinfection	–	75–100		75–99	Ehdaie et al. (2017) Ehdaie et al. (2014)

Sodium hypochlorite	Disinfection	0.20 a month/Variable	>99	>85	ND	Anderson et al. (2010) Wilhelm et al. (2018) CDC (2008)
Titanium dioxide	Disinfection	–	99–100	ND	ND	Alrousan et al. (2009)
Zeolite	Disinfection	–	ND	ND	ND	Rivera-Garza et al. (2000) Top and Ülkü (2004)
Zinc oxide	Disinfection	–	99–100	ND	ND	Pal et al. (2007)

ND not determined

devices by the deprived communities will not occur. It is thus vital for consumer's preferences, choices and aspirations to be understood.

A study by Abebe and co-authors (Abebe et al. 2014) in South Africa on ceramic filters impregnated with silver nanoparticles showed a higher acceptance rate of 84%, with a diarrhoeal reduction of 80% in HIV-infected individuals. The higher diarrhoeal reduction could be attributed to a higher acceptance rate of these HWTS. However, in their study, they did not determine the adoption rate, which is one of the factors that influence the sustainability of the HWTS. Congruently, in another study in South Africa on assessment of the sustainability and the acceptance/adoption rate of BSZ-SICG and SIPP filters by rural communities in Makwane village, it was found that the communities accepted the HWTS (79.5%) enthusiastically and diarrhoeal incidences were reported to be reduced by 92.2% over a period of six months. However, the communities showed no willingness to buy the HWTS (84.3%) in future (Clasen et al. 2008b). Moreover, the authors determined the adoption rate based on the number of HWTS in use during their study, and they found that most of the community members adopted SIPP filters (54.3%) compared to the BSZ-SICG (20.8%). Variation in adoption rates for the two HWTS could be attributed to the users' preferences and the appearance of the systems. Just like the ceramic filters impregnated with silver nanoparticles, the SIPP filters are much lighter than the BSZ-SICG in weight. Therefore, they are effortless for the users to carry and maintain (maintenance in terms of cleaning the system), unlike the BSZ-SICG, which is cumbersome. This further proves that the appeal of HWTS is of crucial importance when implementing these systems. One other factor that could have attributed to a lower adoption rate of the BSZ-SICG and SIPP filters might be the high unemployment rate in the village (60%), which contributed to the community's inability to afford the devices. These findings show that there is a need for governments in African countries to subsidise systems or collaborate with non-governmental organisations and the private sector in manufacturing and implementing HWTS in deprived communities.

Furthermore, in Kenya, a study on the SODIS water treatment method evaluated the acceptance rate of users and found that the acceptance rate was high at first (75%), but reduced to 30% (du Preez et al. 2011). This could have been due to unavailability of proper bottles and the time it takes to treat water. Although SODIS has been proven to be cost-effective and easy to use, the low cost applies to users who are already in possession of bottles. Also, SODIS requires the use of appropriate, clean transparent plastic bottles, which in most cases may not be readily available. Moreover, the time required for disinfection, which is 4–6 h on sunny days and 2 days on cloudy days, may affect the rate of adoption by users (CDC 2008).

Although the removal of pathogenic bacteria in combination with diarrhoea reduction using HWT methods/systems is well documented in the literature, very few studies have evaluated the acceptance and adoption rate of the methods/devices in rural areas of Africa. The acceptance and adoption rates are the determining factors for the sustainability of PoU water treatment methods/technology in rural communities.

Moreover, the longevity of HWT methods/technologies can influence the adoption rate positively. Most of the studies on the determination of the longevity of HWTS did not indicate the volume of water that the systems can treat before they expire. They only indicated that, for example, the biosand filter's lifespan might range from 1 to 12 years (Sisson et al. 2013). However, the volume of water filtered, together with the maintenance time, was not determined. On the contrary, a study by Moropeng and co-authors on BSZ-SICG in South Africa revealed a decrease in removal efficiency of pathogenic *E. coli* from 99.9% in the first month to 79.6% in the third month of being in used in Makwane village (Moropeng et al. 2018). However, they showed that the removal efficiency could be increased with proper maintenance. In their study, the BSZ-SICG showed a rise in *E. coli* removal from 79.6% to 96.9% after being washed in the third month of use (Moropeng et al. 2018). If the longevity of the systems is not known, especially by the users, the health benefits of the users might be compromised, as they will continue to filter the water while the systems are no longer effective in removing waterborne organisms. As a result, vulnerable communities will continue to suffer devastating diseases caused by waterborne organisms. Table 11.5 is a summary of some HWT systems implements in Africa.

11.9 Challenges and Future Perspectives

1. Although some water treatment technologies implemented at PoU have proven to remove waterborne pathogens, with some technologies successfully mitigating diarrhoeal illnesses, it is of paramount importance that future studies should focus on the social acceptability of these methods in combination with waterborne pathogen removal efficiency and health benefits. Most rural communities prefer devices with a higher flow rate simply because of the large quantity of water produced and the fact that it saves time.
2. HWT systems that have been reported to provide a higher flow rate and microbiologically free water, such as SIPP, CCF and BSZ-SICG, may be suitable options for rural African communities. However, monitoring the leaching of silver over a long period of use in rural communities has not been well documented. Therefore, future researchers should focus on monitoring silver leached into treated water, given the health hazards associated with silver.
3. Little information is available on the performance of HWT systems/devices in the removal of pathogenic microorganisms from different water sources with different physical and chemical parameters. Since contamination levels differ depending on the source, future researchers should also investigate the efficiency of HWT systems in removing pathogenic microorganisms and turbidity from different drinking water sources.
4. It is essential to take into consideration the preferences of users when implementing HWTS methods/technology for sustainable use of the devices. One of the essential factors that researchers should look into is the visual attractiveness

Table 11.5 Implemented HWT systems in African regions between 2008 and 2018

Country	Type of system	Health outcomes	Acceptance rate	Period of the study	Reference
Democratic Republic of Congo	LifeStraw filters	15% fewer weeks with diarrhoea	ND	12 months	Boisson et al. (2010)
Ethiopia	LifeStraw filters	Reported 25% reduction in the longitudinal prevalence of diarrhoea.	ND	5 months	Boisson et al. (2009)
Ethiopia	Chlorination	63% of diarrheal reduction	ND		Mengistie et al. (2013)
Ghana	Sodium dichloroisocyanurate (NaDCC) tablets	Reduced diarrheal episodes to 2.2%	ND	12 weeks	Hoekstra et al. (2010)
Ghana	Plastic biosand Filters (BSF)	60% diarrheal reduction in children	ND	6 months	Stauber et al. (2012)
Kenya	Concrete biosand Filters (BSF)	47% reduction in diarrheal diseases	ND	6 months	Tiwari et al. (2009)
Kenya	SODIS	Diarrheal reductions of roughly 50% in the incidence of dysentery and approximately 30% in the incidence of non-dysentery diarrhoea	The decrease from above 75% to 30%		du Preez et al. (2011)
Kenya	Biosand Filters (BSF)	N/A	N/A	6 months	McKenzie et al. (2013)
Kenya	Ceramic filters	71% reduction in diarrheal diseases	ND	6 months	Priest et al. (2018)
Rwanda	LifeStraw filters	50% lower odds of reported diarrhoea among children <5	ND	12–24 months	Kirby et al. (2016)
South Africa	Ceramic water filters impregnated with silver nanoparticles	80% diarrheal reduction	84% acceptability	12 months	Abebe et al. (2014)
South Africa	Low-cost ceramic candle filter system	The results showed that the systems were not satisfactory for home-based water treatment in a remote rural community of southern Africa.	ND	6 months	Lange et al. (2016)

South Africa	Silver-embedded ceramic tablet	N/A	N/A		Ehdaie et al. (2017)
South Africa	Biosand-zeolite silver-impregnated clay granular filter (BSZ-SICG) and silver-impregnated porous pot filter (SIPP)	92.2% reduction in diarrheal diseases	ND	6 months	Moropeng et al. (2018)
Zambia	LifeStraw family filter	The diarrheal reduction was reported to be 53% among children and 54% among all other household members	ND	12 months	Peletz et al. (2012)
Zimbabwe	Biosand filters (BSF)	N/A	ND	Longitudinal study	Kanda et al. (2013)
Zimbabwe and South Africa	Ceramic filters	80% reduction in diarrheal diseases.	ND	6 months trial	Du Preez et al. (2008)

ND not determined

of devices to be implemented, as most rural communities are visually attracted by the device design and this might influence the acceptance and adoption rate. Also, researchers should take into consideration the portability of the device, which also plays a vital role in acceptance by users.

5. Since Africa is considered culturally and religiously diverse, it is, therefore, important for implementers/researchers to consider these factors in order to improve the acceptance and adoption rate to ensure sustainable use of HWTS.

6. Most importantly, educating vulnerable communities about the importance and the use of HWTS is of paramount importance for sustainable use.

7. Above all, the HWT method/technology must be inexpensive, not causing households to stop treating water because they cannot afford to purchase the technology (as most African countries are faced by financial and political challenges).

References

Abebe LS, Smith JA, Narkiewicz S et al (2014) Ceramic water filters impregnated with silver nanoparticles as a point-of-use water-treatment intervention for HIV-positive individuals in Limpopo Province, South Africa: a pilot study of technological performance and human health benefits. J Water Health 12:288–300. https://doi.org/10.2166/wh.2013.185

Adeyemo FE, Kamika I, Momba MNB (2015) Comparing the effectiveness of five low-cost home water treatment devices for Cryptosporidium , Giardia and somatic coliphages removal from water sources. Desalin Water Treat 56:2351–2367. https://doi.org/10.1080/19443994.2014.96 0457

Albert J, Luoto J, Levine D (2010) End-user preferences for and performance of competing POU water treatment technologies among the rural poor of Kenya. Environ Sci Technol 44:4426–4432. https://doi.org/10.1021/es1000566

Alrousan DMA, Dunlop PSM, McMurray TA, Byrne JA (2009) Photocatalytic inactivation of E. coli in surface water using immobilised nanoparticle TiO2 films. Water Res 43:47–54. https://doi.org/10.1016/j.watres.2008.10.015

Anderson BA, Romani JH, Wentzel M, Phillips HE (2010) Awareness of water pollution as a problem and the decision to treat drinking water among rural African households with unclean drinking water: South Africa 2005

Bain R, Cronk R, Hossain R et al (2014) Global assessment of exposure to faecal contamination through drinking water based on a systematic review. Trop Med Int Heal 19:917–927. https://doi.org/10.1111/tmi.12334

Barnes D, Collin C, Ziff S (2009) The biosand filter, siphon filter and rainwater harvesting

Baumgartner J, Murcott S, Ezzati M (2007) Reconsidering 'appropriate technology': the effects of operating conditions on the bacterial removal performance of two household drinking-water filter systems. Environ Res Lett 2:024003. https://doi.org/10.1088/1748-9326/2/2/024003

Berney M, Weilenmann H-U, Simonetti A, Egli T (2006) Efficacy of solar disinfection of Escherichia coli, Shigella flexneri, Salmonella Typhimurium and Vibrio cholerae. J Appl Microbiol 101:828–836. https://doi.org/10.1111/j.1365-2672.2006.02983.x

Boisson S, Kiyombo M, Sthreshley L et al (2010) Field assessment of a novel household-based Water filtration device: a randomised, placebo-controlled trial in the Democratic Republic of Congo. PLoS One 5:e12613. https://doi.org/10.1371/journal.pone.0012613

Boisson S, Schmidt W-P, Berhanu T et al (2009) Randomized controlled trial in rural Ethiopia to assess a portable water treatment device. Environ Sci Technol 43:5934–5939. https://doi.org/10.1021/es9000664

Brown J, Sobsey MD (2009) Ceramic media amended with metal oxide for the capture of viruses in drinking water. Environ Technol 30:379–391. https://doi.org/10.1080/09593330902753461

Budeli P, Moropeng RC, Mpenyana-Monyatsi L, Momba MNB (2018) Inhibition of biofilm formation on the surface of water storage containers using biosand zeolite silver-impregnated clay granular and silver impregnated porous pot filtration systems. PLoS One 13:e0194715

Buzunis JB (2019) Intermittently operated slow sand filtration [microform] : a new water treatment process

CAWST (Centre for affordable water and sanitation technology) (2017) Biosand filter: manual for design, construction, installation and maintenance

CDC (Center of Disease Control) (2008) Household water treatment options in developing countries: Flocculant/disinfectant powder

Chiller T, CE Mendoza, López B, et al (2006) Reducing diarrhoea in Guatemalan children: randomized controlled trial of flocculant-disinfectant for drinking-water

Chu Y, Jin Y, Flury M, Yates MV (2001) Mechanisms of virus removal during transport in unsaturated porous media. Water Resour Res 37:253–263. https://doi.org/10.1029/2000WR900308

Clarke R, Peyton D, Healy MG et al (2017) A quantitative microbial risk assessment model for total coliforms and E. coli in surface runoff following application of biosolids to grassland. Environ Pollut 224:739–750. https://doi.org/10.1016/j.envpol.2016.12.025

Clasen T, Boisson S (2006) Household-based ceramic water filters for the treatment of drinking water in disaster response: an assessment of a pilot programme in the Dominican Republic. Water Pract Technol 1

Clasen T, Haller L, Walker D et al (2007) Cost-effectiveness of water quality interventions for preventing diarrhoeal disease in developing countries. J Water Health 5:599–608. https://doi.org/10.2166/wh.2007.010

Clasen T, McLaughlin C, Boisson S et al (2008b) Microbiological effectiveness and cost of disinfecting water by boiling in semi-urban India. Am J Trop Med Hyg 79:407–413. https://doi.org/10.4269/ajtmh.2008.79.407

Clasen T, Naranjo J, Frauchiger D, Gerba C (2009) Laboratory assessment of a gravity-fed ultrafiltration water treatment device designed for household use in low-income settings. Am J Trop Med Hyg 80:819–823

Clasen TF, Thao DH, Boisson S, Shipin O (2008a) Microbiological effectiveness and cost of boiling to disinfect drinking water in rural Vietnam. Environ Sci Technol 42:4255–4260. https://doi.org/10.1021/es7024802

Colwell RR, Huq A, Islam MS et al (2003) Reduction of cholera in Bangladeshi villages by simple filtration. Proc Natl Acad Sci 100:1051–1055. https://doi.org/10.1073/pnas.0237386100

Conroy RM, Elmore-Meegan M, Joyce T et al (1996) Solar disinfection of drinking water and diarrhoea in Maasai children: a controlled field trial. Lancet 348:1695–1697. https://doi.org/10.1016/S0140-6736(96)02309-4

Conroy RM, Meegan ME, Joyce T et al (1999) Solar disinfection of water reduces diarrhoeal disease: an update. Arch Dis Child 81:337–338. https://doi.org/10.1136/adc.81.4.337

Conroy RM, Meegan ME, Joyce T et al (2001) Solar disinfection of drinking water protects against cholera in children under 6 years of age. Arch Dis Child 85:293–295

Crump JA, Otieno PO, Slutsker L et al (2005) Household based treatment of drinking water with flocculant-disinfectant for preventing diarrhoea in areas with turbid source water in rural western Kenya: cluster randomised controlled trial. BMJ 331:478. https://doi.org/10.1136/bmj.38512.618681.E0

Cutler D, Miller G (2005) Water, Water, Everywhere: Municipal Finance and Water Supply in American Cities. Natl Bur Econ Res Work Pap Ser No. 11096 https://doi.org/10.3386/w11096

Danley-Thomson AA, Huang EC, Worley-Morse T, Gunsch CK (2018) Evaluating the role of total organic carbon in predicting the treatment efficacy of biosand filters for the removal of Vibrio cholerae in drinking water during startup. J Appl Microbiol 125:917–928. https://doi.org/10.1111/jam.13909

Doocy S, Burnham G (2006) Point-of-use water treatment and diarrhoea reduction in the emergency context: an effectiveness trial in Liberia. Trop Med Int Heal 11:1542–1552. https://doi.org/10.1111/j.1365-3156.2006.01704.x

du Preez M, Conroy RM, Ligondo S et al (2011) Randomized intervention study of solar disinfection of drinking water in the prevention of dysentery in Kenyan children aged under 5 years. Environ Sci Technol 45:9315–9323. https://doi.org/10.1021/es2018835

Du Preez M, Conroy RM, Wright JA et al (2008) Use of ceramic water filtration in the prevention of diarrheal disease: a randomized controlled trial in rural South Africa and Zimbabwe. Am J Trop Med Hyg 79:696–701

Duke WF, Nordin RN, Baker D, Mazumder A (2006) The use and performance of BioSand filters in the Artibonite Valley of Haiti: a field study of 107 households. Rural Remote Health 6:570

Earwaker P (2006) Evaluation of household biosand filters in Ethiopia

Ehdaie B, Krause C, Smith JA (2014) Porous ceramic tablet embedded with silver nanopatches for low-cost point-of-use water purification. Environ Sci Technol 48:13901–13908. https://doi.org/10.1021/es503534c

Ehdaie B, Rento CT, Son V et al (2017) Evaluation of a silver-embedded ceramic tablet as a primary and secondary point-of-use water purification technology in Limpopo Province, S. Africa. PLoS One 12:e0169502. https://doi.org/10.1371/journal.pone.0169502

Elliott MA, Stauber CE, Koksal F et al (2008) Reductions of E. coli, echovirus type 12 and bacteriophages in an intermittently operated household-scale slow sand filter. Water Res 42:2662–2670. https://doi.org/10.1016/j.watres.2008.01.016

Farrah S, Preston DR, Toranzos G, et al (1991) Use of modified diatomaceous earth for removal and recovery of viruses in water

Fewtrell L, Kaufmann RB, Kay D et al (2005) Water, sanitation, and hygiene interventions to reduce diarrhoea in less developed countries: a systematic review and meta-analysis. Lancet Infect Dis 5:42–52. https://doi.org/10.1016/S1473-3099(04)01253-8

Figueroa ME, Kincaid DL (2010) Social, cultural and behavioral correlates of household water treatment and storage. Cent Publ HCI 2010-1 Heal Commun Insights

Fritter CF, Netke SP, Scruggs III JE, Gröss SA (2003) Water purifying apparatus

Fulton GP (2000) Diatomaceous earth filtration for safe drinking water. American Society of Civil Engineers, Reston, VA

Gómez-Couso H, Fontán-Sainz M, Fernández-Ibáñez P, Ares-Mazás E (2012) Speeding up the solar water disinfection process (SODIS) against Cryptosporidium parvum by using 2.5l static solar reactors fitted with compound parabolic concentrators (CPCs). Acta Trop 124:235–242. https://doi.org/10.1016/j.actatropica.2012.08.018

Gottlieb MC (2005) Ion exchange application in water treatment plant design, handbook, 4th edn. McGraw-Hill, New York

Guo M, Huang J, Hu H et al (2012) UV inactivation and characteristics after photoreactivation of Escherichia coli with plasmid: health safety concern about UV disinfection. Water Res 46:4031–4036. https://doi.org/10.1016/j.watres.2012.05.005

Gupta V, Pathania D, Agarwal S, Sharma S (2012) Removal of Cr(VI) onto Ficus carica biosorbent from water

Hoekstra RM, Wannemuehler KA, Schmitz A et al (2010) Sodium dichloroisocyanurate tablets for routine treatment of household drinking water in Periurban Ghana: a randomized controlled trial. Am J Trop Med Hyg 82:16–22. https://doi.org/10.4269/ajtmh.2010.08-0584

Huq A, Yunus M, Sohel SS et al (2010) Simple sari cloth filtration of water is sustainable and continues to protect villagers from cholera in Matlab, Bangladesh. MBio 1. https://doi.org/10.1128/mBio.00034-10

Kanda A, Gotosa J, Chagwiza G (2013) Performance of biosand filters in treating source water in post emergency: a case of two rural districts of northern Zimbabwe. J Appl Sci 6:31–38

Kirby MA, Nagel CL, Rosa G et al (2016) Faecal contamination of household drinking water in Rwanda: a national cross-sectional study. Sci Total Environ 571:426–434. https://doi.org/10.1016/j.scitotenv.2016.06.226

Lange J, Materne T, Grüner J (2016) Do low-cost ceramic water filters improve water security in rural South Africa? Drink Water Eng Sci 9:47–55. https://doi.org/10.5194/dwes-9-47-2016

Långmark J (2004) Biofilms and microbial barriers in drinking water treatment and distribution

Lantagne D, Clasen T (2012) Point-of-use water treatment in emergency response. Waterlines 31:30–52

Lantagne DS, Quick R, Mintz ED (2009) Household water treatment and safe: storage options in developing countries

Lantagne SD, Quick R, ED Mintz (2006) Household water treatment and safe storage options in developing countries: a review of current implementation practices

Laurent P (2005) Household drinking water systems and their impact on people with weakened immunity. Geneva

Lea M (2010) Bioremediation of turbid surface Water using seed extract from Moringa oleifera lam. (drumstick) tree. In: Current protocols in microbiology. John Wiley & Sons, Inc., Hoboken, NJ

Légaré-Julien F, Lemay O, Vallée-Godbout U et al (2018) Laboratory efficacy and disinfection by-product formation of a coagulant/disinfectant tablet for point-of-use Water treatment. Water 10:1567. https://doi.org/10.3390/w10111567

Li X (2010) Applications for nanomaterials in critical technologies

Locas A, Demers J, Payment P (2008) Evaluation of photoreactivation of Escherichia coli and enterococci after UV disinfection of municipal wastewater. Can J Microbiol 54:971–975. https://doi.org/10.1139/W08-088

Lule J, Bunnell R, Wafula W et al (2005) Effect of home-based water chlorination and safe storage on diarrhea among persons with human immunodeficiency virus in Uganda. Am J Trop Med Hyg 73:926–933. https://doi.org/10.4269/ajtmh.2005.73.926

Luoto J, Najnin N, Mahmud M et al (2011) What point-of-use water treatment products do consumers use? Evidence from a randomized controlled trial among the urban poor in Bangladesh. PLoS One 6:e26132. https://doi.org/10.1371/journal.pone.0026132

Mattelet CCEHY (2006) Household ceramic water filter evaluation using three simple low-cost methods: membrane filtration, 3m petrifilm and hydrogen sulfide bacteria in northern region, Ghana

Mcguigan K, Joyce T, Conroy R, et al (1998) Solar disinfection of drinking water contained in transparent plastic bottles: Characterizing the bacterial inactivation process

McKenzie ER, Jenkins MW, S-SK T et al (2013) In-home performance and variability of biosand filters treating turbid surface and rain water in rural Kenya. J Water Sanit Hyg Dev 3:189–198. https://doi.org/10.2166/washdev.2013.050

Medina-Valtierra J, Calixto S, Ruiz F (2004) Formation of copper oxide films on fiberglass by adsorption and reaction of cuprous ions. Thin Solid Films 460:58–61. https://doi.org/10.1016/j.tsf.2004.01.107

Méndez-Hermida F, Ares-Mazás E, McGuigan KG et al (2007) Disinfection of drinking water contaminated with Cryptosporidium parvum oocysts under natural sunlight and using the photocatalyst TiO2. J Photochem Photobiol B Biol 88:105–111. https://doi.org/10.1016/j.jphotobiol.2007.05.004

Mengistie B, Berhane Y, Worku A (2013) Household water chlorination reduces incidence of diarrhea among under-five children in rural Ethiopia: a cluster randomized controlled trial. PLoS One 8:e77887. https://doi.org/10.1371/journal.pone.0077887

Mintz ED (1995) Safe water treatment and storage in the home. JAMA 273:948. https://doi.org/10.1001/jama.1995.03520360062040

Mohamed H, Brown J, Njee RM et al (2015) Point-of-use chlorination of turbid water: results from a field study in Tanzania. J Water Health 13:544–552. https://doi.org/10.2166/wh.2014.001

Moher D, Liberati A, Tetzlaff J, et al (2010) Preferred reporting items for systematic reviews and meta-analyses: the PRISMA statement

Mol A (2001) The success of household sand filtration. Waterlines 20:27–30. https://doi.org/10.3362/0262-8104.2001.043

Molelekwa GF, Mukhola MS, Van der Bruggen B, Luis P (2014) Preliminary studies on membrane filtration for the production of potable water: a case of Tshaanda Rural Village in South Africa. PLoS One 9:e105057. https://doi.org/10.1371/journal.pone.0105057

Momba MNB (2013) South African guidelines for the selection and use of appropriate home water-treatment systems by rural households. Report to the Water Research Commission, Pretoria

Moropeng R, Budeli P, Mpenyana-Monyatsi L, Momba M (2018) Dramatic reduction in diarrhoeal diseases through implementation of cost-effective household drinking water treatment systems in Makwane Village, Limpopo Province, South Africa. Int J Environ Res Public Health 15:410

Moropeng RC, Momba MNB (2019) Assessing the sustainability and the acceptance/adoption rate of cost effective household water treatment systems by rural communities of Makwane village, South Africa. Unpublished

Murcott S (2006) Implementation, critical factors and challenges to scale-up of household drinking water treatment and safe storage systems

Mwabi J, Mamba B, Momba M (2013) Removal of waterborne bacteria from surface water and groundwater by cost-effective household water treatment systems (HWTS): a sustainable solution for improving water quality in rural communities of Africa. Water SA 39. https://doi.org/10.4314/wsa.v39i4.2

Mwabi JK, Adeyemo FE, Mahlangu TO et al (2011) Household water treatment systems: a solution to the production of safe drinking water by the low-income communities of Southern Africa. Phys Chem Earth, Parts A/B/C 36:1120–1128. https://doi.org/10.1016/j.pce.2011.07.078

Mwabi JK, Mamba BB, Momba MNB (2012) Removal of Escherichia coli and Faecal coliforms from surface water and groundwater by household water treatment devices/systems: a sustainable solution for improving water quality in rural communities of the Southern African Development Community Region. Int J Environ Res Public Health 9:139–170. https://doi.org/10.3390/ijerph9010139

Nangmenyi G, Li X, Mehrabi S et al (2011) Silver-modified iron oxide nanoparticle impregnated fiberglass for disinfection of bacteria and viruses in water. Mater Lett 65:1191–1193. https://doi.org/10.1016/j.matlet.2011.01.042

Narasiah KS, Vogel A, Kramadhati NN (2002) Coagulation of turbid waters using Moringa oleifera seeds from two distinct sources. Water Sci Technol Water Supply 2:83–88. https://doi.org/10.2166/ws.2002.0154

Oyanedel-Craver VA, Smith JA (2008) Sustainable colloidal-silver-impregnated ceramic filter for point-of-use Water treatment. Environ Sci Technol 42:927–933. https://doi.org/10.1021/es071268u

Pal S, Tak YK, Song JM (2007) Does the antibacterial activity of silver nanoparticles depend on the shape of the nanoparticle? A study of the gram-negative bacterium Escherichia coli. Appl Environ Microbiol 73:1712–1720. https://doi.org/10.1128/AEM.02218-06

Peletz R, Simunyama M, Sarenje K et al (2012) Assessing water filtration and safe storage in households with young children of HIV-positive mothers: a randomized, controlled trial in Zambia. PLoS One 7:e46548. https://doi.org/10.1371/journal.pone.0046548

Peter G (2010) Impact of rural water projects on hygienic behaviour in Swaziland. Phys Chem Earth, Parts A/B/C 35:772–779. https://doi.org/10.1016/j.pce.2010.07.024

Petersen HH, Petersen TB, Enemark HL et al (2016) Removal of Cryptosporidium parvum oocysts in low quality water using Moringa oleifera seed extract as coagulant. Food Waterborne Parasitol 3:1–8. https://doi.org/10.1016/j.fawpar.2016.03.002

Pooi CK, Ng HY (2018) Review of low-cost point-of-use water treatment systems for developing communities. npj Clean Water 1(11). https://doi.org/10.1038/s41545-018-0011-0

Poonia DS, Singh A, Santra P, Mishra D (2017) Design, development and performance evaluation of non-tracking cooker type solar water purifier

Priest JW, Morris JF, Narayanan J et al (2018) A randomized controlled trial to assess the impact of ceramic Water filters on prevention of diarrhea and cryptosporidiosis in infants and young children—Western Kenya, 2013. Am J Trop Med Hyg 98:1260–1268. https://doi.org/10.4269/ajtmh.17-0731

Pritchard M, Mkandawire T, Edmondson A et al (2009) Potential of using plant extracts for purification of shallow well water in Malawi. Phys Chem Earth, Parts A/B/C 34:799–805. https://doi.org/10.1016/j.pce.2009.07.001

Quick RE, Kimura A, Thevos A et al (2002) Diarrhea prevention through household-level water disinfection and safe storage in Zambia. Am J Trop Med Hyg 54:511–516

Quick RE, Venczel LV, Gonzalez O, Mintz ED (1996) Narrow-mouthed water storage vessels and in situ chlorination in a Bolivian community: a simple method to improve drinking water quality. Am J Trop Med Hyg 54:511–516

Ratnayaka DD, Brandt MJ, Johnson M (2009) Water supply, 6th edn. Butterworth-Heinemann, Oxford

Reller ME, Mendoza CE, Lopez MB et al (2003) A randomized controlled trial of household-based flocculant-disinfectant drinking water treatment for diarrhea prevention in rural Guatemala. Am J Trop Med Hyg 69:411–419

Ren G, Hu D, Cheng EWC et al (2009) Characterisation of copper oxide nanoparticles for antimicrobial applications. Int J Antimicrob Agents 33:587–590. https://doi.org/10.1016/j.ijantimicag.2008.12.004

Reyneke B, Cloete TE, Khan S, Khan W (2018) Rainwater harvesting solar pasteurization treatment systems for the provision of an alternative water source in peri-urban informal settlements. Environ Sci Water Res Technol 4:291–302. https://doi.org/10.1039/C7EW00392G

Rivera-Garza M, Olguín M, García-Sosa I et al (2000) Silver supported on natural Mexican zeolite as an antibacterial material. Microporous Mesoporous Mater 39:431–444. https://doi.org/10.1016/S1387-1811(00)00217-1

Rohm and Haas (2008) Ion exchange for dummies, an introduction

Schutte F, Focke W (2006) Handbook for the operation of water treatment works. Water Res Comm Water Inst South Africa

Semenza JC, Roberts L, Henderson A et al (1998) Water distribution system and diarrheal disease transmission: a case study in Uzbekistan. Am J Trop Med Hyg 59:941–946

Sisson AJ, Wampler PJ, Rediske RR et al (2013) Long-term field performance of biosand filters in the Artibonite Valley, Haiti. Am J Trop Med Hyg 88:862–867. https://doi.org/10.4269/ajtmh.12-0345

Sobsey MD, Stauber CE, Casanova LM et al (2008) Point of use household drinking water filtration: a practical, effective solution for providing sustained access to safe drinking water in the developing world. Environ Sci Technol 42:4261–4267. https://doi.org/10.1021/es702746n

Sobsey MD, Water S, Organization WH (2002) Managing water in the home: accelerated health gains from improved water supply. World Health Organization, Geneva

Sondi I, Salopek-Sondi B (2004) Silver nanoparticles as antimicrobial agent: a case study on E. coli as a model for gram-negative bacteria. J Colloid Interface Sci 275:177–182. https://doi.org/10.1016/j.jcis.2004.02.012

Souter P, GD Cruickshank, MZ Tankerville, et al (2003) Evaluation of a New Water Treatment for Point of Use Household Applications to Remove Microorganisms and Arsenic from Drinking Water

Stauber C, Kominek B, Liang K et al (2012) Evaluation of the impact of the plastic BioSand filter on health and drinking water quality in rural tamale, Ghana. Int J Environ Res Public Health 9:3806–3823. https://doi.org/10.3390/ijerph9113806

Stauber CE, Elliott MA, Koksal F et al (2006) Characterisation of the biosand filter for E. coli reductions from household drinking water under controlled laboratory and field use conditions. Water Sci Technol 54:1–7. https://doi.org/10.2166/wst.2006.440

Suthar S (2011) Contaminated drinking water and rural health perspectives in Rajasthan, India: an overview of recent case studies. Environ Monit Assess 173:837–849. https://doi.org/10.1007/s10661-010-1427-2

Tian Y, Hu H, Zhang J (2017) Solution to water resource scarcity: water reclamation and reuse. Environ Sci Pollut Res 24:5095–5097. https://doi.org/10.1007/s11356-016-8331-4

Tiwari S-SK, Schmidt W-P, Darby J et al (2009) Intermittent slow sand filtration for preventing diarrhoea among children in Kenyan households using unimproved water sources: randomized controlled trial. Trop Med Int Heal 14:1374–1382. https://doi.org/10.1111/j.1365-3156.2009.02381.x

Top A, Ülkü S (2004) Silver, zinc, and copper exchange in a Na-clinoptilolite and resulting effect on antibacterial activity. Appl Clay Sci 27:13–19

Van Halem D (2006) Ceramic silver impregnated pot filters for household drinking water treatment in developing countries

van Halem D, van der Laan H, Heijman SGJ et al (2009) Assessing the sustainability of the silver-impregnated ceramic pot filter for low-cost household drinking water treatment. Phys Chem Earth, Parts A/B/C 34:36–42. https://doi.org/10.1016/j.pce.2008.01.005

Varkey A, Dlamini M (2012) Point-of-use water purification using clay pot water filters and copper mesh

Waldman RJ, Mintz ED, Papowitz HE (2013) The cure for cholera — improving access to safe water and sanitation. N Engl J Med 368:592–594. https://doi.org/10.1056/NEJMp1214179

Wang S, Hou Y, Lin S, Wang X (2014) Water oxidation electrocatalysis by a zeolitic imidazolate framework. Nanoscale 6:9930. https://doi.org/10.1039/C4NR02399D

Wegelin M, Canonica S, Mechsner K, et al (1994) Solar water disinfection: Scope of the process and analysis of radiation experiments

WHO/UNICEF (2013) Progress on sanitation and drinking water-2013 update. World Health Organization, Geneva

WHO/UNICEF (2014) Progress on drinking water and sanitation: 2014 update. World Health Organization, Geneva

Wilhelm N, Kaufmann A, Blanton E, Lantagne D (2018) Sodium hypochlorite dosage for household and emergency water treatment: updated recommendations. J Water Health 16:112–125. https://doi.org/10.2166/wh.2017.012

Wolf J, Prüss-Ustün A, Cumming O et al (2014) Systematic review: assessing the impact of drinking water and sanitation on diarrhoeal disease in low- and middle-income settings: systematic review and meta-regression. Trop Med Int Heal 19:928–942. https://doi.org/10.1111/tmi.12331

Wrigley T (2007) Microbial counts and pesticide concentrations in drinking water after alum flocculation of channel feed water at the household level, in Vinh Long Province, Vietnam. J Water Health 5:171–178

Chapter 12
Different Approaches to the Biodegradation of Hydrocarbons in Halophilic and Non-halophilic Environments

David Olugbenga Adetitun

12.1 Introduction

The right to a satisfactorily clean and healthy environment has been enshrined in the constitution of many countries around the world (Ebeku 2003). However, environmental pollution remains an enormous challenge for humanity. This process is aggravated by increasing populations and the demand for more land, and services to meet the needs of this growing population. The pollutants that are introduced in the environment could be chemical or biological. Prominent among these pollutants are hydrocarbons that have become a significant threat to humans and the environment (Kumar and Yadav 2018). While oil pollution has been reported globally, the problem is escalating in the African continent due to the recent discovery of numerous oil reserves in the continent (Adekola et al. 2017) with some countries like Equatorial Guinea depending entirely on oil for sustaining her economy (UNICEF 2020). The adverse human and ecosystem health impacts caused by oil pollution in Africa have been reported by many researchers. For example, the oil spill in the Bonga Oil field resulted in severe adverse effects on fish life in the Gulf of Guinea (Okafor-Yarwood 2018) with similar pollution resulting in decreased fish production (Osuagwu and Olaifa 2018) and increased infant mortality rates (Bruederle and Hodler 2019) in Nigeria. Another study in Nigeria also reported increased disease symptoms and environmental distress such as worry, annoyance, and intolerance, associated with the Niger Delta oil pollution (Nriagu et al. 2016). Similarly, it was reported that oil spills in the Niger Delta resulted in elevated contamination in water used by nearby communities, a situation that could have severe consequences on the health and socioeconomic status of the surrounding populations (Ipingbemi 2009). Similar

D. O. Adetitun (✉)
Department of Microbiology, Faculty of Life Sciences, University of Ilorin, Ilorin, Nigeria
e-mail: adetitun.do@unilorin.edu.ng

© Springer Nature Switzerland AG 2020
A. L. K. Abia, G. R. Lanza (eds.), *Current Microbiological Research in Africa*,
https://doi.org/10.1007/978-3-030-35296-7_12

high concentrations of heavy metals that could be detrimental to aquatic life have been reported in the Mediterranean coastline of Morocco (Er-Raioui et al. 2009) and the Bakassi Peninsula of Cameroon (Abdourahimi et al. 2016). Research on oil pollution in Africa has, however, declined over the years, like in South Africa, for example (O'Donoghue and Marshall 2003), despite the devastating effects of oil pollution in the African continent.

Numerous methods have been used to remediate oil-polluted environments, and these include physicochemical, thermal, and biological processes (Erdogan and Karaca 2011). Among these methods, remediation of oil-polluted environments using microorganisms has gained more considerable attention due to their simplicity, environmental friendliness, and cost-effectiveness (Kumar and Yadav 2018).

12.2 Biodegradation of Hydrocarbons in Non-halophilic Sites

Biodegradation by naturally occurring microorganisms represents one of the major methods by which crude oil and its component hydrocarbon elements can be removed from the ecosystem (Ulrici 2000). This approach is more economical than other remediation methods (Kumar and Yadav 2018). Biodegradation is the biologically catalyzed reduction of complex chemicals to harmless byproducts (Erdogan and Karaca 2011). Biodegradation is the method by which organic substances are broken down into smaller compounds by living microorganisms (Marinescu et al. 2009). The process of biodegradation ultimately generates carbon dioxide and water. The process is termed mineralization, and it involves making elements available in the soil in forms that plants and microbes can utilize.

On the other hand, immobilization involves the assimilation of the mineralized compounds into the protoplasm of the microbes and plants. However, in most cases, the term biodegradation is usually described as nearly any biologically mediated modification of a substrate (Bennet et al. 2002). Comprehending the process of biodegradation requires an understanding of the microorganisms that make the process work. The microorganisms transform the substance through metabolic or enzymatic processes. It is based on the processes of growth and cometabolism.

For growth to occur, an organic waste product is employed as the sole supplier of carbon and energy. This process results in the mineralization of organic pollutants. Cometabolism is defined as the metabolism of an organic compound in the presence of a growth substrate that is used as the primary carbon and energy source (Fritsche and Hofrichter 2008). Numerous organisms, including fungi, bacteria, and yeasts, are involved in the biodegradation process. Algae and protozoa reports are quite scanty concerning their involvement in biodegradation (Das and Chandran 2011). Biodegradation can be achieved in many ways, but often the end product of the process is water and carbon dioxide (Pramila et al. 2012). The biodegradation of organic material can be in the presence of oxygen (aerobic) or absence of oxygen (anaerobically) (Fritsche and Hofrichter 2008; Mrozik et al. 2003). Biodegradable matter is usually an organic material like plant and animal matter and fossil fuels inclusive. Fossil fuels are products of plants and animal remains.

Some microorganisms have the unique, innate, catabolic ability to transform, degrade, or accumulate an extensive range of compounds including hydrocarbons (e.g., oil), polyaromatic hydrocarbons (PAHs), polychlorinated biphenyls (PCBs), radionuclides, and metals (Leitão 2009). Hydrocarbons are organic compounds whose structures comprise hydrogen element and carbon. Hydrocarbons are seen as cyclic, branched, or straight-chain molecules that occur as aromatic or aliphatic hydrocarbons. The aromatics have benzene (C_6H_6) in their structure, while the aliphatics are seen in three forms: alkanes, alkenes, and alkynes (McMurry 2000). Aromatic hydrocarbons are hydrocarbons that contain one (monoaromatic) or many (polyaromatic) benzene rings (Korenaga et al. 2000). They are highly hydrophobic, with low water solubility and biodegradability (Shemer and Linden 2007). Most of these compounds may be formed during incomplete combustion of organic materials, and as food contaminants during food processing.

12.3   Some Hydrocarbons and Other Chemical Pollutants Biodegraded in the Environment

Polycyclic aromatic hydrocarbons (PAHs) are of significant environmental concern because of their toxicity, low volatility, resistance to microbial degradation, and high affinity for sediments. Some of them are also carcinogenic and bioaccumulate in aquatic organisms (Shemer and Linden 2007). They are an important pollutant class of hydrophobic organic contaminants (HOCs) widely found in air, soil, and sediments. The core origin of PAH pollution is industries during the production process (Mrozik et al. 2003). They have been studied with increasing interest for more than two decades because of more findings of their toxic effects, environmental persistence, and prevalence (Okere and Semple 2012). PAHs adsorb to organic-rich soils and sediments, accumulate in fish and other water life forms and may be transferred to humans through the consumption of seafood (Mrozik et al. 2003). Biodegradation of PAHs is assumed on the one hand to be a part of the normal steps of the carbon cycle and on the other as the removal of anthropogenic pollutants from the environment. The use of microorganisms for bioremediation of PAH-contaminated ecosystems is an attractive technology for restoration of polluted sites.

Polychlorinated biphenyls (PCBs) are mixtures of artificial organic chemicals. They are nonflammable, chemically stable, have a high boiling point, and have electrical insulating properties. Hence, they have been used in many industrial and commercial applications. These include electrical, heat transfer and hydraulic equipment and many other industrial applications. It follows naturally that PCBs are toxic compounds that could act as negative endocrine transformers and lead to cancer. Therefore, environmental pollution with PCBs is of increasing concern (Seeger et al. 2010).

Pesticides are substances or mixtures meant for preventing, destroying, repelling, or mitigating any pests. Pesticides that are speedily degraded are known as nonpersistent, whereas those that are recalcitrant to degradation are referred to as persistent (Bradman and Whyatt 2005). The popular type of degradation of the pesticides is carried out in the soil by microorganisms, especially fungi and bacteria that use pesticides as carbon and energy source (Vargas 1975).

Dyes are generally utilized in cloth making and many other industries (Raffi et al. 1997; Armağan et al. 2003). Azo dyes, which are aromatic compounds, are the most vital and most populous class of synthetic dyes used commercially (Chen 2006; Vandevivere et al. 1998). These dyes are poorly biodegradable because of their structures and treatment of wastewater containing dyes usually involves physical and chemical methods such as filtration, adsorption, coagulation-flocculation, oxidation, and electrochemical methods (Verma and Madamwar 2003). The success of a biological process for color removal from a given wastewater is dependent on the use of microorganisms that effectively change the color.

Heavy metals are unlike organic contaminants because they cannot be destroyed but must either be changed to a stable form or removed. Bioremediation of metals is actualized through modification or biotransformation. The principles by which microorganisms act on many metals include biosorption, bioleaching, biomineralization, intracellular accumulation, and enzyme-catalyzed transformation (Lloyd and Lovely 2001).

12.4 Distribution of Hydrocarbon-Degrading Microorganisms

The localization of hydrocarbon-degrading bacteria in natural environments is well known. It has also received good mention in literature because they can easily be harnessed for biodegradation in the treatment of oil pollution. Pollution due to transportation, exploration, and sabotage has made isolation of oil-utilizing bacteria rampant. Many researchers have shown that microbial number increased in areas that suffered from oil pollution. The presence of hydrocarbons in the environment frequently brings about a selective enrichment in situ for hydrocarbon-utilizing microorganisms. The information in Table 12.1 shows the response of microbial populations in oil-polluted soil. Hydrocarbon degraders in the soil increased with time and later slowly decreased.

The data in this table shows that:

(a) Seeding an oil-polluted area with microorganisms is generally not necessary.
(b) Genetically engineered microorganisms (GEMs) are not needed for petroleum biodegradation. Nevertheless, under some circumstances, seeding may be advantageous to ensure uniformity of the hydrocarbon breakdown pattern.

Table 12.1   Changes in the number of hydrocarbon degraders in jet fuel-polluted soil

Treatment	Days after treatment	Microorganisms/g of soil
None	0	4×10^4
	28	2×10^4
	112	2×10^4
50 mg jet fuel/g of soil	28	4×10^8
	112	1×10^6

Source: Song and Bartha (1990)

12.5   Changes in Hydrocarbon Composition After Biodegradation

The products of the total biodegradation process are carbon dioxide and water. However, other ways exist for measuring the occurrence of biodegradation. One of such methods is the disappearance of the components of the particular hydrocarbon compound or molecule. In a study on the biodegradation of jet fuel by Adetitun et al. (2018), the chemical composition of the fuel before and after biodegradation was determined by gas chromatography–mass spectrometry (GCMS) analysis (Table 12.2 and Table 12.3). The specific hydrocarbons after the biodegradation process reduced drastically compared with the composition at the start, showing that there had been a significant change in jet fuel composition.

12.6   Determination of Total Petroleum Hydrocarbon

Another way to measure biodegradation is by determining the total petroleum hydrocarbon at the start and end of the biodegradation process. This technique has been reported in the research work of Adetitun et al. (2016b) on the ability of Gram-negative bacilli isolated from soil contaminated with kerosene to biodegrade hydrocarbons. The biodegradation was confirmed by the change in the Total Petroleum Hydrocarbon (TPH) (Table 12.4) using the gravimetric method described by Matthew (2009). Total petroleum hydrocarbon reduced in all cases where petroleum was present. The highest reduction was recorded in 56.0 mL treatment, with 57.4% percentage change from 78.4 mg/kg to 33.4 mg/kg. The least reduction was recorded in 112.0 mL treatment, with 6.3% change from 109.2 mg/kg to 102.3 mg/kg.

The ability to produce biosurfactants is another good way to know if an organism can carry out biodegradation or not. The higher the quantity of biosurfactant produced, the higher the biodegradation ability. Potential organisms can be screened for biosurfactant production qualitatively using a combination of tests such as drop collapse test, blood agar hemolysis test, oil-spreading test, and quantitatively using the emulsification index test (Samuel-Osamoka et al. 2018).

Table 12.2 Chemical composition of jet fuel before biodegradation

S/N	Retention time	Chemical compound	Area %
1	3.588	Cyclohexane, methyl-	1.00
2	4.832	Toluene	0.88
3	5.170	Cyclohexane, 1,3-dimethyl	1.86
4	5.871	Cyclohexane,1,2-dimethyl	0.38
5	6.083	Octane	0.47
6	7.328	Cyclohexane, ethyl	0.84
7	7.472	Cyclohexane,1,3-trimethyl	0.77
8	8.104	Cyclohexane,1,2,4-trimethyl	0.73
9	8.611	Ethylbenzene	0.63
10	8.836	Oxalic acid	0.71
11	8.967	Benzene,1,3-dimethyl	2.28
12	9.142	Heptane,2,5-dimethyl	0.73
13	9.568	Cyclohexane	0.34
14	9.699	Cis-1-ethyl-3-methyl-cyclohexane	1.81
15	9.780	Cyclohexane,1-ethyl-4methyl	1.68
16	9.974	O-xylene	1.46
17	10.287	3-Decy-2-ol	0.78
18	10.443	Nonane	2.66
19	10.554	1-Ethyl-3-methylcyclohexane	1.21
20	11.019	1H-Indene,octahydro	1.54
21	11.194	1-Octadecyne	0.89
22	11.513	1-Dodecanol,3,7,11-trimethyl	3.57
23	11.713	1-Octanol,2-butyl	0.81
24	11.907	Octane,2,6-dimethyl	1.57
25	12.039	Cyclohexane,1-ethyl-2,3-dimethyl	1.23
26	12.182	2-Nonen-1-ol	1.78
27	12.439	2,3,4-Trimethyl-hex-3-enal	1.35
28	12.633	Cyclohexane,1,1,2,3-tetramethyl	3.40
29	13.071	Benzene,1-ethyl-3-methyl	2.80
30	13.351	Dodecanal	0.72
31	13.809	Benzene,1-ethyl-2-methyl-	1.06
32	13.940	m-Menthane	1.12
33	14.028	1-Nonylcycloheptane	1.88
34	14.453	Benzene,1,2,3-trimethyl	6.58
35	14.660	10-Heneicosene	1.62
36	14.897	Decane	4.78
37	15.648	Trans-P-mentha-1(7),8-dien-2-ol	7.28
38	15.998	Decane,5-cyclohexyl	1.78
39	16.111	Cyclohexanecarboxylic acid	0.90
40	16.830	Cyclooctene,1,2-dimethyl	9.13
41	17.193	Cis-P-Mentha-2,8,dien-1-ol	6.85
42	17.474	2-Piperidinone	2.38

(continued)

Table 12.2 (continued)

S/N	Retention time	Chemical compound	Area %
43	18.592	Methanol	3.62
44	18.988	Undecane	3.46
45	19.451	Spiro[3.5]nona-5,7-dien-1-one,trimrthyl	1.78
46	20.258	Hydrocinnamic acid	2.41
47	36.941	Octadecene	1.02
48	37.785	Tetradecene,2,6,10-trimethyl	2.48

Source: Adetitun et al. (2018)

12.6.1 Hemolysis Test

One of the screening tests for the identification of biosurfactant-producing bacteria is the hemolysis test (Carrillo et al. 1996). In this test, each isolate would be streaked on blood agar medium containing 5% v/v blood and incubated at 37 °C for 24–48 h. The essence of this is to assay for hemolytic activity. Plates should be visually inspected for zones of clearance around the colonies, indicating biosurfactant production (Samuel-Osamoka et al. 2018).

12.6.2 Drop Collapse Test

Another test for the screening of biosurfactant production is the drop collapse assay. It relies on the destabilization of a liquid drop on the hydrocarbon surface by a cell-free extract containing a biosurfactant. The test has previously been described by Jain et al. (1991).

12.6.3 Oil-Spreading Test

The oil-spreading test can be carried out as described by Morikawa et al. (2000). In doing the test, 10 μL of crude oil or any particular hydrocarbon of interest would be added to the surface of 40 mL of distilled water in a Petri dish to form a thin oil layer. Then, 10 μL of the culture supernatant would be dropped gently in the center of the oil layer. After a minute, if a biosurfactant is present in the supernatant, the oil will be displaced, and a clearing zone would be formed. The diameter of the clear zone on the oil surface should be measured in triplicates for each isolate. A drop of water is often utilized as a negative control.

Some examples of results obtained with the three tests mentioned above are presented in Table 12.5.

Table 12.3 Chemical composition of degraded jet fuel, facilitated by Gram-negative bacilli and their consortiums

S/n	Chemical compounds	A	B	C	AB	AC	BC	ABC
1	4H-ThioPyrano[4,3:4,5]Furo[2,3-d] pyridine3(6H)-amine,5,8-dimethyl-	40.24	29.46	27.44	59.90	31.48	27.90	47.34
2	2-Thiazolamine,4-(3,4-dimethoxy-phenyl)-5methyl-	25.62	11.65	25.76	24.81	10.68	14.08	24.22
3	Benzo(b)naphtho(2,3-d)thiophene,10-dihydro7-methyl-	0.00	12.08	4.80	0.00	16.22	17.36	1.00
4	Methaqualone	0.00	3.68	0.00	0.00	2.36	0.00	0.00
5	Officinalic acid, methyl ester	0.00	0.00	0.00	0.00	0.00	0.00	0.29
6	1-Ethanone	0.00	0.00	0.00	0.00	0.00	0.00	0.34
7	Anodendrosite E2,monoacetate	4.50	0.00	0.00	0.00	0.00	0.00	3.46
8	Benzimidazole-5-carboxylic acid,2-methyl-1phenyl-	2.88	0.00	3.15	0.00	10.56	0.00	16.37
9	Benz(c) acridine	0.00	0.00	5.60	0.00	0.00	0.00	0.00
10	2,5,Di-t-butyl-4-methoxy-1,4dihydrobenzaldehyde	1.20	0.00	5.20	0.00	0.00	0.00	0.00
11	3,5,6-Trimethyl-P-quinone,2-(2,5dioxotetrahydrofuran-3-yl)thio-	0.00	0.00	1.20	0.00	0.00	0.00	0.00
12	1methyl-2,5-dichloro-1,6-diazaphenaline	0.00	0.00	0.00	1.84	0.00	1.88	0.00
13	Benzoic acid	0.00	0.00	0.00	8.16	0.00	0.00	0.00
14	Acetic acid	0.00	3.26	0.00	2.24	0.00	4.90	0.00
15	1,2-Benzenedicarboxylic acid,4-methyl-5(1methyl)-dimethyl ester	0.00	0.36	0.00	0.00	0.00	0.00	0.00
16	Pregn-4-en-18-oic acid	0.00	1.81	0.00	0.00	0.00	0.00	0.00
17	Benzothiophene-3 carboxamide,4,5,6,7t etrahydro-2-amino-6-tert-butyl-	0.00	4.10	0.00	0.00	0.00	0.00	0.00
18	5-Beta-hydroxymethyl-3-beta-tosylamino cholestane	0.00	0.00	0.00	0.00	0.00	2.36	0.00
19	Pyrido(1,2,9)benzimidazole-4-carbonitrile,3methyl-1-dimetylamino-	0.00	0.00	0.00	0.00	0.00	2.62	0.00
20	(1-Cyclopentenyl)ferrocene perylene	0.00	1.51	0.00	0.89	0.00	0.00	1.03
21	1,2,4-Methenocyclopentlene	0.00	1.21	0.00	1.64	0.00	0.00	0.00
22	6,7,8,9-Tetrahydro-1,2,3-trimethoxy-9methyl-5H-benzocycloheptane	0.00	2.16	0.00	0.00	0.00	0.00	0.00
23	3-(y-methylaminopropyl)-5(4-bromoophenyl)-2-methyl-2H-pyrazole	0.00	0.00	0.00	0.00	0.00	0.00	2.11
24	Methyl7-(5-(methoxycarbonyl) methyl1,2furyl)heptanoate	0.00	0.00	0.00	0.00	0.00	5.65	0.00
25	Cinnamic acid	0.00	2.16	0.00	1.01	0.86	0.00	1.16

A = *Aeromonas hydrophila*, B = *Vibrio parahaemolyticus*, C = *Actinobacillus* sp. AB = *Aeromonas hydrophila* and *Vibrio parahaemolyticus*, AC = *Aeromonas hydrophila* and *Actinobacillus* sp. BC = *Vibrio parahaemolyticus* and *Actinobacillus*, sp., ABC = *Aeromonas hydrophila*, *Vibrio parahaemolyticus,* and *Actinobacillus* sp.
Source: Adetitun et al. (2018)

Table 12.4 Total petroleum hydrocarbon at the start and end of the experiment

Treatment (mL)	TPH (mg/kg) at week 1	TPH (mg/kg) at week 8	Percentage change (%)
0.0	0.0	0.0	0.0
7.0	28.4	18.4	35.2
14.0	32.4	22.1	31.8
21.0	38.9	27.8	28.5
56.0	78.4	33.4	57.4
112.0	109.2	102.3	6.3
168.0	118.8	108.6	8.6
224.0	210.9	132.4	37.2

Source: Adetitun et al. (2016b)

Table 12.5 Summary of results obtained for some microorganisms using the hemolytic activity, drop Collapse test, and oil-spreading test

S/n	Bacteria	Hemolytic activity	Drop collapse test	Oil-spreading test
1	*Stenotrophomonas maltophilia* T7D7	β-Hemolytic	+	11 ± 1.0^{de}
2	*Pseudomonas aeruginosa* TCSS2	β-Hemolytic	+	10 ± 0.5^{cd}
3	*Pseudomonas aeruginosa* BP4	β-Hemolytic	+	10 ± 0.5^{cd}
4	*Pseudomonas aeruginosa* GB24	β-Hemolytic	+	12 ± 1.0^{e}
5	*Alcaligenes* sp. CIFRID-TSB1	α-Hemolytic	−	0 ± 0.0^{a}
6	*Bacillus subtilis* B-28	β-Hemolytic	+	8 ± 1.0^{b}
7	*Bacillus* sp. AH6	γ-Hemolytic	−	0 ± 0.0^{a}
8	*Alcaligenes faecalis* 138C-1	α-Hemolytic	−	0 ± 0.0^{a}
9	*Bacillus safensis* HKG 214	β-Hemolytic	−	8 ± 1.0^{b}
10	*Alcaligenes faecalis* RAJ4	α-Hemolytic	−	0 ± 0.0^{a}
11	*Bacillus pumilus* TW3	α-Hemolytic	+	9 ± 1.0^{bc}
12	*Bacillus subtilis* Y2	β-Hemolytic	+	9 ± 1.0^{bc}
13	*Bacillus* sp. KYLS-CU05	γ-Hemolytic	−	0 ± 0.0^{a}
14	*Pseudomonas* sp. YA6 16S	β-Hemolytic	+	10 ± 1.0^{cd}
15	*Klebsiella Pneumoniae* NGB-FR75	α-Hemolytic	−	0 ± 0.0^{a}

Key: Each value is a mean of three determinations ± standard deviation. Different superscripts are significantly different ($p < 0.05$). Source: Samuel-Osamoka et al. (2018)

12.6.4 Emulsification Index (E24%)

The emulsification index (E24%) has previously been described by Batista et al. (2006). In brief, this method involves centrifugation at 6000 rpm for 20 min to separate the biosurfactant from microbial cells, yielding a biosurfactant cell-free extract. Equal volumes of the cell-free biosurfactant are then added to a test tube. The mix-

Table 12.6 Emulsification index test

S/n	Isolates	Total height of solution (mm)	Height of emulsion (mm)	Percentage emulsification (%)
1	*Stenotrophomonas maltophilia* T7D7	34 ± 1.00^b	25 ± 0.50^e	73.50 ± 0.50^f
2	*Pseudomonas aeruginosa* TCSS2	34 ± 1.00^b	22 ± 0.10^{cd}	64.70 ± 1.00^b
3	*Pseudomonas aeruginosa* BP4	34 ± 2.00^b	23 ± 1.00^d	67.65 ± 0.01^e
4	*Pseudomonas aeruginosa* GB24	33 ± 1.00^{ab}	22 ± 1.00^{cd}	66.67 ± 0.01^d
5	*Bacillus subtilis* B-28	31 ± 1.00^a	20 ± 1.00^b	64.52 ± 0.02^b
6	*Bacillus subtilis* Y2	32 ± 0.10^{ab}	21 ± 0.80^{bc}	65.63 ± 0.03^c
7	*Pseudomonas* sp. YA616S	32 ± 0.00^{ab}	21 ± 1.00^{bc}	65.63 ± 0.02^c
8	Negative control	33 ± 1.00^{ab}	0 ± 0.00^a	0 ± 0.00^a

Key: Each value is a mean of three determinations ± standard deviation. Different superscripts on the columns are significantly different ($p < 0.05$). Source: Samuel-Osamoka et al. (2018)

ture of the biosurfactant and crude oil (or any other hydrocarbon) (2:2) is agitated for about 2 min using a vortex mixer and then stabilized for 24 h. The E24% is determined by measuring the column height of the emulsified oil against its total height and then multiplying by 100. For example, in the study of Samuel-Osamoka et al. (2018), the percentage emulsification for all the isolates tested using this method was above 50% (Table 12.6).

Another approach to estimating biodegradability by microbes is the use of growth measurements in culture fluids with the hydrocarbon as the sole source of energy and carbon. In an experiment by Adetitun et al. (2016a), the time course of degradation of gasoline was carried out using mineral salts medium (MSM) as described by Adetitun et al. (2014). The increase in cell number of bacterial species on hydrocarbon substrates was done by inoculating each of the bacterial cultures into a 250 mL Erlenmeyer flask containing 99 mL of mineral salts medium (MSM) as previously used by Adetitun et al. (2014) and Oboh et al. (2006). The composition of the MSM in g/L was 0.5 KH_2PO_4; 0.3 KNO_3; 0.2 $MgSO_4.7H_2O$; 1.4 Na_2HPO_4; and 1.0 $(NH_4)2SO_4$. The hydrogen ion concentration of the medium was adjusted to 7.0. Each flask was supplemented with 1 mL of gasoline as the sole carbon supply. Control flasks with no gasoline were set up equally. The culture flasks were incubated for 168 h at 37 °C in a rotatory incubator shaker. The optical density (OD 600 nm), colony forming unit (CFU), and hydrogen ion concentration of the culture fluids were monitored every 24 h as indicators of biodegradation. The results (Fig. 12.1 and Fig. 12.2) show that the active total viable count and optical density increased while the pH dropped. It has been argued that pH plays no role in the measurement of biodegradation. This is because the pH of the reaction mixture will change depending on the metabolites and products released into the medium.

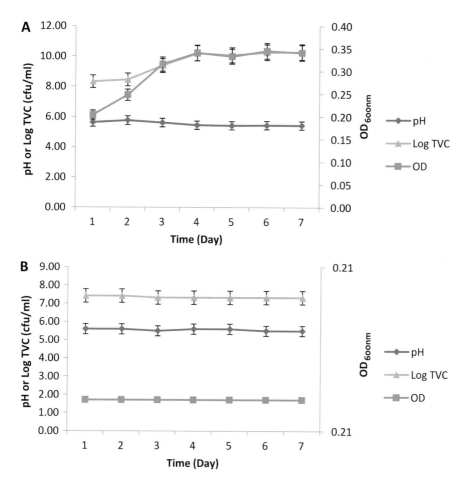

Fig. 12.1 Growth configuration of *Ochrobactrum* sp. (Closest relative—*Ochrobactrum* sp. VH-19) on gasoline. (**a**) With gasoline; (**b**) without gasoline (Source: Adetitun et al. 2016a)

12.7   Biodegradation of Hydrocarbons in Halophilic Ecosystems

Most oil exploration and production activities occur offshore in the marine habitat. This is why it is important to direct attention to microorganisms that can tolerate high salinity in the process of biodegradation and bioremediation. Hypersaline
niches can occur naturally or can be artificially constructed. The artificial construction is usually done by salt industries (De la Huz et al. 2005). Reddening in
the sea is caused by archaea which inhabit high-salinity waters. At low salt concentrations, a more prominent contribution of bacteria and archaea is expected.
Recent studies have nevertheless shown that at high salt concentrations, archaea

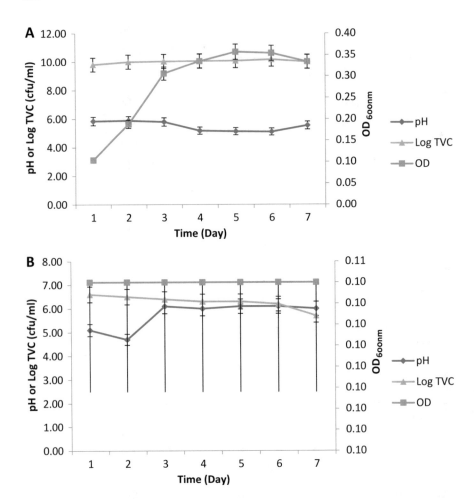

Fig. 12.2 Growth configuration of *Pseudomonas* sp. (Closest relative—*Pseudomonas aeruginosa* strain) on gasoline. (**a**) With gasoline; (**b**) without gasoline (Source: Adetitun et al. 2016a)

are not as predominant as thought. The isolation of *Salinibacter ruber*, a red-pigmented extremely halophilic bacterium, has shown that reddening in seawater is not exclusively caused by halophilic archaea. The most detailed studies of hypersaline habitats are from saline lakes, marine salterns, and saline soils (Bodaker et al. 2010; Pagaling et al. 2009).

Halophiles have solved the problem of coping with salt and other stresses in the environment. Studies of microbial life in high-salt environments can answer many basic questions on the adaptations of microorganisms to their environments. The salinity of salt lakes, for example, differs from place to place. This may be attributed to the amount of fresh water available for dilution. In the Dead Sea, the salinity is in the range of 27% to 29%, while in Lake Assal in Africa it is from 34% to 36%, and that of the Romanian salt lakes range from 6% to 25% salinity (Oren 2002; Enache et al. 2008; Madigan et al. 2009).

The understanding of the genetics, metabolism, ecology, and biochemistry of oil-eating microbes is increasing yearly with the advent of better equipment for studying them. No single microorganism can degrade petroleum. Therefore, a variety of microbes degrade specific hydrocarbons in succession. Biodegradation of oil in the environment depends on the oil type, weather, and other factors. Dominant among the factors is the type of microorganisms involved and whether the components of the hydrocarbons are aliphatics or aromatics. Even among the aromatics, the polycyclic aromatic hydrocarbons (PAHs) are quite challenging to degrade.

Some organisms have been able to grow and even thrive in a hypersaline environment, which many other organisms are unable to tolerate. Those that can survive are termed halophiles (Ma et al. 2010). These organisms use two distinct mechanisms to prevent desiccation via osmotic outflow of water out of their cytoplasm (Weinisch et al. 2018). Both techniques work by increasing the internal osmolality of the cell. The first method is used by most halophilic bacteria, some archaea, algae, molds, and yeasts, and usually involves the accumulation of organic molecules in the cytoplasm. These are osmoprotectants which are known as compatible solutes and can either be synthesized or accumulated from the environment (Santos and Da Costa 2002). The most compatible solutes are neutral ionic. The presence of hydrocarbons in saline aquatic habitats is a great disadvantage to organisms. Biodegradation of hydrocarbons in these environments is, therefore, the most appropriate way to clean up such polluted sites. Le Borgne et al. (2008) reported that the results of the degradation of hydrocarbon in saline or hypersaline sites could differ. Some workers reported the negative influence of salinity on degradation while others report positive effects.

Adding salt to freshwater or soil will affect the biodegradation because such sites had not been previously exposed to salinity. Bacteria initially present in such places cannot utilize salt, and the addition of salt will affect their metabolism (Margesin and Schinner 2001; Minai-Tehrani et al. 2009; Ulrich et al. 2009). This is an example of the negative effect of salinity on biodegradation. Another adverse effect is seen in sites like mangroves and intertidal mats where halotolerant and slightly halophilic microorganisms tend to be dominant (Diaz et al. 2002). In a study on the microbial mats of an Arabian Gulf area that was chronically exposed to oil spills and subject to high daily salinity and temperature changes, such as 150 g/L salts and 40 °C (low tide) to 50 g/L salts and 25 °C (high tide), the degradation rates of some hydrocarbons were evaluated using the following salinities: 0 g/L, 35 g/L, 50 g/L, 80 g/L, 120 g/L, and 160 g/L. It was found that about 100% of added phenanthrene and dibenzothiophene were degraded at 35 g/L while the most efficient degradation results for pristine (about 75%) and n-octadecane (about 85%) occurred between salinities of 35 g/L and 80 g/L (Abed et al. 2006).

On the positive side, a halophilic archaeon degraded hydrocarbons at the highest salinities that was tested more efficiently (Tapilatu et al. 2010). Percentages of biodegradation of eicosane increased from about 9% in medium with 146 g/L NaCl to 65% at 204 g/L NaCl after 30 days incubation. Such haloarchaeon presented optimal growth at 45 °C and capability to degrade a few aliphatic and aromatic hydrocarbons, especially tetradecane whose initial concentration decreased by 88% after 4 weeks.

McKew et al. (2007) discovered that:

(a) *Thalassolituus* was the dominant species when n-alkanes with 12–32 carbons were added. The organism was not found when decane was the only alkane added to seawater.
(b) *Alcanivorax* was predominant when the branched alkane, pristane, was supplied, but not detected when it was not.
(c) *Cycloclasticus* was dominant with most polycyclic aromatic hydrocarbons (PAHs) but was not isolated when fluorine was supplied.

From the preceding, it is clear that different organisms are required for the degradation of specific hydrocarbons. High salinity is required for biodegradation of hydrocarbons, but at extremely high salinity levels, the rate of biodegradation is drastically reduced.

12.8   Enzymes Involved in Biodegradation of Hydrocarbons

Biodegradation of hydrocarbons is mediated by enzymes, which are known to be biological catalysts. Enzymes are specific in action; hence, different enzymes are specifically involved in different aspects of biodegradation and for different hydrocarbons. Some enzymes involved in the biodegradation of petroleum hydrocarbons together with the degrading microorganisms and substrates are shown in Table 12.7.

Table 12.7 Enzymes involved in the biodegradation of petroleum hydrocarbons

S/N	Enzyme	Substrate	Degrading microorganisms	References
1	Soluble monooxygenases	C1-C8 alkanes, alkenes, and cycloalkanes	*Methylococcus* sp., *Methylocella* sp., *Methylosinus* sp., *Methylomonas* sp.	McDonald et al. (2006)
2	Particulate methane monooxygenase	C1-C5 halogenated alkanes and cycloalkanes	*Methylocystis* sp., *Methylococcus* sp., *Methylobacter* sp.,	McDonald et al. (2006)
3	Dioxygenase	C10-C50	*Acinetobacter* sp.	Maeng et al. (1996)
4	Bacterial P450 system	C5-C16	*Acinetobacter* sp., *Caulobacter, Mycobacterium*	van Beilen et al. (2006)
5	Eukaryotic P450	C10-C16	*Candida maltose, Yarrowiali polytica Candida tropicalis*	Lida et al. (2000)
6	AlkB enzyme	C5-C15 alkanes, fatty acids, benzenes	*Pseudomonas* sp.	Zampolli et al. (2014)

12.9 Enrichments in Biodegradation

Microorganisms that are involved in biodegradation can be isolated using enrichment methods (Mancini et al. 2002). Enrichments help to narrow down the microorganisms that can utilize a particular hydrocarbon. Enrichment can involve repeated feeding of a polluted sample with specific hydrocarbon and measuring its disappearance and the total viable counts of the microbes present. When the microorganisms have reduced the hydrocarbon, more hydrocarbon is fed into the reaction mixture (Rios-Hernandez et al. 2003). Spectrophotometer, denaturing gradient gel electrophoresis (DGGE), or gas chromatography can be used to observe the reduction in the hydrocarbon concentration over time. Using nonenrichment methods makes some microorganisms that can degrade hydrocarbons to be lost before the biodegradation process even begins. This is because the process of isolation using nonenrichments makes for competition. Oil-degrading organisms may not be able to compete well with nonoil degraders (Liu et al. 2017). Enrichment of cultures is an excellent way to narrow down potential hydrocarbonoclastic microorganisms. Enrichment also makes it easier to isolate all sorts of microorganisms and not bacteria or fungi alone.

12.10 Conclusion

The effects of oil pollution in Africa are devastating and long-lasting. For example, the United Nations had reported that it would take 30 years for the oil-polluted Niger Delta of Nigeria to be thoroughly cleaned up, assuming that the pollution stops at the point it has reached today. However, as at today, not much has been heard about how far it has gone as is also the case in many other affected African countries mainly due to the lack of adequate research on improved ways to address the situation. Thus, there is a need for an integrated approach involving government entities, the private sector, and funding bodies that would ensure that the applications of the findings from the few existing studies are successfully implemented. Also, it must be stated here that the different methods of biodegradation should involve experts from different fields of engineering as the microbiologists alone cannot do the task. The microbiologist will identify and make available the best cultures for degradation. The engineers are the ones who will apply the cultures and others to make it work. Aside from cultures, the need for venting cannot be overemphasized. Through such an integrated approach the different approaches of biodegradation and bioremediation (enhanced biodegradation) mentioned in this chapter could be explored, harnessed, and upscaled for sustainable development in the Niger Delta and other African countries affected by oil pollution.

References

Abdourahimi S, Fantong WY et al (2016) Environmental pollution by metals in the oil bearing Bakassi Peninsula, Cameroon. Carpathian J Earth Environ Sci 11:529–538

Abed RMM, Al-Thukair A, De Beer D (2006) Bacterial diversity of a cyanobacterial mat degrading petroleum compounds at elevated salinities and temperatures. FEMS Microb Ecol 57:290–301

Adekola J, Fischbacher-Smith M, Fischbacher-Smith D, Adekola O (2017) Health risks from environmental degradation in the Niger Delta, Nigeria. Environ Plan C Gov Policy 35:334–354. https://doi.org/10.1177/0263774X16661720

Adetitun DO, Olayemi AB, Kolawole OM (2014) Hydrocarbon degrading capability of bacteria isolated from a maize planted kerosene contaminated Ilorin alfisol. Biokemistri 26(1):13–18

Adetitun DO, Olayemi AB, Kolawole OM, Fathepure B (2016a) Molecular identification of hydrocarbon-degrading bacteria isolated from alfisol-loam experimentally contaminated with gasoline. Biokemistri 28(3):135–143

Adetitun DO, Awoyemi OD, Adebisi OO, Kolawole OM, Olayemi AB (2016b) Biodegradative activities of some gram-negative bacilli isolated from kerosene treated soil grown with cowpea (Vigna unguiculata). Agrosearch 16(1):41–57

Adetitun DO, Akinmayowa OV, Atolani O, Olayemi AB (2018) Biodegradation of jet fuel by three gram-negative bacilli isolated from kerosene contaminated soil. Pollution 4(2):291–303

Armağan B, Özdemir O, Turan M, Çelik MS (2003) The removal of reactive azo dyes by natural and modified zeolites. J Chem Technol Biotechnol 78:725–732. https://doi.org/10.1002/jctb.844

Batista SB, Mounteer AH, Amorim FR, Totola MR (2006) Isolation and characterization of biosurfactant/bioemulsifier-producing Bacteria from petroleum contaminated sites. Bioresour Technol 97:868–875

Bennet JW, Wunch KG, Faison BD (2002) Use of Fungi biodegradation. In: Manual of environmental microbiology, vol 2. ASM Press, Washington, D.C., pp 960–971

Bodaker II, Sharon MT, Suzuki R, Reingersch M, Shmoish E, Reishcheva ML, Sogin M, Rosenberg S, Belkin A, Oren, Béjà O (2010) The dying dead sea: comparative community genomics in an increasingly extreme environment. ISME J 4:399–407

Bradman A, Whyatt RM (2005) Characterizing exposures to nonpersistent pesticides during pregnancy and early childhood in the National Children's study: a review of monitoring and measurement methodologies. Environ Health Perspect 113:1092–1099. https://doi.org/10.1289/ehp.7769

Bruederle A, Hodler R (2019) Effect of oil spills on infant mortality in Nigeria. Proc Natl Acad Sci 116:5467–5471. https://doi.org/10.1073/pnas.1818303116

Carrillo P, Madaraz C, Pitta-Alvarez S, Giulietti A (1996) Isolation and selection of bio surfactant producing Bacteria. World J Microbiol Biotechnol 12:82–84

Chen H (2006) Recent advances in azo dye degrading enzyme research. Curr Protein Pept Sci 7:101–111

Das N, Chandran P (2011) Microbial degradation of petroleum hydrocarbon contaminants: an overview SAGE-Hindawi access to research biotechnology. Theatr Res Int:13

De la Huz R, Lastra M, Junoy J, Catellanos C, Vieitez JM (2005) Biological impacts of oil pollution and cleaning in the intertidal zone of exposed sandy beaches: preliminary study of the 'prestige' oil spill. Estuarine Coast Shelf Sci 65:19–29

Diaz MP, Boyd KG, Grigson SGW, Burgess JG (2002) Biodegradation of crude oil across a wide range of salinities by an extremely halotolerant bacterial consortium MPD-M, immobilized onto polypropylene fibers. Biotechnol Bioeng 79(2):145–153

Ebeku KSA (2003) The right to a satisfactory environment and the African commission. African Hum Rights Law J 3:149–166

Enache M, Itoh T, Kamekura M, Popescu G, Dumitru L (2008) Halophilic archaea isolated from man made young (200 years) Salt Lakes in Slanic Prahova. Romania. Central Eur J Biol Bucharest 3:388–395

Erdogan EE, Karaca A (2011) Bioremediation of crude oil polluted soils. Asian J Biotechnol 3:206–213

Er-Raioui H, Bouzid S, Marhraoui M, Saliot A (2009) Hydrocarbon pollution of the Mediterranean coastline of Morocco. Ocean Coast Manag 52:124–129. https://doi.org/10.1016/j.ocecoaman.2008.10.006

Fritsche W, Hofrichter M (2008) In: Rehm H-J, Reed G (eds) Aerobic degradation by microorganisms in biotechnology set, 2nd edn. Wiley-VCH Verlag GmbH, Weinheim

Ipingbemi O (2009) Socio-economic implications and environmental effects of oil spillage in some communities in the Niger delta. J Integr Environ Sci 6:7–23. https://doi.org/10.1080/15693430802650449

Jain D, Collins-Thompson D, Lee H (1991) A collapsing test for screening surfactant producing microorganisms. J Microbiol Methods 13(4):271–279

Korenaga T, Liu X, Tsukiyama Y (2000) Dynamics analysis for emission sources of polycyclic aromatic hydrocarbons in Tokushima soils. J Health Sci 46:380–384

Kumar R, Yadav P (2018) Novel and cost-effective Technologies for Hydrocarbon Bioremediation. In: Kumar R, Yadav P (eds) Microb. Action Hydrocarb. Springer, Singapore, pp 543–565

Le Borgne S, Paniagua D, Vazquez-Duhalt R (2008) Biodegradation of organic pollutants by Halophilic Bacteria and Archaea. J Mol Microb Biotechnol 15(2–3):74–92

Leitão AL (2009) Potential of *Penicillium* species in the bioremediation field. Int J Environ Res 6:1393–1417

Lida T, Sumita T, Ohta A, Takagi M (2000) The cytochrome P450 multigene family of an n-alkane-assimilating *Yeast, Yarrowia lipolytica*: cloning and characterization of genes coding for new cyp52 family members. Yeast 16(12):1077–1087

Liu J, Bacosa HP, Liu Z (2017) Potential environmental factors affecting oil-degrading bacterial populations in deep and surface waters of the northern Gulf of Mexico. Front Microbiol 7:1–14. https://doi.org/10.3389/fmicb.2016.02131

Lloyd JR, Lovely DR (2001) Microbial detoxification of metals and radionuclides. Curr Opin Biotechnol 12:248–253

Ma Y, Galinski EA, Grant WD et al (2010) Halophiles 2010: Life in saline environments. Appl Environ Microbiol 76:6971–6981. https://doi.org/10.1128/AEM.01868-10

Madigan MT, Martinko JM, Dunlap PV, Clark DP (2009) Brock biology of microorganisms, 12th edn. Person Benjamin Cummings, New York

Maeng JHO, Sakai Y, Tani Y, Kato N (1996) Isolation and characterization of a novel oxygenase that catalyses the first step of n-alkane oxidation in *Acinetobacter* sp. strain M-1. J Bacteriol 178(13):3695–3700

Mancini SA, Lacrampe-Couloume G, Jonker H et al (2002) Hydrogen isotopic enrichment: an indicator of biodegradation at a petroleum hydrocarbon contaminated field site. Environ Sci Technol 36:2464–2470. https://doi.org/10.1021/es011253a

Margesin R, Schinner F (2001) Biodegradation and bioremediation of hydrocarbons in extreme environments. Appl Environ Microbiol 56:650–663

Marinescu M, Dumitru M, Lacatusu A (2009) Biodegradation of petroleum hydrocarbons in an artificial polluted soil. Res J Agri Sci 41(2)

Matthew M (2009) A Comparison Study of Gravimetric and Ultraviolet Fluorescence Methods for the analysis of Total Petroleum Hydrocarbons in surface water. Civil Engineering Master's Theses

McDonald IR, Miguez CB, Rogge G, Bourque D, Wendlandt KD, Groleau D, Murrel J (2006) Diversity of soluble methane monooxygenase-containing methanotrophs isolated from polluted environments. FEMS Microbiol Lett 255(2):225–232

McKew BA, Coulon F, Osborn AM, Timmis KN, McGenity TJ (2007) Determining the identity and roles of oil-metabolizing marine bacteria from the Thames Estuary, UK. Environ Microbiol 9:165–176

McMurry J (2000) In aromatic hydrocarbons. In: Organic chemistry, vol 5. Thomas learning, New York, pp 120–180

Minai-Tehrani D, Minuoi S, Herfatmanesh A (2009) Effect of salinity on biodegradation of poly-cyclic aromatic hydrocarbons (PAHs) of heavy crude oil in soil. Bull Environ Contam Toxicol 82:179–184

Morikawa M, Hirata Y, Imanaka T (2000) A study on the structure-function relationship of the lipopeptide biosurfactant. Biochim Biophys Acta 1488:211–218

Mrozik A, Piotrowska-Seget Z, Labuzek S (2003) Bacterial degradation and bioremediation of polycyclic aromatic hydrocarbons. Pol J Environ Stud 12(1):15–25

Nriagu J, Udofia EA, Ekong I, Ebuk G (2016) Health risks associated with oil pollution in the Niger Delta, Nigeria. Int J Environ Res Public Health 13:1–23. https://doi.org/10.3390/ijerph13030346

O'Donoghue S, Marshall DJ (2003) Marine pollution research in South Africa: a status report. S Afr J Sci 99:349–356

Oboh BO, Ilori MO, Akinyemi JO, Adebusoye SA (2006) Hydrocarbon-degrading potentials of Bacteria Isolated from a Nigerian Bitumen (Tarsand) Deposit. Nature and Science 4:51–57

Okafor-Yarwood I (2018) The effects of oil pollution on the marine environment in the Gulf Of Guinea-The Bonga oil field example. Transnatl Leg Theory 9:254–271. https://doi.org/10.1080/20414005.2018.1562287

Okere UV, Semple KT (2012) Biodegradation of PAHs in 'Pristine' Soils from Different Climatic Regions. J Bioremed Biodegrad

Oren A (2002) Diversity of halophilic microorganisms: environments, phylogeny, physiology, and applications. J Ind Microbiol Biotechnol 28:56–63

Osuagwu ES, Olaifa E (2018) Effects of oil spills on fish production in the Niger Delta. PLoS One 13:1–14. https://doi.org/10.1371/journal.pone.0205114

Pagaling E, Wang HZ, Venables M, Wallace A, Grant WD, Cowan DA, Jones BE, Heaphy S (2009) Microbial biogeography of six Salt Lakes in Inner Mongolia, China and a Salt Lake in Argentina. Appl Environ Microbiol 75:5750–5760

Pramila R, Padmavathy K, Ramesh KV, Mahalakshmi K (2012) *Brevibacillus parabrevis*, *Acinetobacter baumannii* and *Pseudomonas citronellolis* - potential candidates for biodegrada-tion of low-density polyethylene (LDPE). J Bacteriol Res 4(1):9–14

Raffi F, Hall JD, Cernigila CE (1997) Mutagenicity of azo dyes used in foods, drugs, and cosmetics before and after reduction by Clostridium species from the human intestinal tract. Food Chem Toxicol 35:897–901

Rios-Hernandez LA, Gieg LM, Suflita JM (2003) Biodegradation of an alicyclic hydrocarbon by a sulfate-reducing enrichment from a gas condensate-contaminated aquifer. Appl Environ Microbiol 69:434–443. https://doi.org/10.1128/AEM.69.1.434-443.2003

Samuel-Osamoka FC, Kolawole OM, Adetitun DO (2018) Effects of environmental conditions and nutritional factors on biosurfactant production by bacteria from diesel contaminated soil. Nig J Technol Res 13(1):113–122

Santos H, Da Costa MS (2002) Compatible solutes of organisms that live in hot saline environ-ments. Environ Microbiol 4:501–509

Seeger M, Hernández M, Méndez V, Ponce B, Córdoval M, González M (2010) Bacterial deg-radation and bioremediation of chlorinated herbicides and biphenyls. J Soil Sci Plant Nutr 10(3):320–332

Shemer H, Linden KG (2007) Photolysis, oxidation and subsequent toxicity of a mixture of poly-cyclic aromatic hydrocarbons in natural waters. J Photochem Photobiol A Chem 187:186–195

Song HG, Bartha R (1990) Effects of jet fuel spills on the microbial community in soil. Appl Environ Microbiol 56:646–651

Tapilatu YH, Grossi V, Acquaviva M, Militon C, Bertrand JC, Cuny P (2010) Isolation of hydro-carbon-degrading extremely halophilic archaea from an uncontaminated hypersaline pond (Camargue, France). Extremophiles 14:225–231

Ulrich AC, Guigard SE, Fort JM, Semple KM, Pooley K, Armstrong JE, Biggar KW (2009) Effect of salt on aerobic biodegradation of petroleum hydrocarbons in contaminated groundwater. Biodegradation 20:27–38

Ulrici W (2000) Contaminant soil areas, different countries, and contaminant monitoring of contaminants, in environmental process II. In: Rehm HJ, Reed G (eds) Soil decontamination biotechnology, vol 11, pp 5–42

UNICEF (2020) UNICEF Annual Report 2017 Equatorial Guinea. https://www.unicef.org/publications/index_102899.html

Van Beilen JB, Funhoff EG, van Loon A, Kaysser L, Bouza M, Witholt B (2006) Cytochrome P450 alkane hydroxylases of the CYP153 family are common in alkane-degrading eubacteria lacking integral membrane alkane hydroxylases. Appl Environ Microbiol 7(1):59–65

Vandevivere PC, Bianchi R, Verstraete W (1998) Treatment and reuse of wastewater from the textile wet-processing industry: review of emerging technologies. J Chem Technol Biotechnol 72:289–302

Vargas JM (1975) Pesticide degradation. J Arboric 1(12):232–233

Verma P, Madamwar D (2003) Decolorization of synthetic dyes by a newly isolated strain of *Serratia marcescens*. World J Microbiol Biotechnol 19:615–618

Weinisch L, Kuhner S, Roth R et al (2018) Identification of osmoadaptive strategies in the halophile, heterotrophic ciliate Schmidingerothrix salinarum. PLoS Biol 16:e2003892. https://doi.org/10.1371/journal.pbio.2003892

Zampolli J, Collina E, Lasagni M, Di Gennaro P (2014) Biodegradation of variable-chain-length *n*-alkanes in *Rhodococcus opacus* R7 and the involvement of an alkane hydroxylase system in the metabolism. AMB Express 4(73):1–9

Chapter 13
The Use of Biosurfactants in the Bioremediation of Oil Spills in Water

Leonard Kachienga

13.1 Introduction

The continuous use of petroleum hydrocarbons and their by-products causes the generation of large quantities of insoluble petroleum wastes leading to a conglomeration of hydrocarbon pollutants in the environment (Berthe-Corti and Hopner 2005). According to Mohanty et al. (2013), the persistence of petroleum hydrocarbons in the aquatic environment eventually affects both flora and fauna. There are numerous treatment options of affected oil spill areas such as incineration, solvent extraction, and pump process, which have been used for the remediation of oil-polluted areas. However, bioremediation options have emerged in recent years as better and viable options. This is because they are typically more cost-effective compared to conventional physicochemical options. Although the constituents of petroleum hydrocarbon are structurally complex, they can be completely mineralized or reduced through the activities of some microbes involved (Leahy and Colwell 1990; Johnsen et al. 2005). Georgiou et al. (1990) suggested that biosurfactants could emulsify hydrocarbon-water mixtures, thereby enhancing the degradation of hydrocarbons in the environment. Most biosurfactants have a low adverse effect regardless of the time frame spent in the environment compared to chemical surfactants (Georgiou et al. 1990). It has been reported that biosurfactants are nontoxic, harmless, and ecologically acceptable compared to chemical surfactants (Yu and Huang 2011).

Biosurfactants are amphipathic molecules which possess both polar and nonpolar regions (Fig. 13.1).

These molecules are produced by extracellular or intracellular microorganisms and can reduce surface tension in the air-water interface between two immiscible

L. Kachienga (✉)
Faculty of Science, Department of Environmental, Water and Earth Sciences,
Tshwane University of Technology, Pretoria, South Africa

© Springer Nature Switzerland AG 2020   333
A. L. K. Abia, G. R. Lanza (eds.), *Current Microbiological Research in Africa*,
https://doi.org/10.1007/978-3-030-35296-7_13

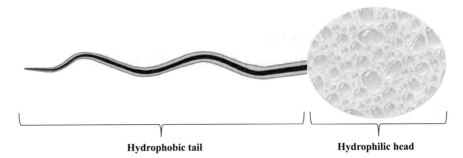

Hydrophobic tail   **Hydrophilic head**

Fig. 13.1 Schematic representation of a biosurfactant molecule

liquids or create an interface of polar and nonpolar pollutants (Yu and Huang 2011). The three primary characteristics of biosurfactant-active agents are the following: (1) enrichment at interfaces, (2) lowering interfacial tension, and (3) micelle formation (Santos et al. 2016). Biosurfactants facilitate the uptake of various hydrocarbons through various processes such as emulsification, pseudo-solubilization, and facilitated transport (Mohanty et al. 2013). The reduction in interfacial tension between the polar phase and nonpolar phase is what is known as emulsification and involves an increase in an interfacial area between the phases resulting in the transfer of hydrocarbons from the nonpolar to the polar region (Volkering et al. 1998). According to Salihu et al. (2009), biosurfactants are also highly effective in breaking down surface tension. Several microbial strains obtained from hydrocarbon-affected water sources are involved in the emulsification process, which increases their biomass.

Volkering et al. (1998) proposed a detailed interaction between the microorganism producing biosurfactants and hydrocarbon pollutants (Fig. 13.2).

Solubilization is defined as an increase in hydrocarbon solubility due to the partitioning within biosurfactants micelles. These micelles usually consist of a circular arrangement of various biosurfactant monomers where the hydrophobic tails of the monomers are pointing toward the center of the whole biosurfactant molecule. Hydrocarbons are, thus, partitioned in the hydrophobic core of the micelles (Mohanty and Mukherji 2007).

The mass transfer of various hydrocarbons from the nonpolar region through the interaction of any oil either by any single biosurfactant or many with sorbed oil is referred to as facilitated uptake (Kubicki et al. 2019). Biosurfactants have recently been found to affect the distribution of oil and the surfaces of microbial cells, hence influencing the rate of biodegradation either positively or negatively as reported by Mohanty and Mukherji (2013). It is also vital to select microorganisms that are resistant to the surfactant and are not adversely affected by its toxicity (Kaczorek et al. 2008). The biodegradability of biosurfactants after oil degradation is a desirable process which usually leads to the sustainable application of this kind of technology, although any accumulation of toxic intermediates during biosurfactants degradation may equally result to additional challenges (Mohanty and Mukherji 2012). According to Haba et al. (2000), higher removal percentage observed in the presence of biosurfactant is attributed to the interaction of surfactant-water and

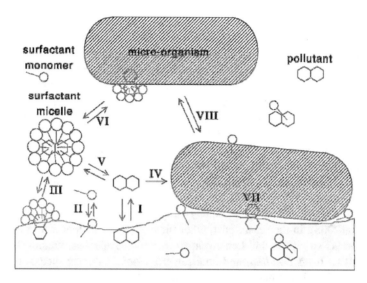

Fig. 13.2 Microbial interactions with pollutant (Volkering et al. 1998) I: sorption of pollutant, II: sorption of surfactant molecules, III: solubilization of pollutant, IV: the uptake of pollutant molecules from the water phase by microorganisms, V: partitioning of the pollutant between the water phase and micelles, VI: sorption of micelles to microorganisms, VII: direct microbial uptake of pollutant from soil, VIII: sorption of microorganisms to soil

surfactant-oil (such as interfacial tension reduction), which dominates the interaction of oil-water and is further attributed to the reduction of surface and interfacial tensions of surfactant solutions. This increases the mobility of oil and consequently enhances the separation of oil from aqueous solution.

Biosurfactants have diverse structures that enhance their usefulness in industrial applications, and their production by microbes ensures both their bioavailability and survival under low moisture conditions (Silva et al. 2014a, b, c). These molecules play a crucial role in the bioremediation processes due to their efficacy as well as their environmentally friendly processes such as low toxicity and high biodegradability (Silva et al. 2014a, b, c). Marchant and Banat (2012) reported that biosurfactants have also been used in various industrial processes of bioremediation and that these molecules could in the future be counted among the most multifunctional materials of the twenty-first century for the bioremediation of oil-polluted environments.

13.2   Biosynthesis and Genetic Regulation of Biosurfactant Production

Biosurfactants are usually synthesized or produced by microorganisms (glycolipids), plants (saponins), or animals (bile salts) (Souza et al. 2014). According to Campos et al. (2013), the majority of biosurfactants are produced by aerobic microorganisms in carbon-containing media such as carbohydrates, hydrocarbons, fats, and oils. They are secreted into carbon-containing media during the bioremediation process to facilitate the translocation of nonpolar molecules such as oil spill substrates across cell membranes. These biosurfactants usually display a range of different amphiphilic structures, and they are composed of hydrophobic and hydrophilic moieties. For the synthesis of these two moieties, two different synthetic pathways must be used: one leading to the hydrophobic and the other to the hydrophilic moiety. The hydrophobic fatty acid components—which may be a long-chain fatty acid, hydroxyl fatty acid, or alpha-alkyl-beta-hydroxy fatty acid—are synthesized by a rather common pathway of lipid metabolism. The hydrophilic moieties, however, exhibit a higher degree of structural complexity.

The first biosurfactant to be discovered were rhamnolipid (a glycolipid-type biosurfactant produced by *Pseudomonas aeruginosa*) (Dobler et al. 2017) and surfactin (a lipopeptide biosurfactant produced by *Bacillus subtilis*) (Sen 2010). According to Das et al. (2008), in the recent past, other biosurfactants such as arthrofactin from *Pseudomonas* sp., iturin and lichenysin from *Bacillus* species, mannosylerythritol lipids (MEL) from *Candida,* and emulsan from *Acinetobacter* species have been discovered. The biosynthetic pathway can be divided into three major steps, mainly (1) synthesis of the hydrophilic part, (2) synthesis of the hydrophobic part, and lastly (3) synthesis of rhamnolipid from these two parts. Most of the precursors for their synthesis are dTDP-L-rhamnose and activated 3-(3-hydroxyalkanoyloxy) alkanoate (HAA), respectively, for hydrophilic and hydrophobic parts (Soberón-Chávez 2004). The rhamnolipid is normally produced by the reaction of two special rhamnosyltransferases catalyzing the sequential rhamnosyl transfer reactions from the precursors over mono- toward di-rhamnolipids. Ochsner et al. (1994) suggested that, several genes that have been found in the biosynthesis of rhamnolipid in *Pseudomonas aeruginosa* are 2 kb fragments. These fragments contain a single open reading frame (rhlR) of 723 bp specifying a putative 28-kDa protein (RhlR). A comprehensive review of the genes regulating rhamnolipid production has been published (Dobler et al. 2016).

13.3   Advantages of Biosurfactants

There are numerous advantages of biosurfactants over well-known chemical surfactants, such as the simplicity of their chemical structure, environmental compatibility, and low toxicity (Kapadia and Yagnik 2013). These traits allow them to be used in various industries such as cosmetic, pharmaceutical, and food industries (Santos et al. 2013). They also have a higher selectivity due to the presence of specific functional groups which allow the detoxification of specific pollutants in any affected environment (Ławniczak et al. 2013). According to Abdel-Mawgoud et al. (2010), biosurfactants usually exhibit low toxicity, potentially high activities, and high tolerability in extreme environmental conditions such as extreme temperatures, pH, and salinity. Some of the advantages of biosurfactants have previously been reviewed (Santos et al. 2016).

13.3.1 Biodegradability

These molecules are naturally biodegradable, and this trait is an essential issue concerning environmental pollution. Given that they can be broken down through natural processes by bacteria, fungi, or other simple organisms into more basic components, they do not create many problems to the environment and are suited for eco-friendly environmental applications like bioremediation and dispersion of oil spills (Mulligan 2005).

13.3.2 Low Toxicity

Biosurfactants have low toxicity, and hence they cause no damage/harm to the biotic ecosystem. They are generally considered as low or nontoxic products and, therefore, appropriate for pharmaceutical, cosmetic, and food use. The synthetic anionic surfactant (Corexit) displayed an LC_{50} (concentration lethal to 50% of test species) against *Photobacterium phosphoreum*, which is ten times lower than rhamnolipids. This demonstrates higher toxicity of the chemically derived surfactant. It has also been demonstrated that biosurfactants had higher EC_{50} (effective concentration to decrease 50% of a test population) values than synthetic dispersants (Poremba et al. 1991).

13.3.3 Biocompatibility and Digestibility

These molecules are biocompatible, which means they are well tolerated by living organisms (Santos et al. 2016). Thus, their interaction with living organisms does not alter the bioactivity of the organisms, hence allowing their use in cosmetics, pharmaceuticals, and as functional food additives.

13.3.4 Availability of Raw Materials

Cheap raw materials which are abundant in the environments such as rapeseed oil, potato process effluents, oil refinery waste, cassava flour wastewater, curd whey and distillery waste, and sunflower oil can be used to produce biosurfactants (Muthusamy et al. 2008). The carbon source may originate from hydrocarbons, carbohydrates, and/or lipids, which may be used separately or in combination with each other. Industrial wastes and by-products are also a significant source of biosurfactants (Martins and Martins 2018), and these are of particular interest for bulk production of economically acceptable biosurfactants.

13.3.5 Specificity

The complex structure of biosurfactants includes specific functional groups that enhance their specificity (Santos et al. 2016). This means that they can readily and effectively be used to target a specific pollutant in the environment.

13.4 Disadvantages of Biosurfactants

A significant drawback in the use of biosurfactants is upscaling their production and the associated (Kumar et al. 2015; Olasanmi and Thring 2018). Their use in the petroleum industry and environmental applications requires considerably large quantities which are associated with extremely high costs (Perfumo et al. 2010). This shortcoming could, however, be curbed by making use of waste substrates. Another challenge is obtaining pure products after their manufacture, mainly when they are intended for use in pharmaceutical, food, and cosmetic applications (Chong and Li 2017).

13.4.1 Limitations of Large-Scale Applications of Biosurfactants

According to Das and Mukherjee (2007) despite the numerous identified advantages of biosurfactants, a common theme in literature is the inability of biosurfactants to compete commercially with their synthetic counterparts due to the high cost of production and sometimes low yield. Biosurfactants are 20–30% more expensive than synthetic surfactants, hence negatively affecting their applications in large scale (Hazra et al. 2011). This can only be addressed by the need for cost-effective substrates in the production of biosurfactants. Reports on the cost implication of substrates vary in the literature. The second aspect after the substrate is the production process which is a considerable part of the production cost. In particular, research into avenues to minimize cost and optimize the production process of biosurfactants is intertwined with their potential as agents of sustainability. The need to highlight the economic feasibility of biosurfactant production for commercial applications in the sustainability discussion is validated by the statement, "sustainability is only sustainable when it is profitable." This specific drawback is among the main reasons why it is still not feasible in the African continent unless these two questions are fully answered or tackled.

This technology must be looked upon despite the above challenges because of the devastating effects of oil spills caused on aquatic life, marine ecosystems, and the environment at large. Another need for biosurfactants lies in the toxicity of chemically synthesized surfactants that have been extensively reported during their bioremediation process.

13.5   Biosurfactant-Producing Microorganisms

Many microorganisms (fungi, yeasts, and bacteria) feed on substances that are immiscible in water, producing and using surface-active substances (biosurfactants) (Table 13.1). The production of these substances has, however, not been reported with any Protozoan microorganisms (Banat et al. 2010).

13.6   Biosurfactant-Enhanced Bioremediation of Oil Spill Environments

Petroleum is among the primary energy sources worldwide. The energy demand from petroleum deposits worldwide has an increase of 1.7% in the number of barrels of oil produced annually for the past three decades, with an annual consumption rate of 15.3 billion tons (BP P.l.c. 2019). According to Bachmann et al. (2014), most of the oil reserves are still able to meet the world's demand for the next four decades if current levels of consumption are maintained. It is, therefore, essential to develop technologies that will allow efficient use of this resource. As such, the current drive in petroleum production is toward unconventional oil alternatives like heavy and extra-heavy oils. These kinds of crude oils are widely recoverable from oil resources in Europe, Far East Asia (China), and South and North American countries such as Canada, Mexico, Venezuela, and the USA as reported by Cerón-Camacho et al. (2013). The growing global demand for oil is not without consequences.

According to Luna et al. (2013), there is an increase in the release of petroleum and petroleum by-products into the environment, and this has become a significant cause of global pollution and great concern in both industrialized and developing countries. These pollutions usually occur in existing water bodies through fuel transportation accidents either by road or water and continuous sipping from numerous surface and underground storage tanks that are subject to corrosion, i.e., filling stations, oil extraction, and processing operations (Sarubbo et al. 2014). The hydrophobic nature of petroleum or its by-products usually has structural cell membrane effects (Sobrinho et al. 2013).

There are various reports regarding numerous potential applications of biosurfactants in any affected environment (Sarubbo et al. 2014). According to the study done by Sobrinho et al. (2008) where a biosurfactant produced by the yeast *Cycas sphaerica* was tested for the removal of oil spills from both land and aquatic environment, they found out that there was a removal rate of between 75% and 92%, respectively. Two years later, a study by Batista et al. (2010) found that biosurfactants produced by *Candida tropicalis* had a removal rate of 78% to 97%, which was slightly higher for a similar oil spill-affected environment. A higher removal rate of over 90% of an oil spill in both soil and water by crude oil biosurfactants secreted by *C. glabrata* UCP1002 has also been reported (Gusmão et al. 2010). In a different study done by Luna and co-workers (2011, 2012), the removal rate of over 95% of

Table 13.1 Major classes of biosurfactant, microorganisms involved in their production, and their economic importance

Biosurfactant			
Group	Class	Microorganism	Economic importance
Glycolipids	Rhamnolipids	*Pseudomonas aeruginosa, Pseudomonas* sp., *Burkholderia glumae, Burkholderia plantarii, Burkholderia thailandensis*	Antimicrobial activity against *Mycobacterium tuberculosis*, anti-adhesive activity against several bacterial and yeast strains isolated from voice prostheses, enhancement of the degradation and dispersion of different classes of hydrocarbons; emulsification of hydrocarbons and vegetable oils; removal of metals from soil.
	Trehalose lipids	*Rhodococcus erythropolis, Nocardia erythropolis, Mycobacterium* sp., *Arthrobacter* sp.	Enhancement of the bioavailability of hydrocarbons, antiviral activity against HSV and influenza virus
	Sophorolipids	*Torulopsis bombicola, Torulopsis apicola, Torulopsis petrophilum*	Recovery of hydrocarbons from dregs and muds; removal of heavy metals from sediments; enhancement of oil recovery
	Mannosylerythritol lipid	*Candida antartica*	Antimicrobial, immunological and neurological properties
	Cellobiolipids	*Ustilago zeae, Ustilago maydis*	–
Lipopeptides and lipoproteins	Surfactin/iturin/fengycin	*Bacillus subtilis, Bacillus licheniformis*	Enhancement of the biodegradation of hydrocarbons and chlorinated pesticides; removal of heavy metals from a contaminated soil, sediment and water; antimicrobial activity and antifungal activity against profound mycosis effect on the morphology and membrane structure of yeast cells increase in the electrical conductance of biomolecular lipid membranes nontoxic and non-pyrogenic immunological adjuvant.
	Viscosin	*Pseudomonas fluorescens*	Antimicrobial activity.
	Lichenysin	*Bacillus licheniformis*	Enhancement of oil recovery, antibacterial activity chelating properties that might explain the membrane-disrupting effect of lipopeptides.
	Serrawettin	*Serratia marcescens*	Chemorepellent.
	Subtilism	*Bacillus subtilis*	Antimicrobial activity
	Gramicidin	*Brevibacterium brevis*	Antibiotic, disease control.
	Polymyxin	*Bacillus polymyxa*	Bactericidal and fungicidal activity
	Antibiotic TA	*Myxococcus xanthus*	Bactericidal activity, chemotherapeutic applications.

Category	Biosurfactant	Microorganism	Function
Fatty acids/neutral, lipids/phospholipids	Corynomycolic acid	*Corynebacterium lepus*	Enhancement of bitumen recovery.
	Spiculisporic acid	*Penicillium spiculisporum*	Removal of metal ions from aqueous solution; dispersion action for hydrophilic pigments; preparation of new emulsion-type organogels, superfine microcapsules (vesicles or liposomes), heavy metal sequestrants.
	Phosphatidylethanolamine	*Acinetobacter sp., Rhodococcus, erythropolis Mycococcus sp.*	Increasing the tolerance of bacteria to heavy metals.
Polymeric surfactants	Emulsan	*Acinetobacter calcoaceticus*	Stabilization of the hydrocarbon-in water emulsions.
	Alasan	*Acinetobacter radioresistens*	Stabilization of the hydrocarbon-in water emulsions.
	Biodispersant	*Acinetobacter calcoaceticus A2*	Dispersion of limestone in water.
	Polysaccharide protein complex	*Acinetobacter calcoaceticus*	Bioemulsifier.
	Liposan	*Candida lipolytica*	Stabilization of hydrocarbon in- water emulsions.
	Mannoprotein	*Saccharomyces cerevisiae*	Stabilization of hydrocarbon in water emulsions.
	Protein PA	*Pseudomonas aeruginosa*	Bioemulsifier.
Particulate biosurfactants	Vesicles	*Acinetobacter calcoaceticus, Pseudomonas marginalis*	Degradation and removal of hydrocarbons.
	Whole microbial cells	*Cyanobacteria*	Degradation and removal of hydrocarbons.

oil in motor oil wastewater and soil was achieved by a new biosurfactant known as
Lunasan. This biosurfactant was produced by *C. sphaerica* UCP 0995, which
showed a great potential use in any bioremediation of oil-polluted water sources and
land. Table 13.2 demonstrates a list of different types of biosurfactants and their
producing microorganisms with potential applications in the bioremediation of oil-
polluted environments (Sarubbo and Campos-Takaki 2010; Makkar et al. 2011).

Table 13.2 Microorganisms, various types of biosurfactants produced, and their application in the
bioremediation processes

Microorganisms	Type of biosurfactant	Applications
R. erythropolis 3C-9	Glucolipid and trehalose lipid	Oil spill cleanup operations
P. aeruginosa S2	Rhamnolipid	Bioremediation of oil-contaminated sites
C. sphaerica UCP0995	Protein-carbohydrate-lipid complex	Oil removal
C. lipolytica UCP0988	Sophorolipids	Removal of petroleum and motor oil adsorbed to sand
C. tropicalis UCP0996	Protein-carbohydrate-lipid complex	Removal of petroleum and motor oil adsorbed to sand
C. guilliermondii UCP0992	Glycolipid complex	Removal of petroleum derivate motor oil from sand
C. glabrata UCP1002	Protein-carbohydrate-lipid complex	Oil removal
C. sphaerica UCP0995	Protein-carbohydrate-lipid complex	Bioremediation processes
C. lipolytica UCP0988	Sophorolipids	Control of environmental oil pollution
C. sphaerica UCP0995	Protein-carbohydrate-lipid complex	Removal of oil from sand
C. lipolytica UCP0988	Sophorolipids	Oil removal
C. lipolytica UCP0988	Sophorolipids	Oil recovery
C. glabrata UCP1002	Protein-carbohydrate-lipid complex	Oil recovery from sand
P. aeruginosa BS20	Rhamnolipid	Bioremediation of hydrocarbon-contaminated sites
P. cepacia CCT6659	Rhamnolipid	Bioremediation of marine and soil environments
P. hubeiensis	Glycolipid	Bioremediation of marine oil pollution
C. soyoae	Mannosylerythritol	Lipid bioremediation of marine environment
Pseudoxanthomonas sp. PNK-04	Rhamnolipid	Environmental applications
P. alcaligenes	Rhamnolipid	Environmental applications
A. chroococcum	Lipopeptide	Environmental applications
B. subtilis BS5	Lipopeptide	Bioremediation of hydrocarbon-contaminated sites

13.7 Some Studies on Biosurfactants Production and Application in Oil Pollution Abatement

Several countries in Africa have oil reserves, with oil production being a considerable contributor to the economy of some, like Equatorial Guinea (Pariona 2017). These countries, like many oil-producing countries, are faced with the problem of pollution arising from these oil producing facilities, oil transportation systems, and damage of oil infrastructure like pipelines. The adverse effects from these pollution events have been reported. For example, several events and their associated consequences on the environment, animals and humans, have been reported in Nigeria (Nriagu et al. 2017; Kanu and Achie 2011; Okafor-Yarwood 2018; Bruederle and Hodler 2019), Gabon (Vinson et al. 2008), Egypt (Ramadan et al. 2010; Hamed et al. 2013; El-Borai et al. 2016), Moroco (Er-Raioui et al. 2009), and Cameroon (Abdourahimi et al. 2016), among others. In most instances, there is a complete lack of adequate facilities to remediate such polluted areas. The studies on the use of biosurfactants for bioremediation of oil-polluted environments are limited in Africa, although some studies conducted in the continent have isolated biosurfactants from different environments (Table 13.3).

Table 13.3 Some studies on the isolation/synthesis of biosurfactants in different African countries

Country	Source of isolation	Reference
Algeria	Bacteria from uncontaminated fertile surface soil	Zohra et al. (2014)
	Bacteria from contaminated soil	Mesbaiah et al. (2016)
	Bacterium from salt lake	Salah (2013)
Burkina Faso	*Bacteria from Adansonia digitate* seeds	Kaboré et al. (2018)
Cameroon	*Bacteria from fermented milk*	Hippolyte et al. (2018)
	Bacter from Cocoyam rhizosphere	Oni et al. (2014)
Egypt	Bacteria from oil-polluted water from the Red Sea	Barakat et al. (2017)
	Egyptian sunflower seeds	Diab et al. (2016)
	Rhizosphere soil of an Egyptian salt marsh plant	Diab et al. (2013)
	Bacteria from drain sediments	Gomaa and El-Meihy (2019)
	Bacteria from oil-contaminated soil	Ramadan et al. (2010)
Ethiopia	Selected fruit peel waste	Murugesan et al. (2019)
Libya	Bacteria from oil-contaminated soil	Hakima and Ian (2017)
Namibia	Bacteria from a wastewater treatment plant	Ndlovu et al. (2017)
Nigeria	Bacteria from oil-polluted river water	Ndibe et al. (2019)
	Soil bacteria	Nwaguma et al. (2016)
	Palm oil mill effluent	Hope and Gideon (2015)
	Bacteria from diesel-contaminated soil	Salam et al. (2016)
South Africa	Bacteria from diesel-contaminated soil	Ganesh and Lin (2009)
	Bacteria from petroleum-contaminated soil	Bezza and Chirwa (2017)
	Bacteria from petroleum-contaminated soil	Lutsinge and Chirwa (2018)
Tunisia	Soil	Ghribi and Ellouze-Chaabouni (2011)

Studies on biosurfactants and their potential benefits are very limited in Africa. The few ones that have been conducted have mainly focused on the isolation of specific organisms and testing them for their potential to produce biosurfactants, mostly through the investigation of genes responsible for such action. Also, the studies have been entirely laboratory based and large-scale production has not been investigated. These limitations are probably influenced by the lack of adequate instrumentation to perform such studies. While these studies have been conducted, much is still to be done to fully explore the biosurfactant-producing potentials of the African environment, the dynamics, and kinetics of the produced molecules, and finally the possibility for mass production and industrial application.

13.8 Biosurfactants Toxicity During the Bioremediation of Oil Spill in Both Aquatic and Soil Environments

The effect of biosurfactants in any aquatic and surface environment in terms of toxicity is still not well studied. However, suspected lower toxicity levels and higher biodegradability are observed for these biosurfactants compared to their chemical counterparts making them highly acceptable for bioremediation purposes (Sarubbo et al. 2014). Franzetti et al. (2012) reported that the toxicity features of various biosurfactants produced by microbes are often assumed to have a direct impact on their origin. There is a need for thorough studies on the level of toxicity of these biosurfactants before their release to facilitate any biodegradation of oil spills (Franzetti et al. 2012). According to Van Hamme and Ward (1999), the concentration of biosurfactants closer to critical micelle concentration (CMC) is usually not toxic to any microbe driving the bioremediation process. However, if it is above the CMC, then there is a possibility of bioremediation inhibition. According to Prince et al. (2003), the use of dispersants/biosurfactants has been thought to have stimulated the natural process of biodegradation, due to the formation of an oil-water interface and the dispersion of the oil, which will dramatically increase the surface area for microbial attack. The results of a study carried out by Kachieng'a and Momba (2017) also confirmed the effectiveness of a dispersant in improving the bioavailability of hydrocarbons and showed the biodegradation capability of a consortium of *Aspidisca* sp., *Trachelophyllum* sp., and *Peranema* sp.

Their toxicity level might have a more significant impact toward specific microbes during the bioremediation process in comparison to little or no inhibitory effect on a consortium of microbes, which are more diverse, performing similar bioremediation process (Singh et al. 2007). According to Lima et al. (2011), the general proposal of toxicity tests is to establish the potential effect which chemicals could have on the biota of a specific environment. The results of such tests could aid in the design of regulation for the use and disposal of these chemicals. The toxicity is usually measured using the LC_{50} (the lethal dose capable of killing 50% of a given population) (Zhang et al. 2007). The higher the LC_{50} of a surfactant, the lower its toxicity. Another method that has been used to determine the toxicity of biosurfactants is the seed germination test through the calculation of the germination index (GI); this method is one of the most commonly used methods (Sobrinho et al. 2013). Usually, a GI value of 80% indicates the disappearance of phytotoxicity. The toxicity values of biosurfactants, dispersants, crude oils, and dispersant/crude oil mixtures to vegetables and organisms are summarized in Table 13.4.

Table 13.4 Results of toxicity tests of biosurfactants, dispersants, crude oils, and dispersant/crude oil mixtures to vegetables and organisms (adapted from Silva et al. 2014a, b, c)

Test compound	Organisms	Toxicity
Biosurfactants		
Emulsan	*Mysidopsis bahia*	LC50 (200 mg/L)
Candida sphaerica UCP 0995 biosurfactant	*Brassica oleracea*	86% GI
Candida sphaerica UCP 0995 biosurfactant	*Artemia salina*	LC50 (600 mg/L)
Candida sphaerica UCP 0995 biosurfactant	*Brassica oleracea*	No toxicity
Pseudomonas aeruginosa UCP 0992 biosurfactant	*Brassica oleracea*	80% GI
Pseudomonas aeruginosa UCP 0992 biosurfactant	*Artemia salina*	LC50 (525 mg/L)
Emulsifiers/dispersing agents		
Dodecylbenzene sulfonate/LAS	*Dugesia japonica*	LC50 (1.45 mg/L)
Lauryl sulfate/SDS	*Dugesia japonica*	LC50 (0.36 mg/L)
Triton X-100	*Mysidopsis bahia*	LC50 (3.3 mg/L)
Triton X-100	*Menidia beryllina*	LC50 (2.5 mg/L)
Lauryl sulfate/SDS	*Americamysis ahia*	LC50 (18–23 mg/L)
Lauryl sulfate/SDS	*Menidia beryllina*	LC50 (10 mg/L)
Oil spill dispersants		
Corexit 9500	*Mysidopsis bahia*	LC50 (13.4 mg/L)
Corexit 9500	*Menidia beryllina*	LC50 (75.7 mg/L)
Corexit 9500 *Porites astreoides*	*Porites astreoides*	13% surviving
Corexit 9500	*Montastraea faveolata*	0% surviving
Corexit 9500	*Brachionus plicatilis*	LC50 (0.447 (mg/L)
Corexit 9500	*Brachionus manjavacas*	LC50 (14.2 mg/L)
Crude oils		
BP Horizon source oil	*Porites astreoides*	67% surviving
BP Horizon source oil	*Montastraea faveolata*	27% surviving
Louisiana sweet crude oil	*Americamysis bahia*	LC50 (2.7 mg/L)
Louisiana sweet crude oil	*Menidia beryllina*	LC50 (3.5 mg/L)
Macondo sweet crude oil	*Brachionus plicatilis*	LC50 (2.47 mg/L)
Macondo sweet crude oil	*Brachionus* sp.	LC50 (19.3 mg/L)
Dispersant/oil mixtures		
Corexit 9500/BP Horizon source oil	*Porites astreoides*	67% surviving
Corexit 9500/BP Horizon source oil	*Montastraea faveolata*	20% surviving
Corexit 9500/Louisiana sweet crude oil	*Americamysis bahia*	LC50 (5.4 mg/L)
Corexit 9500/Louisiana sweet crude oil	*Menidia beryllina*	LC50 (7.6 mg/L)
1:10 Corexit 9500/Macondo sweet crude oil	*Brachionus manjavacas*	0.21 (mg/L)
1:50 Corexit 9500/Macondo sweet crude oil	*Brachionus manjavacas*	0.23 (mg/L)

GI germination index, *LC$_{50}$* concentration lethal to 50% of the test species

13.9 Conclusion

This review examined the use of biosurfactants as promising facilitators of biore-mediation of oil spill mostly from the petroleum industry. Little is known about the potential of biosurfactant production by microorganisms, and the majority of studies or application of biosurfactants are simulated mainly at laboratory scale. More information is still required regarding the structures of different biosurfactants, their interaction with soil, contaminants, scaling up, and cost-effectiveness during production. Since biosurfactants are costly compared to chemical surfactants, alternative sources such as agro-industrial waste need to be investigated for their production. Also, further research is needed to develop production processes to increase yield and minimize the cost of purification when purification is a requirement for application. The discovery of novel low-temperature biosurfactants also poses an exciting area for research and industry, both in the potential role and application of the biosurfactants in cold regions and the development of low-energy production processes. While these challenges surround the use of these promising substances for oil-pollution abatement, the challenge is even greater in Africa where resources are limited, the machinery for their production is lacking, and research on the continents' potentials as a rich source for the isolation of useful species is extremely limited.

References

Abdel-Mawgoud AM, Lépine F, Déziel E (2010) Rhamnolipids: diversity of structures, microbial origins and roles. Appl Microbiol Biotechnol 86:1323–1336

Abdourahimi S, Fantong WY et al (2016) Environmental pollution by metals in the oil bearing Bakassi Peninsula, Cameroon. Carpathian J Earth Environ Sci 11:529–538

Bachmann RT, Johnson AC, Edyean RJ (2014) Biotechnology in the petroleum industry: an overview. Int Biodeter Biodegr 86:225–237

Banat IM, Franzetti A, Gandolfi I et al (2010) Microbial biosurfactants production, applications. Appl Microbiol Biotechnol 87:427–444

Barakat KM, Hassan SWM, Darwesh OM (2017) Biosurfactant production by haloalkaliphilic Bacillus strains isolated from Red Sea, Egypt. Egypt J Aquat Res 43:205–211

Batista RM, Rufino RD, Luna JM et al (2010) Effect of medium components on the production of a biosurfactant from Candida tropicalis applied to the removal of hydrophobic contaminants in soil. Water Environ Res 82:418–425

Berthe-Corti L, Hopner T (2005) Geo-biological aspects of coastal oil pollution. Palaeogeogr Palaeocl 219(1–2):171–189

Bezza FA, Chirwa EMN (2017) Biosurfactant-assisted bioremediation of polycyclic aromatic hydrocarbons (PAHs) in liquid culture system and substrate interactions. Polycycl Aromat Compd 37:375–394

BP P.l.c. (2019) BP Statistical Review of World Energy Statistical Review of World 2019, 68th edition. https://www.bp.com/content/dam/bp/business-sites/en/global/corporate/pdfs/energy-economics/statistical-review/bp-stats-review-2019-full-report.pdf

Bruederle A, Hodler R (2019) Effect of oil spills on infant mortality in Nigeria. Proc Natl Acad Sci 116:5467–5471

Campos JM, Stamford TLM, Sarubbo LA et al (2013) Microbial biosurfactants as additives for food industries. Biotechnol Prog 29:1097–1108

Cerón-Camacho R, Martínez-Palou R, Chávez-Gómez B et al (2013) Synergistic effect of alkyl-O-glucoside and -cellobioside biosurfactants as effective emulsifiers of crude oil in water. A proposal for the transport of heavy crude oil by pipeline. J Fuel 110:310–317

Chong H, Li Q (2017) Microbial production of rhamnolipids: opportunities, challenges and strategies. Microb Cell Factories 16:1–12

Das K, Mukherjee AK (2007) Comparison of lipopeptide biosurfactants production by Bacillus subtilis strains in submerged and solid state fermentation systems using a cheap carbon source: some industrial applications of biosurfactants. Process Biochem 42:1191–1199

Das P, Mukherjee S, Sen R (2008) Improved bioavailability and biodegradation of a model polyaromatic hydrocarbon by a biosurfactant producing bacterium of marine origin. Chemosphere, 72(9):1229–1234

Diab A, Gamal S, Din E (2013) Application of the biosurfactants produced by Bacillus spp. (SH 20 and SH 26) and P. aeruginosa SH 29 isolated from the rhizosphere soil of an Egyptian salt marsh plant for the cleaning of oil - contaminataed vessels and enhancing the biodegradat. Afr J Environ Sci Technol 7:671–679

Diab A, Sami S, Diab AA (2016) Production, characterization and application of a new biosurfactant derived from Egyptian sunflower seeds. Int J Sci Res 5:602–612

Dobler L, De Carvalho BR, De Sousa Alves W et al (2017) Enhanced rhamnolipid production by Pseudomonas aeruginosa overexpressing estA in a simple medium. PLoS One 12:1–12. https://doi.org/10.1371/journal.pone.0183857

Dobler L, Vilela LF, Almeida RV, Neves BC (2016) Rhamnolipids in perspective: gene regulatory pathways, metabolic engineering, production and technological forecasting. New Biotechnol 33:123–135

El-Borai AM, Eltayeb KM, Mostafa AR, El-Assar SA (2016) Biodegradation of industrial oil-polluted wastewater in Egypt by bacterial consortium immobilized in different types of carriers. Polish J Environ Stud 25:1901–1909

Er-Raioui H, Bouzid S, Marhraoui M, Saliot A (2009) Hydrocarbon pollution of the Mediterranean coastline of Morocco. Ocean Coast Manag 52:124–129

Franzetti A, Gandolfi I, Raimondi C et al (2012) Environmental fate, toxicity, characteristics and potential applications of novel bioemulsifiers produced by Variovorax paradoxus 7bCT5. Bioresour Technol 108:245–251

Ganesh A, Lin J (2009) Diesel degradation and biosurfactant production by gram-positive isolates. Afri J Biotechnol 8:5847–5854

Georgiou G, Lin SC, Sharma MM (1990) Surface active compounds from micro-organisms. J Biotechnol 10:60–65

Ghribi D, Ellouze-Chaabouni S (2011) Enhancement of Bacillus subtilis Lipopeptide biosurfactants production through optimization of medium composition and adequate control of aeration. Biotechnol Res Int 2011:1–6. https://doi.org/10.4061/2011/653654

Gomaa EZ, El-Meihy RM (2019) Bacterial biosurfactant from Citrobacter freundii MG812314.1 as a bioremoval tool of heavy metals from wastewater. Bull Natl Res Cent 43:69. https://doi.org/10.1186/s42269-019-0088-8

Gusmão CAB, Rufino RD, Sarubbo LA (2010) Laboratory production and characterization of a new biosurfactant from Candida glabrata UCP1002 cultivated in vegetable fat waste applied to the removal of hydrophobic contaminant. World J Microbiol Biotechnol 26:1683–1692

Haba E, Espuny MJ, Busquets M, Manresa A (2000) Screening and production of Rhamnolipids by Pseudomonas aerogirosa from waste frying oils. J Appl Microbiol 88:379–387

Hakima A, Ian S (2017) Isolation of indigenous hydrocarbon transforming Bacteria from oil contaminated soils in Libya: selection for use as potential inocula for soil bioremediation. Int J Environ Bioremed Biodegrad 5:8–17

Hamed YA, Abdelmoneim TS, Elkiki MH et al (2013) Assessment of heavy metals pollution and microbial contamination in water, sediments and fish of Lake Manzala. Egypt Life Sci J 10:86–99

Hazra C, Kundu D, Ghosh P et al (2011) Screening and identification of *Pseudomonas aeruginosa* AB4 for improved production, characterization and application of a glycolipid biosurfactant using low-cost agro-based raw materials. J Chem Technol Biotechnol 86:185–198

Hippolyte MT, Augustin M, Hervé TM et al (2018) Application of response surface methodology to improve the production of antimicrobial biosurfactants by lactobacillus paracasei subsp. tolerans N2 using sugar cane molasses as substrate. Bioresour Bioprocess. https://doi.org/10.1186/s40643-018-0234-4

Hope N, Gideon A (2015) Biosurfactant production from Palm Oil Mill Effluent (POME) for applications as oil field chemical in Nigeria. In: SPE Nigeria Annual International Conference and Exhibition. Society of Petroleum Engineers

Johnsen AR, Wick LY, Harms H (2005) Principles of microbial PAH-degradation in soil. Environ Pollut 133(1):71–84

Kaboré D, Gagnon M, Roy D et al (2018) Rapid screening of starter cultures for maari based on antifungal properties. Microbiol Res 207:66–74

Kachieng'a L, Momba MNB (2017) Kinetics of petroleum oil biodegradation by a consortium of three protozoan isolates (*Aspidisca* sp., *Trachelophyllum* sp. and *Peranema* sp.). Biotechnol Rep 15:125–131

Kaczorek E, Chrzanowski Ł, Pijanowska A, Olszanowski A (2008) Yeast and bacteria cell hydrophobicity and hydrocarbon biodegradation in the presence of natural surfactants: rhamnolipids and saponins. Bioresour Technol 99(10):4285–4291

Kanu I, Achie OK (2011) Industrial effluents and their impact on water quality of receiving rivers in Nigeria. J Appl Technol Environ Sanit 1:75–86

Kapadia SG, Yagnik BN (2013) Current trend and potential for microbial biosurfactants. Asian J Exp Biol Sci 4:1–8

Kubicki S, Bollinger A, Katzke N et al (2019) Marine biosurfactants: biosynthesis, structural diversity and biotechnological applications. Mar Drugs 17:1–30

Kumar AP, Janardhan A, Radha S et al (2015) Statistical approach to optimize production of biosurfactant by Pseudomonas aeruginosa 2297. 3 Biotech 5:71–79

Ławniczak L, Marecik R, Chrzanowski L (2013) Contributions of biosurfactants to natural or induced bioremediation. Appl Microbiol Biotechnol 97:2327–2339

Leahy JG, Colwell RR (1990) Microbial degradation of hydrocarbons in the environment. Microbiol Rev 54(3):305–315

Lima TMS, Procópio LC, Brandão FD et al (2011) Evaluation of bacterial surfactant toxicity towards petroleum degrading microorganisms. J Bioresour Technol 102:2957–2964

Luna JM, Rufino R, Sarubbo LA et al (2011) Evaluation antimicrobial and anti-adhesive properties of the biosurfactant lunasan produced by *Candida sphaerica* UCP 0995. Curr Microbiol 62:1527–1534

Luna JM, Rufino RD, Campos-Takakia GM, Sarubbo LA (2012) Properties of the biosurfactant produced by Candida Sphaerica cultivated in low-cost substrates. J Chem Eng Transact 27:67–72

Luna JM, Rufino RD, Sarubbo LA, Campos-Takaki GM (2013) Characterisation surface properties and biological activity of a biosurfactant produced from industrial waste by *Candida sphaerica* UCP0995 for application in the petroleum industry. Colloid Surface B 102:202–209

Lutsinge TB, Chirwa EMN (2018) Biosurfactant-assisted biodegradation of fluoranthene in a two-stage continuous stirred tank bio-reactor system using microorganism. Chem Eng Trans 64:67–72

Makkar RS, Cameotra SS, Banat IM (2011) Advances in utilization of renewable substrates for biosurfactant production. Appl Microbiol Biotechnol 1(1):5

Marchant R, Banat IM (2012) Microbial biosurfactants: challenges and opportunities for future exploitation. Trend Biotechnol 11:558–565

Martins PC, Martins VG (2018) Biosurfactant production from industrial wastes with potential remove of insoluble paint. Int Biodeterior Biodegrad 127:10–16

Mesbaiah FZ, Eddouaouda K, Badis A et al (2016) Preliminary characterization of biosurfactant produced by a PAH-degrading Paenibacillus sp. under thermophilic conditions. Environ Sci Pollut Res 23:14221–14230

Mohanty G, Mukherji S (2007) Effect of an emulsifying surfactant on diesel degradation by cultures exhibiting inducible cell surface hydrophobicity. J Chem Technol Biotechnol 82(11):1004–1011

Mohanty S, Jasmines J, Mukherji S (2013) Practical considerations and challenges involved in surfactants enhanced bioremediation of oil. Review article. J Biomed Res Int. https://doi.org/10.1155/2013/328608

Mohanty S, Mukherji S (2012) Alteration in cell surface properties of *Burkholderia* spp. during surfactant-aided biodegradation of petroleum hydrocarbons. Appl Microbiol Biotechnol 94(1):193–204

Mohanty S, Mukherji S (2013) Surfactant-aided biodegradation of NAPLs by *Burkholderia multivorans*: comparison between triton X-100 and rhamnolipid JBR 515. Colloid Surface B 102:644–652

Mulligan CN (2005) Environmental applications for biosurfactants. Environ Pollut 133(2):183–198

Murugesan K, Tesfaye Y, Mahmmud A et al (2019) A comparative preliminary analysis of selected fruit peel waste fermented solutions: impact of shorter fermentation in biosurfactant production. Appl Biotechnol Rep 6:69–72

Muthusamy K, Gopalakrishnan S, Ravi TK, Sivachidambaram P (2008) Biosurfactants: properties, commercial production and application. Curr Sci 94(6):00113891

Ndibe T, Eugene W, Usman J (2019) Screening of biosurfactant-producing Bacteria isolated from river Rido, Kaduna, Nigeria. J Appl Sci Environ Manag 22:1855. https://doi.org/10.4314/jasem.v22i11.22

Ndlovu T, Rautenbach M, Vosloo JA et al (2017) Characterisation and antimicrobial activity of biosurfactant extracts produced by Bacillus amyloliquefaciens and Pseudomonas aeruginosa isolated from a wastewater treatment plant. AMB Express. https://doi.org/10.1186/s13568-017-0363-8

Nriagu J, Udofia EA, Ekong I, Ebuk G (2017) Health risks associated with oil pollution in the Niger Delta, Nigeria. Environ Plan C 35(2):334–354

Nwaguma IV, Chikere CB, Okpokwasili GC (2016) Isolation, characterization, and application of biosurfactant by Klebsiella pneumoniae strain IVN51 isolated from hydrocarbon-polluted soil in Ogoniland, Nigeria. Bioresour Bioprocess 3:40. https://doi.org/10.1186/s40643-016-0118-4

Ochsner UA, Fiechter A, Reiser J (1994) Isolation, characterization, and expression in Escherichia coli of the Pseudomonas aeruginosa rhlAB genes encoding a rhamnosyltransferase involved in rhamnolipid biosurfactant synthesis. J Biol Chem 269(31):19787–19795

Okafor-Yarwood I (2018) The effects of oil pollution on the marine environment in the Gulf of Guinea-the Bonga oil field example. Transnatl Leg Theory 9:254–271

Olasanmi IO, Thring RW (2018) The role of biosurfactants in the continued drive for environmental sustainability. Sustain 10:1–12. https://doi.org/10.3390/su10124817

Oni FE, Geudens N, Onyeka J, et al (2014) Diversity and bioactivity of biosurfactant-producing pseudomonads isolated from cocoyam rhizosphere in Nigeria and Cameroon. Crop Protection, 66th International symposium, Abstracts. Presented at the 66th International symposium on Crop Protection

Pariona A (2017) "Top 10 Oil Producing Countries in Africa." WorldAtlas, Oct. 10, 2017. worldatlas.com/articles/top-10-oil-producing-countries-in-africa.html

Perfumo A, Rancich I, Banat IM (2010) Possibilities and challenges for biosurfactants use in petroleum industry. Adv Exp Med Biol 672:135–145

Poremba K, Gunkel W, Lang S, Wagner F (1991) Marine biosurfactants, III. Toxicity testing with marine microorganisms and comparison with synthetic surfactants. Z Naturforsch C 46(3–4):210–216

Prince RC, Lessard RR, Clark JR (2003) Bioremediation of marine oil spills. Oil Gas Sci Technol Rev 58:463–468

Ramadan EM, Kheiralla ZM, Foaad MA et al (2010) Biosurfactant producing bacteria from oil contaminated Egyptian soil. Egypt J Microbiol 45:1–13. https://doi.org/10.21608/ejm.2010.273

Salah A (2013) Production of biosurfactant on crude date syrup under saline conditions by entrapped cells of Natrialba sp. strain E21, an extremely halophilic bacterium isolated from a solar saltern, pp 981–993

Salam LB, Obayori OS, Hawa O (2016) Hydrocarbon degradation and biosurfactant production by an acenaphthene-degrading *Pseudomonas* species. Soil Sediment Contam An Int J 25:837–856

Salihu A, Abdulkadir I, Almustapha MN (2009) An investigation for potential development on biosurfactants. A review. Biotechnol Mol Biol Rev 3(5):111–117

Santos DKF, Rufino RD, Luna JM et al (2013) Synthesis and evaluation of biosurfactant produced by *Candida lipolytica* using animal fat and corn steep liquor. J Pet Sci Eng 105:43–50

Santos DKF, Rufino RD, Luna JM et al (2016) Biosurfactants: multifunctional biomolecules of the 21st century. Int J Mol Sci 17:1–31. https://doi.org/10.3390/ijms17030401

Sarubbo LA, Campos-Takaki GM (2010) *Candida* biosurfactant-enhanced removal hydrophobic organic pollutants. In: Mason AC (ed) Bioremediation: biotechnology, engineering and environmental management. Nova Publishers, New York, NY, pp 435–448

Sarubbo LA, Santos VA, Luna JM, Rufino RD, Almeida DG, Silva FS (2014) Applicants of biosurfactants in the petroleum industry and the remediation of oil spills. Review. Int J Mol Sci 15:12523–12542

Sen R (2010) Biosynthesis, genetics and potential applications. Adv Exp Med Biol 672:316–323

Silva EJ, Rocha e Silva NMP, Rufino RD et al (2014a) Characterization of a biosurfactant produced by *Pseudomonas cepacia* CCT6659 in the presence of industrial wastes and its application in the biodegradation of hydrophobic compounds in soil. Colloid Surface B 117:36–41

Silva R, Almeida D, Rufino R et al (2014b) Applicants of biosurfactants in the petroleum industry and the remediation of oil spills. Int J Mol Sci 15(7):12523–12542

Silva RFS, Almeida DG, Rufino RD et al (2014c) Applications of biosurfactants in the petroleum industry and the remediation of oil spills. Int J Mol Sci 15:12523–12542

Singh A, van Hamme JD, Ward OP (2007) Surfactants in microbiology and biotechnology: part 2. Application aspects. Biotechnol Adv 25:99–121

Soberón-Chávez G (2004) Biosynthesis of rhamnolipids. In: Ramos J-L (ed) Pseudomonas. Kluwer Academic/Plenum Publishers, New York, pp 173–189

Sobrinho HB, Rufino RD, Luna JM, Salgueiro AA, Campos-Takaki GM, Leite LF, Sarubbo LA, (2008) Utilization of two agroindustrial by-products for the production of a surfactant by Candida sphaerica UCP0995. Process Biochemistry, 43(9):912–917

Sobrinho HB, Luna JM, Rufino RD et al (2013) Biosurfactants: classification, properties and environmental applications. In: Recent developments in biotechnology, vol 11, 1st edn. Studium Press LLC, Houston, TX, pp 1–29

Souza EC, Vessoni-Penna TC, Souza Oliveira RP (2014) Biosurfactant-enhanced hydrocarbon bioremediation: an overview. Int Biodeterior Biodegradation 89:88–94

Van Hamme JD, Ward OP (1999) Influence of chemical surfactants on the biodegradation of crude oil by a mixed-bacterial culture. Canadian J Microbiol 45:130–137

Vinson MR, Dinger EC, Kotynek J, Dethier M (2008) Effects of oil pollution on aquatic macroinvertebrate assemblages in Gabon wetlands. African J Aquat Sci 33:261–268

Volkering F, Breure AM, Rulkens WH (1998) Microbiological aspects of surfactant use for biological soil remediation. J Biodegr 8(6):401–417

Yu H, Huang GH (2011) Isolation and characterization of biosurfactant and bioemulsifier-producing bacteria from petroleum contaminated sites in Western Canada. Soil Sed Contam 20(3):274–288

Zhang M, Aguilera D, Das C et al (2007) Measuring cytotoxicity: a new perspective on LC50. Anticancer Res 27:35–38

Zohra F, Mnif S, Badis A, Rebbani S (2014) Biodeterioration & Biodegradation Naphthalene and crude oil degradation by biosurfactant producing Streptomyces spp . Isolated from Mitidja plain soil (north of Algeria). Int Biodeterior Biodegradation 86:300–308

Index

© Springer Nature Switzerland AG 2020
A. L. K. Abia, G. R. Lanza (eds.), *Current Microbiological Research in Africa*,
https://doi.org/10.1007/978-3-030-35296-7

Printed in the United States
By Bookmasters